PRINCIPLES OF

ANIMAL GROWTH AND DEVELOPMENT

DAVID E. GERRARD

ALAN L. GRANT

Purdue University

KENDALL/HUNT PUBLISHING COMPANY

4050 Westmark Drive Dubuque, Iowa 52002

Book Team

Chairman and Chief Executive Officer Mark C. Falb
Senior Vice President, College Division Thomas W. Gantz
Director of National Book Program Paul B. Carty
Editorial Development Supervisor Georgia Botsford
Vice President, Production and Manufacturing Alfred C. Grisanti
Assistant Vice President, Production Services Christine E. O'Brien
Prepress Project Coordinator Angela Puls
Permissions Editor Colleen Zelinsky
Designer Deb Howes

Revised Printing 2006

ISBN13: 978-0-7575-2986-3

Printed in the United States of America

10 9 8

CONTENTS

Figure List v
Table List xiii
Preface xvii
About the Authors xix

Chapter 1
An Introduction to Animal Growth and Development 1

INTRODUCTION 1
What Is Animal Growth and Development? 1
Why Study Animal Growth and Development? 1
Historical Contributions 4

Chapter 2
Tissue Structure and Organization 7

INTRODUCTION 7
Epithelium 7
Connective Tissue 9
Nervous Tissue 21
Muscle Tissue 21

Chapter 3
Embryonic Growth 43

INTRODUCTION 43
Prenatal Development 45
Maternal Membrane Development 58
Measurements of Prenatal Development 60

Chapter 4
Myogenesis 63

INTRODUCTION 63
Molecular Aspects of Myogenesis 63
Morphological Aspects of Myogenesis 68
Myofibrillogenesis 73

Chapter 5
Muscle Growth 77

INTRODUCTION 77
Muscle Fiber Number 77
Muscle Fiber Size 86

Chapter 6
Muscle Fiber Plasticity and Growth 103

INTRODUCTION 103
Muscle Fiber Type Organization 103
Muscle Fiber Types and Classification 105
Determination of Muscle Fiber Type 105
Distinguishing Features of Muscle Fiber Types 109
Development of Muscle Fiber Types 110
Muscle Fiber Types and Growth Performance 111

Chapter 7
Adipose Tissue Growth and Development 119

INTRODUCTION 119
Anatomical Location of Adipose Tissue 119
Adipogenesis 121
Molecular Control of Adipogenesis 123
Adipose Tissue Metabolism 125
Composition of Adipose Tissue 129
Factors Affecting Adipose Tissue Composition and Lipid Metabolism 129

Chapter 8
Growth and Development of Bone 137

INTRODUCTION 137
Origin of Cells Involved in Bone Formation and Resorption 137
Endochondral Ossification 138
Termination of Long Bone Growth 141
Intramembranous Ossification 141
Bone Resorption 142
Bone Remodeling 144
Factors That Affect Bone Growth and Remodeling 145

Chapter 9
Whole Body Animal Growth 149

INTRODUCTION 149
Growth Curves 149
Changes in Body Form 153
Changes in Body Measurements 154
Changes in Body Components 156
Changes in Carcass Components 158
Changes in the Chemical Composition 159
Differential Allometric Growth 161
Factors Affecting Carcass Composition 163

Chapter 10
Hormonal Regulation of Growth 179

INTRODUCTION 179
Hormone Actions 179
Chemical Nature of Hormones 180
Hypothalamic-Pituitary-Peripheral Gland Axes 180
Androgens 184
Estrogens 186
Progestins 186
Synthetic Hormone Combinations 186
Growth Hormone 189
Insulin and Glucagon 192
Leptin 192
Glucocorticoids 193
Catecholamines 194
Thyroid Hormones 196
Parathyroid Homone 198
Other Endocrine Glands 198

Chapter 11
Nutritional Regulation of Growth 199

INTRODUCTION 199
Nutrient Partitioning 199
Undernutrition 202
Compensatory Growth 204
Dietary Protein 208
Dietary Energy 210
Antibacterial Feed Additives 212
Regulation of Protein Accretion 213

Chapter 12
Relationships among Growth, Composition, and Meat Quality 215

INTRODUCTION 215
The Grading System 215
Changing Muscle into Meat 228
Augmenting Growth and Altering Quality 233

Glossary 241
References 251
Index 255

FIGURE LIST

Chapter 1

Figure 1.1 Photographs of market steers illustrating the vast differences in body type between 1950 (top) and present (bottom). *2*

Figure 1.2 Photographs of market barrows illustrating the vast differences in body type between 1950 (top) and present (bottom). *2*

Figure 1.3 Photographs of market wethers illustrating the vast differences in body type between 1950 (top) and present (bottom). *3*

Chapter 2

Figure 2.1 Diagram of various types of epithelium. *8*

Figure 2.2 Illustration of loose connective tissue. *10*

Figure 2.3 Synthesis of tropocollagen and formation of a collagen fibril. *11*

Figure 2.4 Amino acid sequence molecular structure of collagen and tropocollagen molecules, and of collagen fibril formation. *11*

Figure 2.5 Collagen synthesis and maturation. *12*

Figure 2.6 Diagrammatic illustration of the chicken wire-like network of Type IV collagen. *12*

Figure 2.7 Micrograph of elastic fibers. *14*

Figure 2.8 Micrograph of elastic connective tissue. *15*

Figure 2.9 Loose connective tissue. *16*

Figure 2.10 Micrograph of dense, irregular connective tissue. *16*

Figure 2.11 Diagram and photomicrographs of fat cells. *17*

Figure 2.12 Micrograph of cartilage showing the location of chondrocytes (thin arrows) in lacunae (heavy arrows). *18*

Figure 2.13 Micrograph of hyaline cartilage. *18*

Figure 2.14 Micrograph of fibrocartilage at the insertion of tendon into bone. *19*

Figure 2.15 Longitudinal structure of the long bone *19*

Figure 2.16 Photographs showing the different types of bone. *20*

Figure 2.17 Micrograph showing the structure of an osteon. *21*

Figure 2.18 Diagrammatic illustration of a neuron and the motor end plates associated with it. *22*

Figure 2.19 Location of oligodendrocytes. *23*

Figure 2.20 Micrograph of a nerve bundle cross section. *23*

Figure 2.21 Photomicrograph of skeletal muscle fibers (× 630). *25*

Figure 2.22 Drawing of a skeletal muscle in cross section showing muscle fibers, bundle arrangement, supporting connective tissues, nerves, and blood vessels. *25*

Figure 2.23 Photomicrograph of a cross-section of skeletal muscle fibers of the pig, showing their polygonal shape and the peripheral position of nuclei (× 630). *26*

Figure 2.24 Drawing of skeletal muscle fibers showing structural features and their longitudinal orientation. *27*

Figure 2.25 Micrograph of a cardiomyocte showing the elaborate structure of cytoskeleton. Arrows identify the position of the nuclei. *27*

Figure 2.26 Skeletal muscle organization. *28*

Figure 2.27 Ultrastructure of the sarcomere. *29*

Figure 2.28 Diagram of the myosin molecule and its subunits. *30*

Figure 2.29 Electron photomicrograph of a cross-section of a myofibril, showing the thick and thin filaments and their hexagonal arrangement. *31*

Figure 2.30 Diagrammatic representation of the construction and fine structure of an actin filament, showing the relation of tropomyosin and troponin to its (A) G-actin molecules, (B) two chains of polymerized actin (F-actin) coiled about each other, (C) long thin molecules of tropomyosin, (D) pellet-shaped troponin molecules, and (E) an actin filament, showing how two coiled chains of F-actin and polymerized tropomyosin molecules form two long strands of tropomyosin, one strand being associated with each of the F-actin chains. The periodic location of troponin along the actin filament every seven G-actin molecules, and its association with the tropomyosin strand, are also illustrated. *32*

Figure 2.31 Illustration of the location of desmin around myofibrils. *33*

Figure 2.32 Illustration showing of the importance of costameric attachments in facilitating muscle shortening. *34*

Figure 2.33 Diagram showing the structures participating in the organization of the costameres in skeletal muscle in relation to the Z-line. *34*

Figure 2.34 Diagrammatic representation of the sarcoplasmic reticulum and T tubules, and their relation to the myofibrils of mammalian skeletal muscle. *36*

Figure 2.35 Organization of the sarcoplasmic reticulum, transverse tubules and associated structures. *37*

Figure 2.36 Photomicrographs of cardiac muscle. A longitudinal section (A), showing fiber arrangements and nuclei locations and, (B) an electron photomicrograph showing the banding pattern, intercalated disks, mitochondria, and glycogen granules. *38*

Figure 2.37 Organization of smooth muscle elements. *39*

Chapter 3

Figure 3.1 Overview of mammalian development. *44*

Figure 3.2 Differentiation of tissues during prenatal growth and development. *45*

Figure 3.3 Relationship between gestation length and body weight gain. *46*

Figure 3.4 Movement and development of human embryos during early embryo development. *46*

Figure 3.5 Change in embryo cell number during early development. *47*

Figure 3.6 Relative change in cell size and number during early frog development. *48*

Figure 3.7 Blastocyst formation in the human embryo. *48*

Figure 3.8 Types of cell movements during gastrulation. *49*

Figure 3.9 Illustration of hypoblast formation during avian development. *50*

Figure 3.10 Migration of cells during early avian development. *51*

Figure 3.11 Formation of various structures during avian neurulation. *52*

Figure 3.12 Changes in cell morphology during the formation of the neural tube. *53*

Figure 3.13 Illustration showing the changes in somite morphology during chick embryo development. *54*

Figure 3.14 Formation and organization of newly formed somites adjacent to the neural tube. *55*

Figure 3.15 Maturation of a somite into the dermatome, myotome, and scleratome. *55*

Figure 3.16 Migration of sclerotomal cells in preparation for formation of the vertebral column. *56*

Figure 3.17 Migration of mesodermal cells and formation of a limb bud. *56*

Figure 3.18 Organizing centers during limb bud formation. *57*

Figure 3.19 The relative location of limb buds development during early development (A) and the location of the apical ectoderm ridge (B) (AER), the progress zone (PZ) and the zone of polarizing activity (ZPA). *57*

Figure 3.20 Pig fetal membrane organization. *58*

Figure 3.21 Illustration showing the association between maternal and fetal tissues. *59*

Figure 3.22 The unique interlocking maternal and fetal membranes in the pig. *60*

Figure 3.23 Cotyledon structure in cattle and sheep. *61*

Figure 3.24 Microcotyledon structure in the horse. *61*

Figure 3.25 Relationship between crown-rump length (CRL) and gestational age. *62*

Chapter 4

Figure 4.1 Schematic illustrating the events of mitosis in animal cells. *64*

Figure 4.2 Stages of the typical cell cycle. G1 (Gap 1), S (synthetic), G2 (Gap 2), M (Mitosis) and G0 phases vary in length. *64*

Figure 4.3 Illustration showing the action of growth factors on a cell. *65*

Figure 4.4 Major events occurring during myogenesis. *66*

Figure 4.5 Action of myogenic regulatory factors (MRFs) in muscle cells during myogenesis. *67*

Figure 4.6 Process of myoblast fusion. *68*

Figure 4.7 Structures of the dermamyotome. *69*

Figure 4.8 Migration of muscle precursor cells from the dermomyotome. *69*

Figure 4.9 Change in the relative size and frequency of primary and secondary muscle fibers during fetal development. *70*

Figure 4.10 Micrographs of developing primary and secondary muscle fibers. *70*

Figure 4.11 Illustration of embryonic and fetal development of muscle fibers. *71*

Figure 4.12 Evolution of mucle fiber number and size in pig muscle. *73*

Figure 4.13 Formation of myofibrils and their alignment during myofibrillogenesis. *73*

Figure 4.14 Model for the formation of myofibrils from stress fibers (nm, non-muscle). *74*

Figure 4.15 Model showing the organization of nascent myofibrils during myofibrillogenesis. *75*

Chapter 5

Figure 5.1 Effect of muscle cell hyperplasia and hypertrophy on whole muscle hypertrophy. *78*

Figure 5.2 Various types of muscle in pigs. *78*

Figure 5.3 Effect of muscle fiber arrangement on assessing muscle fiber characteristics. *78*

Figure 5.4 Location and type of muscle present in economically important meat cuts. *79*

Figure 5.5 Organization of muscle fibers in a whole muscle. *79*

Figure 5.6 Effect of intrafascicularly terminating muscle fibers on muscle fiber parameters. *79*

Figure 5.7 Effect of restricted nutrient intake (closed boxes) in gestating guinea pigs on fetal muscle fiber number in the biceps femoris muscle. *81*

Figure 5.8 Changes in muscle fiber size (top) and number (bottom) in growing rainbow trout. *82*

Figure 5.9 Relative contribution of muscle fiber hyperplasia versus hypertrophy to overall growth of white (top) and red (bottom) muscle in growing rainbow trout. *83*

Figure 5.10 Relationship between muscle fiber number and performance traits (average daily gain, ADG; and feed efficiency, gain/feed ratio) in growing (25–80 kg) pigs. *84*

Figure 5.11 Muscle fiber number in domestic and wild pigs. *85*

Figure 5.12 Muscle fiber number in developing double-muscled (■) and normal (●) bovine fetuses. *86*

Figure 5.13 Bull and carcass that exhibit double muscling. *87*

Figure 5.14 Comparison of pork carcasses from pigs possessing (right) and lacking (left) the halothane or "stress" gene. *88*

Figure 5.15 Comparison of callipyge and normal genotype lambs. *89*

Figure 5.16 Change in longissimus muscle fiber diameter with increasing age in growing pigs. *90*

Figure 5.17 Micrographs showing longitudinal myofibrillar splitting in adult muscles. *91*

Figure 5.18 Drawing illustrating the role of sarcomere addition in muscle lengthening. *91*

Figure 5.19 Molecular events occurring during protein production in a muscle cell. *92*

Figure 5.20 Effects of the caplain proteases on muscle degradation postmortem. *93*

Figure 5.21 Ubiquitin-proteosome pathway for protein degradation. *94*

Figure 5.22 Fractional protein synthesis, degradation and accretion rates and absolute protein accretion rate in growing animals. *95*

Figure 5.23 Location of satellite cells in relation to a mature muscle fiber. *96*

Figure 5.24 Origin of embryonic, fetal and adult myogenic lineages in developing birds. *96*

Figure 5.25 Sequential appearance of embryonic, fetal and adult myoblasts in developing muscle. *96*

Figure 5.26 Participation of adult satellite cells in regenerating muscle fibers. *97*

Figure 5.27 Estimation of the contribution of satellite cells to muscle growth in growing mice. *98*

Figure 5.28 Changes in muscle weight and nuclei acid content in growing pigs. *98*

Figure 5.29 Changes in nuclei acid concentrations in muscle and fractional synthesis rate in muscle of growing pigs. *98*

Figure 5.30 Percentage increase in the area of muscle fibers in various muscles of pigs treated with somatotropin. *100*

Chapter 6

Figure 6.1 Micrograph of myofibrillar ATPase staining showing the heterogeneity of muscle fibers in the cross section of bovine longissiumus muscle. *104*

Figure 6.2 Myofibrillar ATPase stained muscle cross sections showing the vast differences in muscle fiber type composition in the superficial and deep semitendinosus muscle of pigs. *104*

Figure 6.3 Micrograph of a myofibrillar ATPase stained muscle cross section showing the unique arrangement of muscle fiber type in porcine muscle. *105*

Figure 6.4 Location of the S1 myosin head that is the target of ATPase-mediated muscle fiber typing. *105*

Figure 6.5 Representation of the myofibrillar ATPase staining reaction and fiber typing of various different muscle fiber types. *106*

Figure 6.6 Relationship between the pH sensitivity of different muscle fiber types. *107*

Figure 6.7 Micrographs of bovine muscle cross sections stained with a classical ATPase staining protocol. *107*

Figure 6.8 Photomicrograph of cross-section of porcine longissimus muscle, showing red and white fibers (dark and light, respectively). (× 175). *108*

Figure 6.9 The relationship between type of energy metabolism and muscle fiber type. Note that energy metabolism is predominately either glycolytic or oxidative metabolisms. However, note that all fibers have a combination of both types of metabolisms. *109*

Figure 6.10 Different types of motor nerves lead to specific changes in cytosolic calcium concentrations. *111*

Figure 6.11 Evolution of muscle fiber types in pigs. *112*

Figure 6.12 Disappearance of fetal myosin

heavy chain isoforms in the muscle of various animals. *112*

Figure 6.13 Relative muscle fiber type composition in four cattle breeds. *113*

Figure 6.14 Changes in the abundance of lactate dehydrogenase (LDH) and isocitrate dehydrogenase (ICDH) in growing pig muscle. *115*

Chapter 7

Figure 7.1 Distribution of fat between the subcutaneous, intermuscular, intramuscular, and kidney (and pelvic) fat in beef, lamb, and pork carcasses. *120*

Figure 7.2 Anatomical locations of various fat depots. Transverse section through the lumbar region of a pork carcass depicting six locations of adipose tissue. *121*

Figure 7.3 Adipocyte development. *122*

Figure 7.4 Distribution of adipocyte diameter in pigs with different backfat thickness. BF – average backfat thickness (cm). *123*

Figure 7.5 Age-related changes in carcass adipose tissue weight, volume of adipose cells in the subcutaneous depot and estimated adipose cell number in growing pigs. *124*

Figure 7.6 Molecular mechanisms controlling adipocyte differentiation. *124*

Figure 7.7 Glycerolipid biosynthesis. *126*

Figure 7.8 The role of fatty acids in adipose tissue. *126*

Figure 7.9 *De novo* synthesis of fatty acids from glucose and acetate. *127*

Figure 7.10 Mechanism for lipid mobilization from cells. *128*

Figure 7.11 Location and mechanism of leptin action. *133*

Figure 7.12 Changes in adipocyte volume in the subcutaneous (SQ), perirenal (PR) and intermuscular (IM) fat depots of lambs fed either ad libitum or maintenance diets and slaughtered at 3 different times. *134*

Figure 7.13 Effect of ambient temperature on animal performance traits. *135*

Chapter 8

Figure 8.1 Origin and location of bone cells. *138*

Figure 8.2 Phases of skeletal development. *138*

Figure 8.3 Long bone formation and growth by endochondral ossification. *139*

Figure 8.4 Organization of the growth plate. *140*

Figure 8.5 Intramembranous ossification. *142*

Figure 8.6 Appositional bone growth. *143*

Figure 8.7 Micrograph (transmission microscopy) of a bone resorbing osteoclast. *143*

Figure 8.8 Action of a bone resorbing osteoclast. *144*

Figure 8.9 Regional changes in cartilage and bone that create increased length and width in long bones. *145*

Figure 8.10 Bone remodeling by osteoblastic and osteoclastic cell activity. *146*

Chapter 9

Figure 9.1 Typical growth curve. *150*

Figure 9.2 Non-smoothed growth curve illustrating the actual "stair-stepped" growth occurring in most animals. *150*

Figure 9.3 Prenatal weight growth of the bovine fetus. *150*

Figure 9.4 Identification of major points along a typical growth curve for growth in sheep: conception (a), birth (b), self-accelerating (c), inflection point or puberty (d), self-retarding phase (e) and maturity (f). *150*

Figure 9.5 Alternative means of segmenting a typical growth curve. *151*

Figure 9.6 Growth curves of fast and slow growing animals. *151*

Figure 9.7 Absolute growth rate (weight/unit time). *152*

Figure 9.8 Relative growth rate (growth rate/unit weight). *152*

Figure 9.9 Relative changes in animal weight (cumulative growth), absolute growth rate and relative growth rate over the lifetime of an animal. *153*

Figure 9.10 Relative changes in body proportions during growth. *153*

Figure 9.11 Growth rates of various tissues and proportions of the body of animals fed for rapid or slow growth. *154*

Figure 9.12 A skeletal diagram showing the anatomical locations used to evaluate carcass length. *156*

Figure 9.13 Postnatal growth curves of bone, muscle and fat. *158*

Figure 9.14 Changes in the percentage of bone, muscle, and fat in beef carcasses during growth. *159*

Figure 9.15 Chemical component changes with empty body weight of growing steers. *159*

Figure 9.16 Changes in empty body composition in growing steers. *160*

Figure 9.17 Theoretical representation of monophasic growth patterns. *162*

Figure 9.18 Theoretical representation of diphasic growth patterns. *163*

Figure 9.19 Growth and distribution of fat in growing pigs. *167*

Figure 9.20 Growth and distribution of fat in growing cattle. *168*

Figure 9.21 Topographical maps illustrating the location of fat deposition and relative growth impetus of various muscle groups in cattle. *169*

Figure 9.22 Schematic illustrating the difference in frame size and type among cattle breeds. *170*

Figure 9.23 Relationship between frame size, mature weight, growth rate and fat thickness in cattle. *171*

Figure 9.24 Growth curves of early and late maturing animals. *172*

Figure 9.25 Differences in frame scores, cumulative growth and composition of growing cattle. *172*

Figure 9.26 Differences between muscle to bone ratios of Holstein-Friesian and Hereford steers. *174*

Figure 9.27 Effect of breed type on fat distribution. *174*

Figure 9.28 Growth curves of the various sexes of pigs, sheep, cattle and poultry. *176*

Figure 9.29 Effect of sex on muscle distribution in cattle. *177*

Chapter 10

Figure 10.1 Different mechanisms of hormonal signaling. *180*

Figure 10.2 Hormone action on cells. *181*

Figure 10.3 Potential mechanisms of negative feedback. *181*

Figure 10.4 Relationship between the hypothalamus and the pituitary gland. *182*

Figure 10.5 Gross location of the major endocrine glands in a cow. *183*

Figure 10.6 Potential negative feedback loops of a classical hypothalamus-pituitary-peripheral tissue axis. *184*

Figure 10.7 Effects of various hormones and growth factors on bone resorption and formation. *185*

Figure 10.8 Regulation of growth hormone secretion. *189*

Figure 10.9 Growth hormone-insulin-like growth factor axes. *190*

Figure 10.10 Receptors, ligands and binding proteins of the insulin-like growth factor signaling axes. *190*

Figure 10.11 Hormonal and metabolic changes in response to hypoglycemia. *192*

Figure 10.12 Regulation and action of leptin. *193*

Figure 10.13 Illustration of the β_2-adrenergic receptor and its critical regions. *194*

Figure 10.14 Intracellular signaling by the β-adrenergic receptors. *195*

Figure 10.15 Structure of various adrenergic hormones and their agonists. *196*

Chapter 11

Figure 11.1 Energy partitioning in a pig (60 kg) fed different levels of metabolizable energy (ME). *200*

Figure 11.2 Priority and nutrient flow from the circulation to various tissues during growth. *200*

Figure 11.3 Efficiency of tissue storage and mobilization in lactating (LACT) and non-lactating (DRY) dairy cattle. ***201***

Figure 11.4 Compensatory growth curve: recovering weight-for-age. ***205***

Figure 11.5 Compensatory growth curve: extended growth period. ***205***

Figure 11.6 Permanent stunting of growth. ***205***

Figure 11.7 Effects of restricted nutrient intake (top, left) and compensatory gain (bottom) in twin cattle. ***206***

Figure 11.8 Cumulative growth curves of pigs fed levels of nutrition. ***207***

Figure 11.9 Effects of plane of nutrition on growth of pigs at different ages. ***207***

Figure 11.10 Nitrogen metabolism in ruminants. ***209***

Figure 11.11 Effects of exogenous porcine somatotropin (pST) on protein requirements of growing (25 to 55 kg) pigs. ***211***

Figure 11.12 Theoretical protein and fat deposition rates in an animal with different body weight gains. ***211***

Figure 11.13 Effects of digestible energy (DE) intake on protein deposition in control and porcine somatotropin (pST)-treated gilts and boars. ***212***

Chapter 12

Figure 12.1 Identifying marks used for the U.S. Department of Agriculture grades for meat; (A) beef, lamb, and mutton yield grades; (B) beef, lamb, and mutton quality grades; (C) poultry grades; and (D) pork carcass grades. ***216***

Figure 12.2 Indices of maturity in meat animal carcasses. ***218***

Figure 12.3 Relationships among flank fat streaking, maturity, and quality as used to establish lamb, older lamb, and yearling mutton carcass quality grades. ***219***

Figure 12.4 Marbling in beef longissimus muscle. ***220***

Figure 12.5 Illustrations showing high, intermediate and low degrees of flank fat streakings in the primary (P) and secondary (S) flank. ***221***

Figure 12.6 Relationship between conformation and leg scores in lamb carcasses. ***222***

Figure 12.7 Front view of stewing chickens illustrating from left to right A, B, and C quality grades. ***222***

Figure 12.8 Relationships among marbling, maturity, and quality as used to establish beef carcass quality grades. ***223***

Figure 12.9 Color-texture-exudation standards for ham muscles (top), color standards for *longissimus* muscle (center), and marbling scores for *longissimus* muscle (bottom) for pork carcass evaluation. ***inside front cover***

Figure 12.10 Photograph illustrating the method for assessing loin muscle area in a pork carcass. ***225***

Figure 12.11 Degrees of muscle development for pork carcasses. ***226***

Figure 12.12 Photograph illustrating the method for assessing fat thickness at the 10th rib in a pork carcass. ***226***

Figure 12.13 Beef carcasses showing the outline (in black) of the kidney, pelvic, and heart fat. ***227***

Figure 12.14 Relationship between beef yield grades and cutability. ***227***

Figure 12.15 A simplified diagram of the sites of production, and the general actions of the major hormones related to stress reactions. ***229***

Figure 12.16 Schematic illustrating the metabolism and movements of metabolites in living tissues (left) and muscle (right) immediately postmortem. ***229***

Figure 12.17 A diagrammatic illustration of the pathways that supply energy for muscle function. ***230***

Figure 12.18 Postmortem pH decline curves. ***230***

Figure 12.19 Micrograph illustrating the importance of the connective tissue framework in beef muscle (E, epimysium; P, perimysium). ***231***

Figure 12.20 Effect of ageing on reducible (- - - -) and mature (——) cross-link formation in skeletal muscle. *232*

Figure 12.21 Relationship between meat tenderness (shear force) and sarcomere length. *232*

Figure 12.22 Effect of postmortem ageing on beef tenderness (boxes) and sacromere length (circles). *232*

Figure 12.23 Relationship between beef quality and yield grades. *233*

Figure 12.24 Electron micrographs of (A) relaxed and (B) cold-shortened beef muscle. *235*

Figure 12.25 Role of enhanced sarcoplasmic calcium release on various cellular functions in muscle cells. *236*

TABLE LIST

Chapter 1

Table 1.1 Market Hog Performance. *2*

Table 1.2 Market Steer Performance. *3*

Table 1.3 Market Lamb Performance. *3*

Table 1.4 Tom Turkey Performance. *3*

Table 1.5 Broiler Performance. *3*

Chapter 2

Table 2.1 Examples of the Types of Epithelium. *9*

Table 2.2 Collagen Types and Their Location in Animal Tissues. *13*

Table 2.3 Connective Tissue in Different Tissues. *14*

Table 2.4 Selected Myofibrillar and Other Cytoskeletal Proteins Located in Muscle Fibers. *35*

Table 2.5 Approximate Composition of Mammalian Skeletal Muscle (Percent Fresh Weight Basis). *40*

Chapter 3

Table 3.1 Implantation in a Variety of Domestic Farm Animals. *44*

Table 3.2 Anatomical Features, Sizes and Gestational Ages of Various Species During Prenatal Growth. *62*

Chapter 5

Table 5.1 Muscle Fibre Number in Cattle of Different Breeds (In Millions). *85*

Table 5.2 Effect of Breed on Muscle Fiber Number in Pigs. *85*

Table 5.3 Growth and Muscle Fiber Diameter in Rapid and Slow Growing Chickens. *90*

Table 5.4 Accumulation of DNA in Postnatal Muscle of Different Species. *97*

Table 5.5 Effect of Sex on Muscle Fiber Size (μm^2). *98*

Table 5.6 Effect of Feed Intake on Muscle Fiber Size (μm^2). *99*

Table 5.7 Effect of Dietary Protein on Muscle Fiber Size (μm^2). *99*

Table 5.8 Effect of Beta-Adrenergic Agonists on Muscle Fiber Size (μm^2) in Various Meat Animal Species. *99*

Table 5.9 Effect of Exogenous Growth Hormone (pST) Administration on Muscle Fiber Size (μm^2) in Boars, Gilts and Barrows. *100*

Chapter 6

Table 6.1 Muscle Fiber Type Classification. *108*

Table 6.2 Characteristics of Muscle Fibers in Domestic Meat Animals and Birds. *109*

Table 6.3 Proportion of Fiber Types in *M. Gracilis* of Wild and Domestic Pigs. *114*

Table 6.4 Effects of Major Genes on Histochemical Traits of Valuable Muscles in Different Species. *115*

Table 6.5 Relative Change (percentage of control) in the Amount of Type 1, 2A, 2X and 2B Myosin Heavy Chain (MyHC) Isoform in the Longissimus Muscle of Pigs Fed Varying Levels of Paylean® Over a 6 Week Period. *117*

Table 6.6 Effect of Cimaterol on Muscle Fiber Type Composition in Young Friesian Bulls. *117*

Table 6.7 Effect of Implants (Ralgro® and Synovex-S®) on Muscle Fiber Type Composition in Cattle. *118*

Chapter 7

Table 7.1 Fatty Acid and Triglyceride Composition of Some Animal Fat Depots (Percent by Weight). *130*

Table 7.2 Characteristics of Adipose Tissue from Seven Anatomical Locations in the Pig. *131*

Table 7.3 Characteristics of Porcine Adipose Tissue from Three Locations. *131*

Table 7.4 Effects of Breed, Sex, and Age on Subcutaneous Adipose Tissue Traits in Sheep. *132*

Table 7.5 Influence of Dietary Regime on Diameter of Adipocytes from Three Locations in Pigs Studied at Constant Age and Weight. *133*

Chapter 9

Table 9.1 Comparison of Linear Measurements, Lean Cuts, and Carcass Fat of Pigs at Various Liveweights. *155*

Table 9.2 Distribution of Body Components of Cattle During Growth and Development. *157*

Table 9.3 Dressing Percentages of Various Kinds of Livestock by Grades. *158*

Table 9.4 Changes in Concentration of Proteins in Mammalian Skeletal Muscle During Growth. *160*

Table 9.5 Percentage Chemical Composition of Fat from Depots of Steers. *161*

Table 9.6 Major Muscle Groupings. *162*

Table 9.7 Growth Patterns of Standard Muscle Groups of Steers Using Growth Coefficients. *162*

Table 9.8 Heritability Estimates of Carcass Characteristics of Cattle, Poultry, Sheep and Swine. *164*

Table 9.9 Empty Body Composition. *164*

Table 9.10 Comparison of Muscle Distribution (%) between Adult Cattle and Deer. *165*

Table 9.11 Comparison of Muscle Distribution (%) among Intact Male Bison, Domestic Cattle and Sheep. *165*

Table 9.12 Muscle Weight Distribution of Several Species Expressed Relative to *Bos taurus*. *166*

Table 9.13 Muscle Weight Distribution of Steers and Barrows. *166*

Table 9.14 Fat Distribution among Carcass Depots (% Total Fat). *167*

Table 9.15 Separation of Abdominal Leaf Fat (ALF) in the Broiler Carcass. *168*

Table 9.16 Effect of Frame Size and Breed on Beef Carcass Composition at a Constant Age (473 Days). *172*

Table 9.17 Effect of Frame Size and Breed on Beef Carcass Composition at a Constant Weight (634 lb. Carcass Wt.). *172*

Table 9.18 Effect of Frame Size and Breed on Beef Carcass Composition at a Constant Maturity (19% Fat Trim). *173*

Table 9.19 Effect of Frame Size and Breed on Beef Carcass Compostion at a Constant Quality Grade (5% Fat in Longissimus). *173*

Table 9.20 Effect of Cattle Breed on Carcass Composition. *173*

Table 9.21 Muscle Distribution among Breed Types. *174*

Table 9.22 Muscle to Bone Ratios and Distribution of Muscle Groups in Meat and Mothering Type Pigs (100 kg Live Wt.). *175*

Table 9.23 Effect of Genetic Strain on Carcass and Abdominal Fat in Broilers. *175*

Table 9.24 Distribution of Muscle among Cattle Breeds Selected for Different Conformations. *176*

Table 9.25 Distribution of Muscle, Fat and Bone in Bull and Steer Carcasses. *177*

Table 9.26 Muscle Weight Distribution in Bull, Steer and Heifer Carcass (% of Total Carcass Muscle). *178*

Chapter 10

Table 10.1 History of the FDA Approval Dates and Status for Anabolic Agents. *187*

Table 10.2 Performance and Carcass Characteristics of Implanted Cattle with Estradiol (E_2) and Trenbolone Acetate (TBA) Alone or in Combination. *187*

Table 10.3 Concentrations of Estradiol (E_2) and Estrone (E_1) in Tissues of Steers Following Estrogen Implantation. *188*

Table 10.4 Daily Human 17-β Estradiol (E_2) and Estrone (E_1) Production. *188*

Table 10.5 Major Effects of Growth Hormone. *189*

Table 10.6 Effects of Porcine Somatotropin (pST) on Growth and Carcass Performance in Growing Pigs. *191*

Table 10.7 Effects of Human Somatotropin (pST) Releasing Factor (hSRF) on Growth and Carcass Performance in Growing Lambs. *191*

Table 10.8 Typical Responses to Phenethenolamine Adminstration in Food Producing Animals. *196*

Table 10.9 Effect of Zilpaterol on the Growth Performance of Steers. *197*

Table 10.10 Effect of Zilpaterol on Beef Carcass Performance. *197*

Chapter 11

Table 11.1 Body Weight, Organ Weight, and Cellularity of Normal and Runt Pigs. *202*

Table 11.2 Carcass Composition of Runt and Normal Pigs at 96 kg Body Weight. *202*

Table 11.3 Fat Content (g/100g Body Weight) of Various Newborns. *203*

Table 11.4 Effect of Feed Intake on Muscle and Fat Growth in Pigs. *203*

Table 11.5 Comparison of Live Weights and Live Weight Gains of Cattle Fed a Low Plane of Nutrition Followed by a High Plane of Nutrition versus Those Fed a High Plane of Nutrition Throughout the Study. *204*

Table 11.6 Effects of Dietary Energy on Growth and Composition of Twin Cattle. *206*

Table 11.7 Effects of Plane of Nutrition (From Figure 11.9) on Carcass Composition of Pigs at 90 kg Body Weight. *208*

Table 11.8 Carcass Composition of Male Broilers Fed Finishing Diets Varying in Protein and Amino Acid Composition. *208*

Table 11.9 Effect of Dietary Protein (Energy Held Constant) on Broiler Composition at 49 Days of Age. *209*

Table 11.10 Effect in Pigs of Dietary Protein. *209*

Table 11.11 Amino Acid Supplementation of Low Protein Diets for Pigs. *211*

Table 11.12 Total Body Protein Turnover in Pigs Fed Low and High Carbohydrate Diets. *213*

Table 11.13 Differential Effects of Protein Quality and Quantity on Protein Balance and Turnover in Growing Pigs. *213*

Chapter 12

Table 12.1 United Stated Department of Agriculture Grades for Meat. *217*

Table 12.2 Summary of Specifications of Quality for Individual Carcasses of Ready-To-Eat Poultry. *224*

Table 12.3 Expected Retail Yield for Beef and Lamb Yield Grades and Expected Percent Lean Cuts for Pork Carcass Grades. *225*

Table 12.4 Relationship among Different Porcine Muscle Fiber Types and Resulting Meat Quality Attributes. *237*

Table 12.5 Correlation between Muscle Fiber Type and Meat Quality Characteristics. *237*

Table 12.6 Meat Palatability and Muscle Properties of Cattle Grown at Different Rates. *238*

Table 12.7 Meat Quality Parameters of Animals Treated with Somatotropin and Beta-Adrenergic Agonists. *239*

Table 12.8 Effect of Implants on Quality Grade. *239*

Table 12.9 Effect of Type of Implant on Meat Tenderness (Warner-Bratzler Shear Force). *240*

Table 12.10 Consumer Sensory Perception of Meat from Implanted and Non-Implanted Cattle. *240*

PREFACE

Animal growth and development is not considered a traditional discipline in the animal sciences. Rather, it is an emerging area of interest to many scientists across a broad range of mature disciplines. There is great interest in understanding the processes by which animals grow and develop because of the ramifications on animal productivity. Although several textbooks and numerous scientific journal articles are devoted to this subject matter, those resources are generally more suited for individuals seeking to further advance the science of developmental biology rather than for undergraduate students desiring to learn the basic principles and concepts of meat-animal growth and development. Class materials used in many undergraduate animal growth courses must be gleaned from the literature, a task that may be difficult due to the vast amount of information. This has led us to develop an undergraduate textbook that condenses and compiles the principles and concepts of animal growth and development.

For over forty years, an undergraduate course in Animal, Growth, Development and Evaluation (ANSC 301) has been taught at Purdue University in the Department of Animal Sciences. During this time, several different professors, actively studying animal growth, have compiled and updated an extensive amount of information on this subject. We have attempted to capture much of this information in this textbook with a focus on the major meat-animal species. In preparing the text, we have assumed that students already will have completed introductory courses in biology, chemistry, physiology, and nutrition. A glossary also has been included to assist students with the terminology used in describing the processes involved.

Although we do not claim that all the information contained in this textbook is without error, we have attempted to restrict the contents of this book to principles that have been well-accepted from a scientific standpoint. As with most biology texts, however, it is possible that some of the information contained herein may be challenged, and even determined to be incorrect, in the future. At any rate, it was our intent to focus only on well-established principles that undergraduate students could continue to build on as they make future contributions to the animal industry.

The utility of this material is best delivered in conjunction with an effective laboratory. The Purdue course has included live animal and carcass evaluation exercises which have been useful in illustrating many of the principles discussed in Chapter 9 (Whole Body Animal Growth) and Chapter 12 (Relationships among Growth Composition and Meat Quality). We also have included experiential studies in the laboratory sections to illustrate the effects of compensatory growth, hormone implants, and other metabolism modifiers on growth and composition of mice, swine, cattle, and poultry. These exercises allow the effective integration of different disciplines related to animal growth and offer students experiences that are not possible in a course delivered solely as lectures.

A significant number of individuals have contributed to the data illustrated in this textbook and it is impossible to list the contributions of each. Two individuals, however, deserve a great deal of recognition for providing us with a base set of materials. Elton D. Aberle (Dean of the College of Agriculture and Life Sciences, University of Wisconsin–Madison) and Max

D. Judge (Professor Emeritus, Purdue University) first developed and organized this material far ahead of our arrival at Purdue University. Their hard work and commitment to academic excellence is greatly appreciated and recognized by many. Special thanks are extended to the latter, a former colleague and mentor, for his thoughtful editing of the final manuscript. We also thank the graduate students that continued their research patiently while we were preoccupied with this assignment. Finally, we thank our families for their understanding and encouragement during this monumental undertaking.

David E. Gerrard
Alan L. Grant

ABOUT THE AUTHORS

David E. Gerrard is a Professor of Animal Sciences at Purdue University. He is a native of Indiana and received his education at Purdue University. His teaching responsibilities include an undergraduate growth and development course and a graduate course in muscle biology. Research activities focus on the development and maturation of muscle fibers during pre- and postnatal phases of growth. Furthermore, he conducts research in the area of postmortem metabolism and its role in modulating meat quality. He has been awarded the distinction of University Faculty Scholar at Purdue University and has received research awards from the American Meat Science Association and American Society of Animal Science.

Alan L. Grant is a Professor and Head of Department of Animal Sciences at Purdue University. He is a native of northern New York and received his degrees from Cornell University and Michigan State University. He teaches at the undergraduate and graduate levels. He has conducted research and published in the areas of animal nutrition and muscle biology with major emphasis on cellular processes regulating muscle development. He has received awards for his accomplishments and was named a University Faculty Scholar at Purdue University in recognition of his outstanding academic distinction.

Chapter 1

An Introduction to Animal Growth and Development

Introduction

First and foremost, the intent of this chapter is to introduce the area of animal growth and development and to justify the need for understanding these processes by those interested in careers related to the animal industry. In addition, we will provide some historical background and perspectives regarding advances that have been gained through application of this crucial knowledge. Furthermore, we will introduce early pioneers in the area and describe some early work that forms the basis for continued studies in this area.

What Is Animal Growth and Development?

Growth and development is fundamental to every living creature. As humans, we too, have an intimate understanding and perspective of these processes because we have experienced them first hand. As infants, our parents nurtured us as they provided for our every need. As we grew and developed, we became functionally self-reliant and became larger. Most dictionary definitions of growth use terms such as: "expansion," "increase" and "production." *Growth,* as it pertains to this subject matter, is defined as the general and normal expansion of size as produced by the accretion of tissues similar in composition to that of the original tissue or organ. In the living animal, the process of growth is accomplished by one or more of the following cellular processes: *hypertrophy, hyperplasia* or *accretion.* Hypertrophy is the process by which cells increase in size. Alternatively, growth can occur through the addition of more cells, i.e. hyperplasia. And finally, cells can contribute to the overall growth of an organism by accreting material between cells. Of particular importance to those studying growth is the difference between *true growth,* which results from an increase in the amount of muscle and bone, and *fattening,* which is the accumulation of fat.

Whole body growth does not occur independently. It is accompanied by individual cellular changes allowing tissues and organs to assume different roles and functions as growth proceeds. This age-related adaptation allows and facilitates more complex functions. The term used to describe these changes collectively is *development.* Therefore, development, as used in conjunction with growth, is the gradual progression from a lower to a higher state of complexity.

Why Study Animal Growth and Development?

One of the most compelling reasons for studying animal growth and development is to gain a better understanding of the processes related to animal growth on the part of those interested in careers in the animal industry or allied industries. Subtle or substantial changes in growth and development of animals can directly impact animal productivity. One of the most useful measures of animal performance is *average daily gain,* or the daily increase in weight. This performance trait is a direct reflection of the animal's ability to grow. Increasing the rate of these processes means that producers can raise more livestock in a given amount of time, which usually increases production efficiency. However, the composition of this gain, fat versus lean, can have a dramatic effect on production efficiency.

Study closely Figures 1.1, 1.2 and 1.3 and Tables 1.1, 1.2, 1.3, 1.4 and 1.5. Notice the striking changes that have occurred in the type of livestock reared over the past 50 years. These changes reflect the culmination of improvements made by many facets of the animal industry. Even though these figures provide quite a contrast between the type of animals used in the industry 50 years ago and those currently populating most production systems today, the real differences lie in performance data (Tables 1.1, 1.2, 1.3, 1.4 and 1.5). In particular, the amount of feed required to produce a given unit of animal mass has decreased dramatically over the past several decades. Furthermore, *growth rate,* or the speed at which animals grow, has improved greatly and in some cases, it has nearly doubled. As a result, the amount of time required to produce animals has decreased significantly, and therefore,

TABLE 1.1 MARKET HOG PERFORMANCE

	1950	Present
Market weight (pounds)	200–210	250–280
Average daily gain (pounds)	1.3–1.6	1.7–2.1
Feed efficiency (pounds feed/pound gain)	3.7–4.2	2.6–3.3
Fat thickness (in.)	1.6–1.8	0.5–0.9
Loin muscle area (sq. in.)	3.3–3.7	5.5–9.0
Days to 230	200–210	140–170

Figure 1.1 Photographs of market steers illustrating the vast differences in body type between 1950 (top) and present (bottom). Courtesy of American Hereford Association.

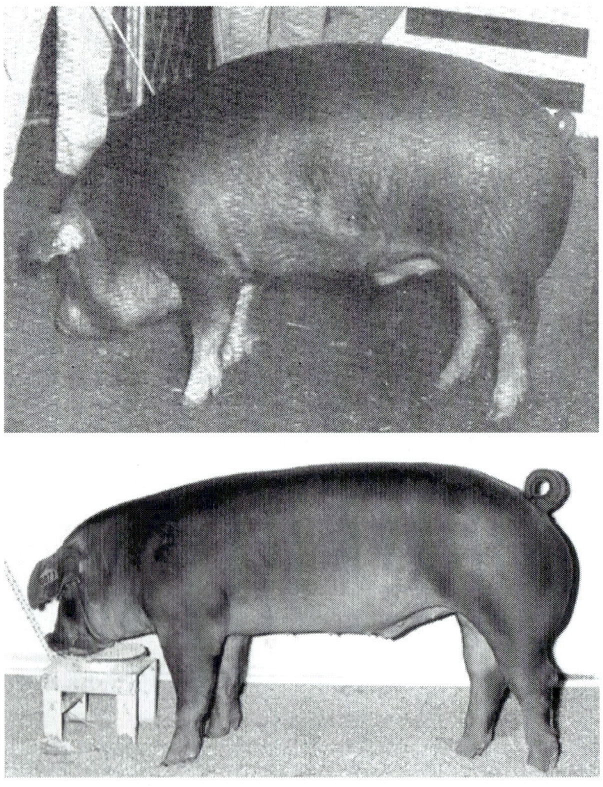

Figure 1.2 Photographs of market barrows illustrating the vast differences in body type between 1950 (top) and present (bottom). Courtesy of National Swine Registry.

TABLE 1.2 MARKET STEER PERFORMANCE

	1950	Present
Market weight (pounds)	850–1100	1100–1300
Average daily gain (pounds)	1.8–2.5	2.5–3.5
Feed efficiency (pounds feed/pound gain)	8.5–10.5	6.0–8.5
Fat thickness (in.)	0.6–1.0	0.3–0.6
Loin muscle area (sq. in.)	8–10	11–14
Time to market (months)	24–36	14–24

TABLE 1.3 MARKET LAMB PERFORMANCE

	1950	Present
Market weight (pounds)	85–90	120–140
Average daily gain (pounds)	0.25–0.40	0.60–0.75
Feed efficiency (pounds feed/pound gain)	8.0–9.5	5.5–6.5
Fat thickness (in.)	0.30–0.50	0.15–0.30
Loin muscle area (sq. in.)	1.8–2.4	2.0–2.7
Days to market	200–210	150–180

TABLE 1.4 TOM TURKEY PERFORMANCE

	1966	Present
Market weight (pounds)	35	35
Average daily gain (pounds)	0.16	0.26
Feed efficiency (pounds feed/pound gain)	3.5–3.0	2.8–2.5
Days to market	220	136

improved overall production efficiency for the industry. From a carcass standpoint, without exception, current market animals possess less fat than those produced earlier. Although much of this improvement is a direct result of selection programs that favor improved lean growth, the reduced demand for fat in the diet by consumers has likely helped the industry move away from extremely fat animals.

Given the dynamics and changes in profitability of the animal industry in recent years, implementation of any technology that may improve animal performance can mean the difference between

TABLE 1.5 BROILER PERFORMANCE

	1955	Present
Market weight (pounds)	3.0–3.5	5.0–6.0
Average daily gain (pounds)	0.05	0.12
Feed efficiency (pounds feed/pound gain)	3.0	1.9
Days to market	70	45

Figure 1.3 Photographs of market wethers illustrating the vast differences in body type between 1950 (top) and present (bottom). Courtesy of the American Hampshire Sheep Association.

operational success and failure. Of particular importance to those interested in animal agriculture is the advent of greater and greater implementation of sophisticated biotechnological advances. For example, only a few years ago, many swine producers relied heavily on seedstock producers for improvement of their stock. At present, however, seedstock is provided by huge genetic companies that have invested great amounts of resources into developing genetic lines with extensively documented performance and are poised to take control of this segment of the industry. The use of sophisticated selection strategies coupled with the latest developments in biotechnology have led to development of more uniform breeding stock, greater uniformity of animals produced, and greater product uniformity for consumers. Such agriculturally-based companies provide many individuals with challenging and rewarding careers. Therefore, it is important for students interested in a career of this nature to have a good understanding of animal growth and development.

Other technologies emerge routinely that are quickly adopted by industry. New compounds that modify animal growth rate and feed efficiency are researched and produced by many agricultural companies. A recent technology adopted by the industry is the use of Paylean®, a compound added to swine diets to improve production efficiency. This compound has stimulating effects on growth of some animal tissues and blunts growth of other tissues. Utilization of this compound in production systems, however, must occur simultaneously with changes in other management practices. Those pigs fed Paylean® require changes in the type and amount of dietary protein provided to maximize benefits. Understanding the basis for this improved performance is, therefore, important for those involved in production agriculture as well as those desiring to work in an industry that supports production agriculture.

Students interested in careers in animal agriculture must understand animal growth processes and they must also have an appreciation for the science underpinning those processes. Future advances depend on discovery of knowledge and investigation of strategies to improve production efficiency. Students that are sufficiently armed with knowledge of growth processes will likely continue to push the limits of these biological systems.

Over many centuries, the meat industry has evolved into a major agricultural segment. Meat is a highly nutritious, coveted food item that contributes substantially to prosperity of a population. Meat is a rich source of high-quality protein, iron and essential B-vitamins, and serves as the major source of dietary protein in industrialized nations. As a result, per capita consumption of meat tends to be the greatest in countries enjoying great economic success. Even though all edible tissue arising from animal origin is classified as meat, muscle tissue represents the majority of those tissues and, therefore, is intimately tied to animal growth and development.

Animals reared for the sole purpose of meat production represent a significant segment of total agriculture production. Taking into account the resources used to produce feed for animal production, the effect of animal production on the proportion of agriculture devoted to meat production is impressive. In the US alone, meat animals represent 50% of total farm receipts and nearly 40% of farm income. Therefore, the economic contribution of meat animal production significantly impacts a major segment of the national economy.

Historical Contributions

It is virtually impossible to credit any one individual or group of individuals for being the first to study animal growth and development. Scientific interest could have developed centuries ago in some remote location where a shepherd first recognized that male offspring grow faster and to larger mature sizes than do females. Moreover, it was well accepted that males yield a greater amount of meat than their female counterparts. Therefore, in this regard, one could argue that interest in animal growth and development has existed since the beginning of time.

In recent history, the pioneering work of Sir John Hammond is considered to be one of the earliest organized efforts to understand animal growth and development. Sir John Hammond was an esteemed animal scientist in England during the early to mid 1900's and conducted some of the earliest documented research showing how animals grow and develop under different environmental and nutritional conditions. Even though many of the studies designed and directed by Hammond were targeted at understanding the basic physiology of domestic animals, he was one of the earliest pioneers in the area of growth biology. In particular, his studies documented that growth of animals produces tissues that accumulate in a sigmoidal fashion. Furthermore, the early growth biologists showed that different tissues and organs of animals grow at different rates and mature at different times.

Subsequently, with help from C. P. McMeeken, they showed that muscle, bone and fat develop at different times. These same researchers also showed that animal growth is related to the level of nutrient intake and that depression of nutrient intake has a differential effect on various tissues and organs. All of these changes may be visualized graphically. These researchers spent much of their academic careers explaining animal growth via changes in the typical growth curve. In more recent years, Rex Butterfield and Roy Berg added to this work by demonstrating growth and development patterns in tissue subsets. Contributions of these early investigators have been paramount and form the foundation to our understanding of how a variety of animals grow and develop during various stages of life.

Over the past half century, hundreds of scientists throughout the world have added to the foundations built by early pioneers. Scientific advances in nutrition, genetics, reproduction, and management have contributed greatly to our understanding of the growth process. In particular, many scientists have fine-tuned the early studies by documenting the contribution of each tissue to overall growth of the animal. As returns from these studies diminished, researchers turned their attention to more fundamental processes controlling whole body metabolism. In particular, advances in measuring hormonal profiles provided great insights into mechanisms by which the endocrine system modulates growth. Significant breakthroughs in this area have led to commercialization of *growth hormone* to promote milk production in dairy cattle. Endocrinologists have discovered a novel circulating factor known as *leptin,* which is capable of modulating whole body fat accretion. Researchers have moved toward studying molecular mechanisms controlling tissue growth. Of particular significance is the discovery of a group of gene products capable of turning many different cell types into muscle. Along those same lines, advances have been made in our ability to "track down" and identify those genes responsible for aberrant growth of muscle in genetically predisposed animals.

The molecular mechanism controlling *double muscling,* a genetic condition in cattle that results in greatly exaggerated muscle growth, has been elucidated recently. Scientists have identified the gene that controls expression of this phenotype. Harnessing the development of this and other novel genes in the future will likely result in exponential gains in animal production efficiency. However, the ramification of implementing such control into production schemes likely will require additional changes in the way animals are produced and marketed. Furthermore, as will be developed in the last chapter of this text, changes cannot occur without affecting final product quality. Therefore, it is not enough that growth biologists have appreciation for the physiology underlying growth, but also must have an appreciation for the kind of products emanating from such changes.

Understanding processes by which living tissue is transformed into highly valued food must occur if the ultimate benefactor, the consumer, of highly efficient animal production is to be served. The ultimate goal of the meat animal industry is to provide a wholesome and consistent product to the consumer. This aspect cannot be overlooked, nor can the consumer's impact on the overall health of the industry be underestimated. In particular, with the likely advent of *transgenic* animals, it is the responsibility of all to understand the benefits and limitations of this type of extensive and aggressive use of science to improve animal production. Even though this aspect of production agriculture will not be covered in detail, it is incumbent on the student to understand the targets of these types of developments so that an educated appraisal of any new developments may be formulated.

Chapter 2

Tissue Structure and Organization

INTRODUCTION

Cells are the smallest living structural units within the body. Contained within the cell membrane is a collection of proteins that enable the cell to remain viable. Each cell type of the body contains a set of properties that allow it to function in a specific manner. These cells do not, however, remain isolated in the body, but rather interact and sometimes aggregate with other cells to form *tissues*. Tissues, by definition, are groups of closely associated cells and cell products that act synergistically to perform specific functions. In most living organisms, there are four primary tissues: *muscle tissue, connective tissue, nervous tissue* and *epithelial tissue*. The primary goal of this chapter is to generate a fundamental understanding of the structure and organization of tissues and their corresponding cell types. Information contained within this chapter should provide the basis for more in depth discussions regarding the growth of economically important tissues in growing food animals.

Epithelium

The *epithelium* is a thin layer of tissue that essentially covers all free surfaces of the body. Its principle function is to act as a barrier between the underlying tissue it covers and its environment. The environment may be the external environment or the internal cavities of the body. Because of its primary function, cells of the epithelium are positioned tightly together and possess minimal extracellular substance. Epithelium covers virtually every exposed surface of the body including the epithelial lining of the oral and nasal cavities and the linings of the digestive and respiratory tracts.

In addition to protecting exposed surfaces, the epithelium controls nutrient flow into underlying tissues. This function is best illustrated by the epithelium covering the innermost surfaces of capillaries known as the *endothelium.* As blood flows through capillary beds, valuable nutrients cross the epithelium (endothelium) and enter the adjacent tissues while cellular metabolites easily diffuse across the epithelium and re-enter the circulation for removal from the body. Movement of small molecules in and out of the capillaries occurs without loss of larger blood proteins that cannot permeate through the epithelium.

The epithelium also produces secretions via specialized gland cells contained within the tissue. These *secretory cells* contain an extensive endoplasmic reticulum that is responsible for packaging and transporting various products to the cell surface for release. The composition of these cell products can vary depending on the substance needed at the cell surface. Many substances are *proteinaceous* in nature, such as enzymes released by the epithelial lining of the digestive tract. Other substances may contain higher proportions of carbohydrates, and may act as lubricants to reduce the friction caused by substances traveling in close opposition to the epithelial surface. One example of such a secretion is *mucous.*

Lastly, the epithelium can aid in sensory communication to the nervous system. Intuitively, the epithelium changes not only with the underlying tissues but also with the environment that it opposes. As a result, nerves closely associated with the surface epithelium can be stimulated or activated when the epithelium is distorted or experiences subtle changes in its environment. The best examples of such specialized epithelia are those associated with taste and sight perceptions which contain specialized cells dispersed throughout the tissue that are especially sensitive to changes in chemicals or illumination in the microenvironment, respectively.

Epithelial tissue is classified according to the structure of the cells and its thickness (Figure 2.1). *Simple* epithelium, as the name implies, consists of a single layer of cells. The nuclei of these cells, when viewed microscopically, appear to line up in a straight line. This type of epithelium is located in places where maximal secretion or absorption is needed. Because of its delicate structure, this type of epithelium is found in well-protected areas, far removed from abrasive forces.

In contrast, *stratified* epithelium is often found where there is much friction with the outer environment. This type of hostile environment can be brought about by both chemical and mechanical perturbations. Because of the necessity to protect underlying tissues in such a harsh environment, stratified epithelium normally consists of many layers of cells, which are routinely shed when challenged by the environment. Skin and the extremities of the digestive tract are examples of such epithelium.

In addition to classification according to cell layers (Table 2.1), epithelium consists of different cell types. *Squamous* cells are very thin and flat in one dimension but rather broad in the other two dimensions. *Cuboidal* cells, as suggested by the name, possess a box-like structure, where all dimensions are relatively equal. In contrast, *columnar* cells are slender, elongated cells. The nucleus of these cells is positioned toward the base of the cell, or that aspect of the cell closest to the underlying *basement membrane,* a layer of fibers and filaments that serve to attach epithelium to under-lying connective tissues.

Combining the aforementioned thicknesses and the two types of cell structures possible, six distinctly different types of epithelia can be found in various places throughout the body (Figure 2.1 and Table 2.1). One epithelium that deserves special attention is *transitional* epithelium that is found in the bladder and other parts of the urinary tract. This type of epithelium resembles a stratified cuboidal epithelium, but changes dramatically with distention of the underlying tissue, in this case, filling of the bladder.

One additional characteristic that merits a brief discussion is a specialization of the epithethial cell membrane, namely *ciliation.* These extensions of

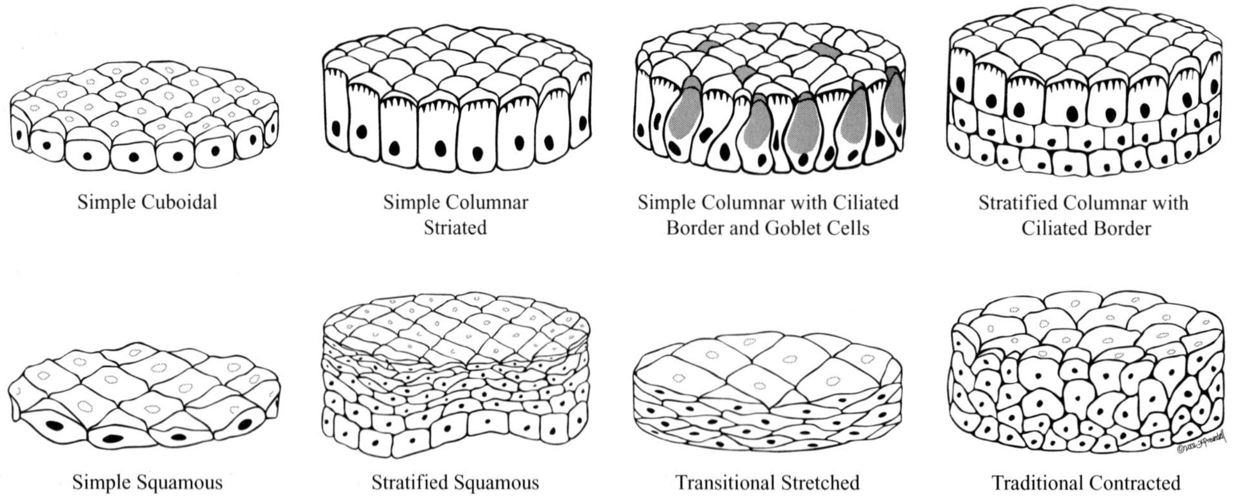

Figure 2.1 Diagram of various types of epithelium. Courtesy of Faye C. Prevedell.

TABLE 2.1 EXAMPLES OF THE TYPES OF EPITHELIUM

Type	Squamous	Cuboidal	Columnar
Simple	Endothelium of blood vessels and lymphatics Epithelium of lung Mesothelium of pleural, peritoneal, and pericardial cavities	Some kidney tubules Thyroid gland Sweat gland	Digestive tract
Pseudostratified	None	None	Trachea Epididymis
Stratified	Epidermis Oral mucosa Vagina Anal mucosa	Sweat gland duct (rare)	Trachea Epiglottis (rare)
Transitional		Bladder Ureter	

the cell membrane provide the cell with two important functions, i.e., improving absorptive and secretory abilities, and aiding in movement of materials along their border. Cells containing *microvilli* have over 20 times more surface area than those lacking these specialized structures. As a result, the ability of the epithelium to absorb nutrients is greatly improved. In some cases, microvilli movement can be orchestrated, which allows for movement across the surface of the epithelium. This characteristic is especially important for the epithelium lining the respiratory tract, as it helps aide removal of particulate matter from the airways.

Connective Tissue

Intuitively, one might regard connective tissue as the structural framework in all organisms on which all other systems are placed. However, connective tissue varies greatly in its physical and functional characteristics. In contrast to the aforementioned epithelium, where the bulk of the tissue consists of cells, the hallmark of connective tissue is the existence of a rich, extensive extracellular matrix that contains relatively few active cells. The underlying structure of all connective tissues is conveyed in the illustration shown in Figure 2.2.

Connective tissue proper is the extracellular substance, which ranges from a semi-solid, glue like substance to a rigid, amorphous material. Within the connective tissue proper, various types of proteinaceous fibers are imbedded in a viscous solution called *ground substance.* Dispersed throughout the connective tissues are *fixed* and *wandering* cells. Therefore, types of connective tissue differ in the relative contribution of each of the aforementioned components. Blood, for example, contains a very fluid ground substance with little to no extracellular fibers. In addition, there is a very specific type of cells associated with this type of connective tissue even though it can vary dramatically under normal physiological conditions. Conversely, cartilage and bone have immensely more extracellular fibers that are densely packed in an amorphous, viscous gel that is capable of withstanding many forces that would normally reconfigure most tissues. In an effort to provide further strength, some connective tissues, the fibrous matrix of bone, for example, deposit various salts to form a calcified, rigid complex. Connective tissues play an important role in the growth of an animal as well as impacting the ultimate quality of the meat produced by numerous species.

Connective Tissue Proper

As mentioned previously, connective tissue proper contains an amorphous or shapeless structure that contains various fibers and numerous cells. Even though this substance may seem rather impenetrable and inert, many cells routinely travel through this

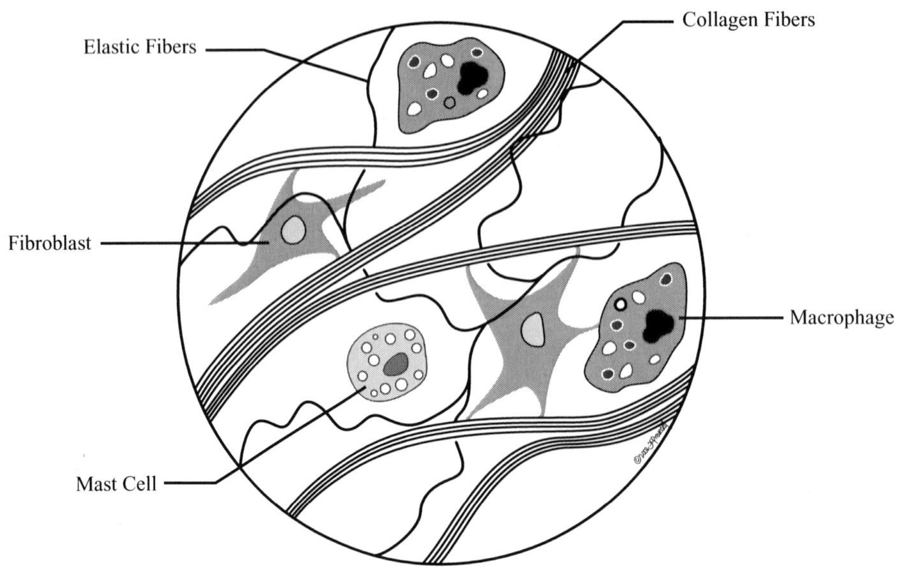

Figure 2.2 Organization of loose connective tissue. Courtesy of Faye C. Prevedall.

substance and on occasion, reside permanently in this locale.

Cell Types

Macrophages are small *ameboid* cells, which move around within the tissue consuming damaged cells and foreign substances that may have gained access to the body. These cells are particularly abundant at sites of injury. *Mast cells* also move freely within connective tissues and are primarily responsible for release of *histamines* that are instrumental in conveying the inflammation response. Because these cells literally move about the body seeking foreign substances, they often are referred to as wandering cells. In contrast, the predominant fixed cells, more restricted in their movements, are *fibroblasts.* These cells are somewhat anchored in the connective tissue matrix and are responsible primarily for production of the fibrous proteins that reinforce the amorphous structure (Figure 2.3). The amount, type and arrangement of these fibers in the extra-cellular matrices are used to classify the various types of connective tissue. *Collagenous, reticular* and *elastic* fibers are the primary reinforcement proteins in connective tissue.

Ground Substance

The fluid-like structure of the connective tissue proper consists of soluble *proteoglycans* (proteins linked to carbohydrates) and is referred to as *ground substance.* The most prominent carbohydrates attached to these proteins are *chondroitin sulfates* and *hyaluronic acid.* The chondroitin sulfates are abundant in cartilage and tendons while hyaluronic acid is quite viscous and is more important in supplying lubrication in the form of *synovial fluid* to most *articulations* (joints) of the skeletal system. The extent of polymerization of hyaluronic acid is important in determining its viscosity. Depolymerization reduces viscosity and results in increased friction in articulations.

Connective tissue fibers

Collagen Fibers

Collagen is the most abundant protein found in animals representing 20 to 30% of the protein in the body. Organization of the collagen molecule is shown in Figure 2.4. The polypeptide or primary sequence of collagen is known as the *alpha chain,* which consists of 1014 to 1023 amino acid residues, depending on the type of collagen. The amino acid composition of collagen is quite unique and contains multiple repeating sequences (X-Y-glycine), where X and Y represent proline and hydroxyproline about two-thirds of the time. The side chain of these amino acids contains pyrole rings, which

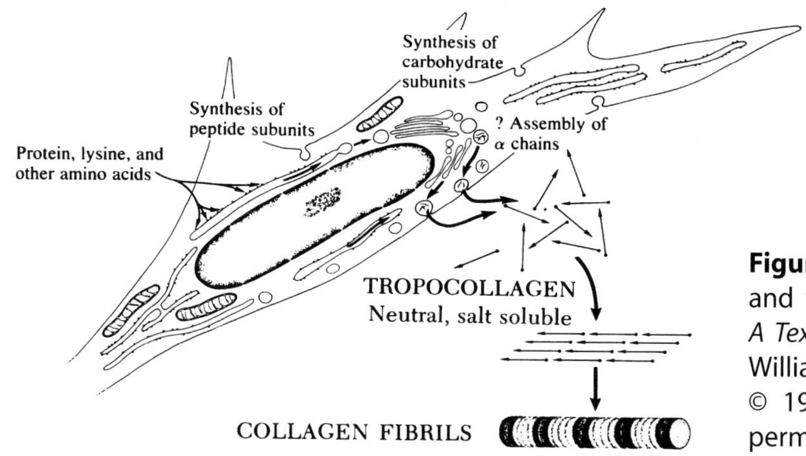

Figure 2.3 Synthesis of tropocollagen and formation of a collagen fibril. From *A Textbook of Histology, Ninth Edition* by William Bloom & Don Fawcett. Copyright © 1968 by Don Fawcett. Reprinted by permssion.

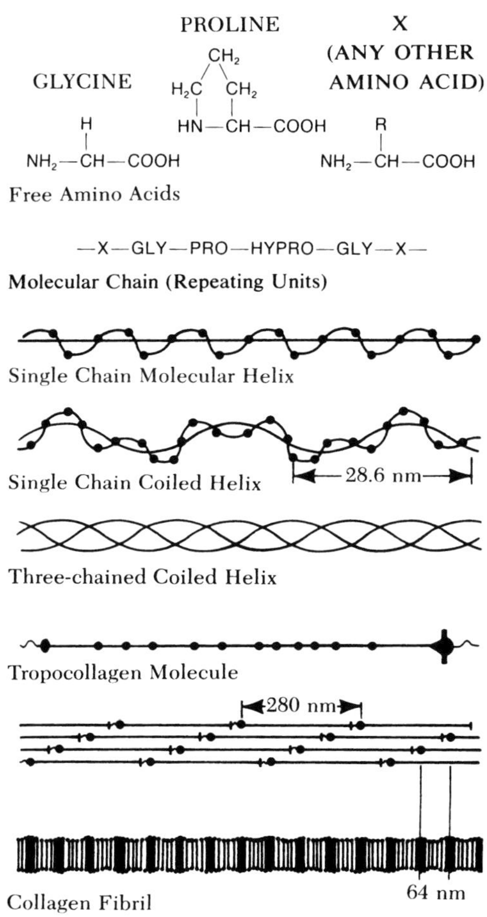

Figure 2.4 Amino acid sequence molecular structure of collagen and tropocollagen molecules, and of collagen fibril formation. Recent advances in collagen biochemistry. *Proc 36th Ann. Recip. Meat Conf.* p. 125–130.

cause the entire polypeptide to coil into a left-handed helix. The presence of the glycine is critical because glycine has a very small side chain, which fits nicely in the center of the helix. The entire molecule is not helical, however, and contains portions that are non-helical, many at the ends of the molecule. The helical portions of the molecule help maintain the stability of the molecule, while the non-helical portions help align the molecule with other alpha chains. Immediately after translation (assembly of amino acids from the mRNA signaling) of the alpha chains (Figure 2.5), disulfide bonds form in the non-helical portion of the molecule and the three alpha helices form a right-handed super helix that is about 1.4 nm in diameter and roughly 300 nm in length. At this point, the collagen precurser, *procollagen* is secreted from the fibroblast. Further processing to remove non-helical portions of the molecule occur via specialized *peptidases* (Figure 2.5).

The ultimate structure of collagen is known as the *pentafibrils,* where collagen fibrils are arranged in a staggered formation that yields highly organized fibers. A micrograph showing the native structure of these fibrils is shown in Figure 2.6. These fibers measure about 75 μm in length. This process of fiber formation occurs outside the cell. At this point, the structure of collagen appears complete, however, the strength of collagen at this point is rather low. The increase in tensile strength associated with collagen comes from cross-linkages between collagen molecules.

The importance of collagen crosslinking is best illustrated by a condition known as *lathyrism,* which

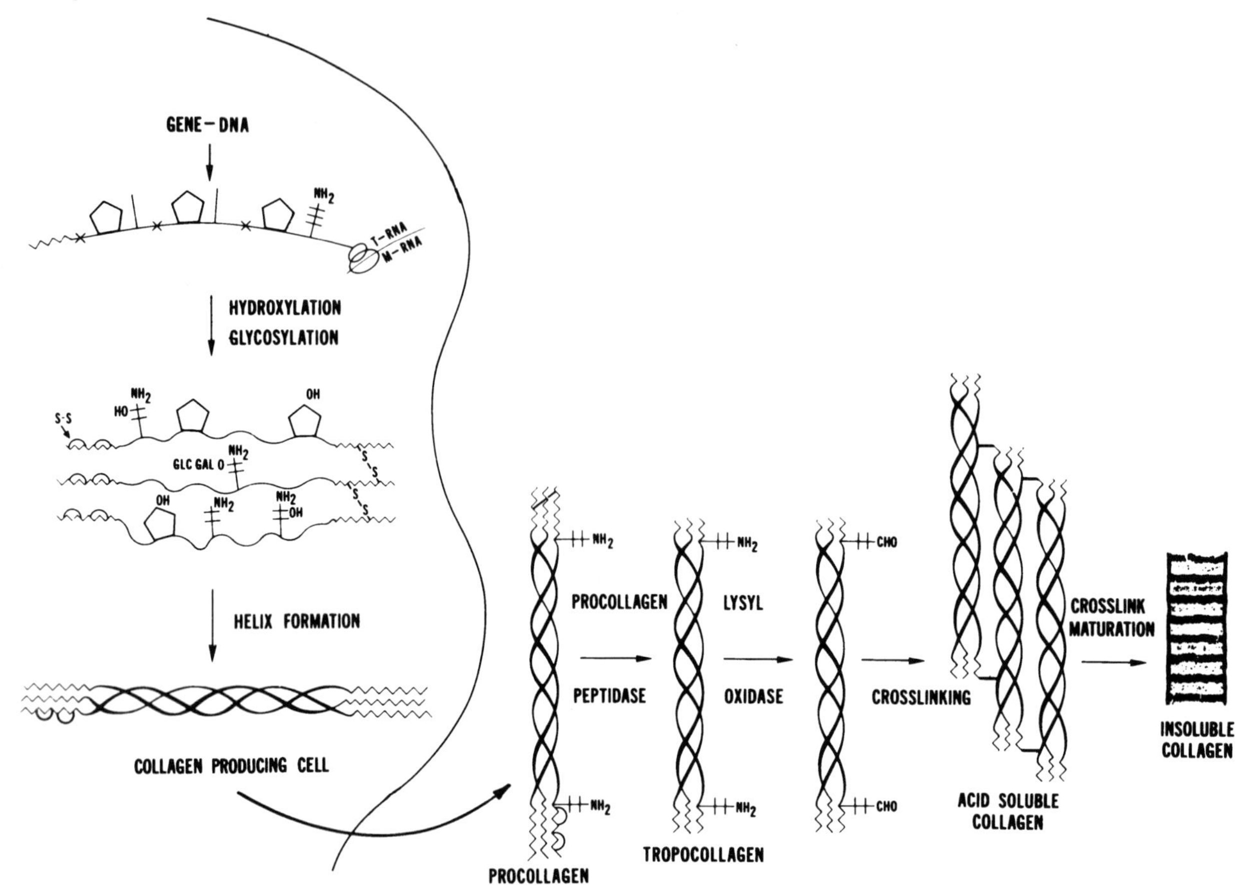

Figure 2.5 Collagen synthesis and maturation. From A. J. Bailey and N. D. Light, Connective Tissue in Meat and Meat Products, Elsevier Applied Science, London, 1989.

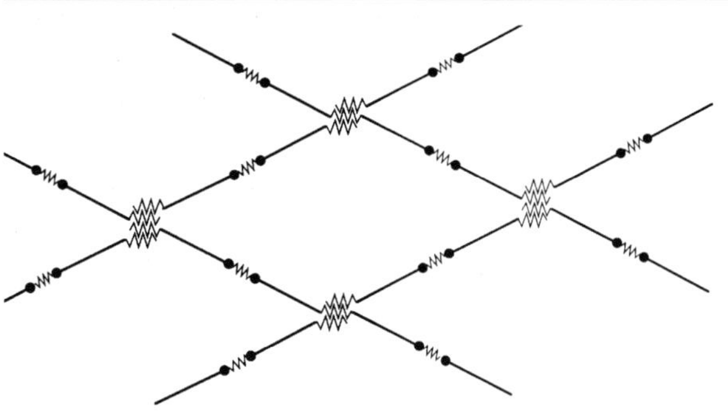

Figure 2.6 Diagrammatic illustration of the chicken wire-like network of Type IV collagen. From *Principles of Meat Science, Fourth Edition* by Aberle et al. Copyright © 2001 by Kendall/Hunt Publishing Company. Reprinted by permission.

results from feeding animals sweet peas containing various nitrile compounds. These nitriles affect the enzyme *lysyl oxidase,* which is partially responsible for the covalent binding or crosslinking of collagen fibrils. As a result, animals ingesting high levels of contaminating feedstuffs suffer from collagen-related maladies, such as spinal deformities and joint degeneration and ultimately succumb to aortic aneurysms or ruptures. The severity of this condition illustrates that the strength associated with connective tissue comes from crosslinks formed within collagen fibrils (Figure 2.5).

To date, there are several types of collagen known to exist in animal tissues (Table 2.2). The type of collagen designates which alpha chains comprise the triple helical collagen molecule. Subtle differences in amino acid composition among these alpha chains affect the structural strength and functionality of different types of col-

TABLE 2.2 COLLAGEN TYPES AND THEIR LOCATION IN ANIMAL TISSUES

Type	Molecular Length (nm)	Retained Terminal Globular Regions	Molecular Composition	Tissue Location
I	300	Processed	$\alpha 1(I)[\alpha 2(I)]_2$	Intramuscular, skin, tendon, bone, dentine
II	300	Processed	$\alpha 1(II)_3$	Cartilage, disc, vitreous humour
III	300	Processed	$\alpha 1(III)_3$	Intramuscular, skin, vascular, intestine
IV	420	C-terminal	$\alpha 1(IV)[\alpha 2(IV)]_2$	Basement membranes
V	300	N-terminal	Various combinations of $\alpha 1(V)[\alpha 2(V)]_2$ and $\alpha 3(V)$	Intramuscular, skin, embryonic tissues
VI	105	N- and C-terminal	$\alpha 1(VI)\alpha 2(VI)\alpha 3(VI)$	Vascular system
VII	450	Not known	Not known	Skin, amniotic membrane
VIII	Not known	Not known	Not known	Aortic endothelium
IX	200	N-terminal	$\alpha 1(IX)\alpha 2(IX)\alpha 3(IX)$	Cartilage
X	150	N- and C-terminal	$\alpha 1(X)_3$	Cartilage
XI	300	Not known	$\alpha 1(XI)\alpha 2(XI)\alpha 3(XI)$	Cartilage

From A.J. Bailey and N.D. Light, Connective Tissue in Meat and Meat Products, Elsevier Applied Science, London, 1989.

lagens in different tissues. Type I collagen is the most abundant connective tissue in the body, comprising 80% of the total collagen in skin and located extensively in muscle tissue. Type I collagen, for example, consists of one alpha 1 chain and two alpha 2 chains and forms large fibrils. In contrast, collagen associated with cartilage consists of three identical alpha 1 chains and forms rather thin, delicate fibers. Yet another collagen, Type IV, is found in the basement membrane and is nonfibrous. This collagen is rather unique because if forms a chicken wire-like structure (Figures 2.6). This type of structure is particularly useful because it provides additional support to tissues which are prone to various physiological distortions.

Reticular Fibers

Reticular fibers are formed in essentially the same manner as collagen fibers; however reticular fibers do not form large bundles of fibers. In fact, reticular fibers are quite small and delicate measuring only 35 nm in diameter. In those connective tissues with high reticulin content, there is a high proportion of ground substance in the connective tissue. Abundant ground substance coupled with reticular fibers arranged randomly throughout the tissue results in a somewhat loose type of connective tissue such as that found in the delicate connective tissue or *parenchyma* surrounding many organs.

Elastin

Elastin differs greatly from collagen in its functional properties (Figure 2.7 and 2.8). This protein contains two unique amino acids, *desmosine* and *isodesmosine,* and possesses rather unique qualities. As stated previously, collagen fibers are relatively inelastic, whereas elastic fibers are capable of being reversibly stretched to nearly twice their length. In a similar manner to collagen, elastin is secreted from fibroblasts as a precursor molecule, *tropoelastin.* Elastin is particularly abundant in blood vessels, ligaments and other compliant tissue such as the lungs and skin (Table 2.3).

TYPES OF CONNECTIVE TISSUE

Loose Connective

Loose connective tissue (Figure 2.9), as the name implies, takes on a porous, highly flexible structure. This type of connective tissue provides structure to blood vessels and nerves. Loose connective tissue

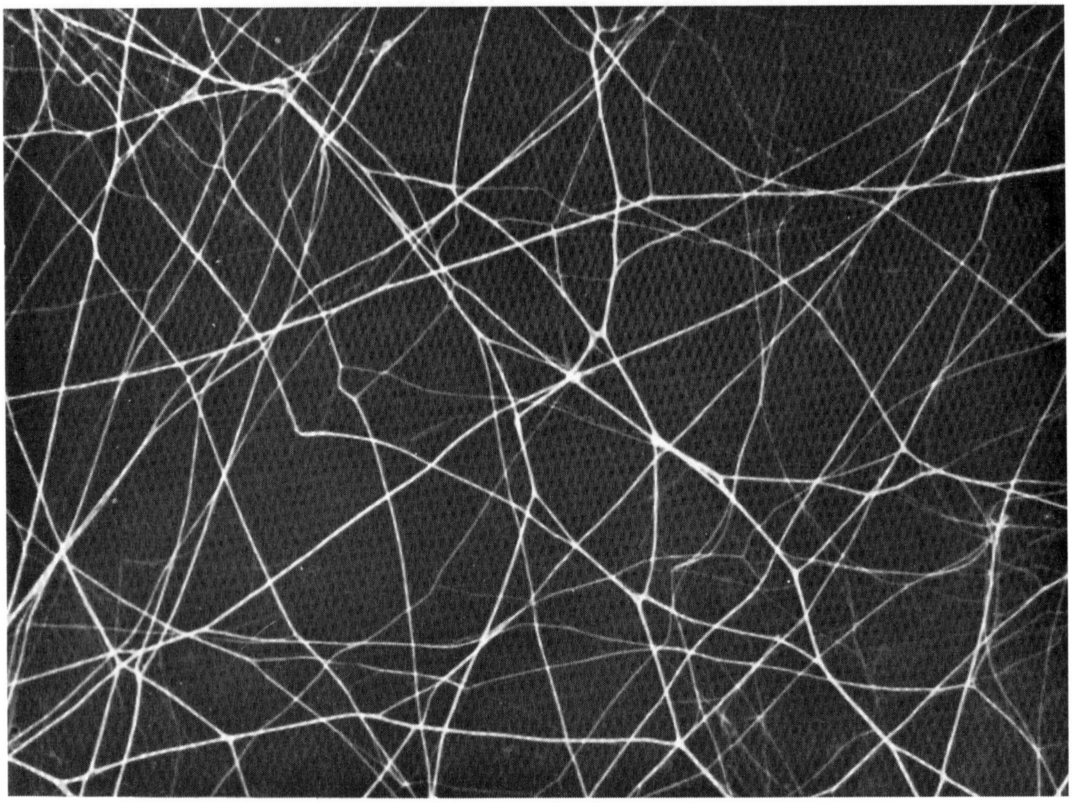

Figure 2.7 Micrograph of elastic fibers. From D. W. Fawcett, Histology, R. O. Greep, ed. McGraw-Hill, New York.

TABLE 2.3 CONNECTIVE TISSUE IN DIFFERENT TISSUES

Tissue	Collagen (%)	Elastin (%)
Tendon	~95	1
Skin	~50	5
Vascular system	~40	50
Bone	25	—
Lung	15	20
Kidney	4	—
Liver	2	—
Muscle	2	0.1

From A.J. Bailey and N.D. Light, Connective Tissue in Meat and Meat Products, Elsevier Applied Science, London, 1989.

contains relatively few collagen and elastic fibers, which are loosely associated providing maximal flexibility and little overall strength and rigidity. Loose connective tissue is highly vascularized and contains numerous cells.

Dense Connective Tissue

In contrast to loose connective tissue, dense connective tissue contains numerous fibers and provides maximal strength with little flexibility. There are two types of dense connective tissue, i.e., *dense regular* and *dense irregular* (Figure 2.10) connective tissues. Again, as the name implies, a high proportion of these tissues consist of collagen, elastin and reticular fibers. Dense regular connective tissue is best illustrated by visualizing the function of a tendon at the end of a muscle. Forces placed on this

Figure 2.8 Micrograph of elastic connective tissue. Courtesy of D. R. Swartz, Department of Anatomy and Cell Biology, Indiana University.

tendon are mainly restricted to one direction. As a result, fibers align with the long axis of the tendon. In this manner, maximal tensile strength can be developed by these fibers against a force. In contrast, connective tissues located in areas where forces are not exerted from one direction but rather from multiple directions, dense irregular connective is better suited to resist these forces. In such connective tissue, fibers constitute the bulk of the mass but are arranged in a random manner. In this way, regardless of the direction of tension on this tissue, some longitudinal fibers will be in opposition.

Specialized Connective Tissue

Adipose Tissue

One type of loose connective tissue that deserves special consideration in this text is *adipose tissue* (Figure 2.9). Adipose tissue is a specialized type of loose connective tissue that contains, in addition to the normally present fibroblasts, numerous *adipocytes* or fat cells (Figure 2.11), which are uniquely capable of storing triglycerides. Adipose tissue is highly vascularized and is often located in areas around critical organs (Figure 2.11, c). In the latter situation, this

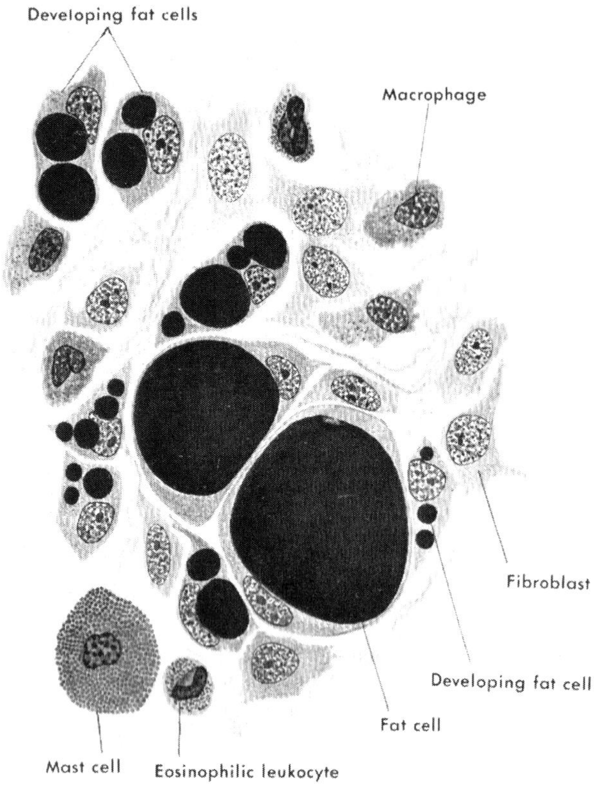

Figure 2.9 Loose connective tissue. From W. Bloom and D. W. Fawcett, A Textbook of Histology, 9th ed. W. B. Saunders Co. Philadelphia, 1968.

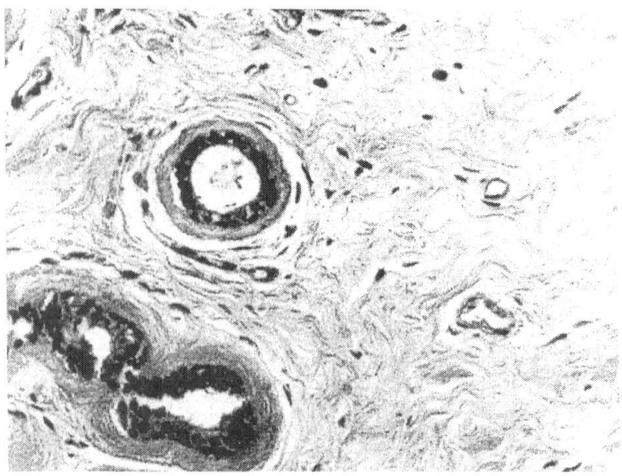

Figure 2.10 Micrograph of dense, irregular connective tissue. Courtesy of D. R. Swartz, Department of Anatomy and Cell Biology, Indiana University.

tissue is an excellent cushion because of its soft, pliable nature. Adipose tissue also serves as an energy storage system and insulates the body from excessive heat loss. Because adipose tissue plays an important role in the growth of animals, it will be covered in much greater detail in Chapter 7.

Blood and Lymph

Hematopoietic tissue, such as *blood* and *lymph,* is a specialized connective tissue. Of course, the cell populations in this form of connective tissue are distinctive and specialized. The fluid phase is viscous and contains a myriad of soluble proteins. Unlike other forms of the connective tissues, blood and lymph have very little filamentous proteins.

Red blood cells represent 50% of the total volume of blood in most animals and serve to exchange oxygen for carbon dioxide to peripheral tissues. The remainder of the blood volume consists of *plasma, white blood cells* and *platelets.* The population of white blood cells consists of *lymphocytes, macrophages* and *microphages,* which constitute the first line of defense during an immune challenge. Platelets, on the other hand, are small anuclear (without nuclei) cells that aid in the formation of clots during blood letting.

The extra-cellular fluid of blood consists of plasma, lymph and *interstitial fluid.* Plasma is mainly confined to the circulatory system and is constantly being transported throughout the body delivering vital nutrients to various tissues via small capillaries. Interstitial fluid forms lymph as it enters the lymphatic system, a vast system of ducts and passageways that deliver interstitial fluids back to the circulatory system.

Supportive Connective Tissue

Cartilage

Cartilage is a form of connective tissue that provides structural support for other tissues. The ground substance or matrix of cartilage is a rather rigid substance consisting of *hyaluronic acid, chondroitin 4-sulfate, chondroitin 6-sulfate, keratan sulfate,* and sparsely dispersed cartilage cells or *chondrocytes.* Chondrocytes (Figure 2.12, white arrows),

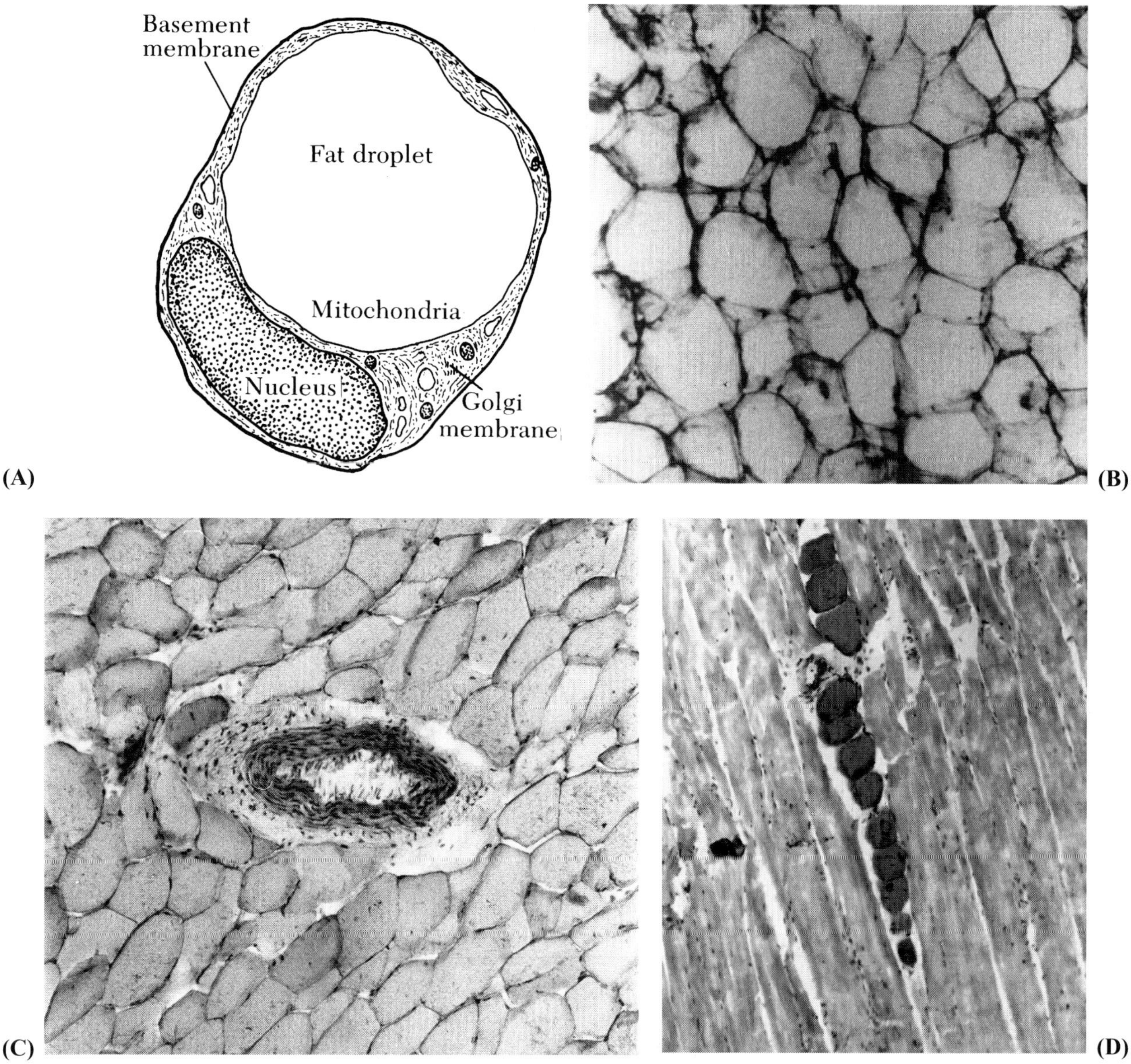

Figure 2.11 Diagram and photomicrographs of fat cells. (A) Diagram of a typical fat cell. (B) Photomicrograph of bovine subcutaneous fat cells. (C) Intermuscular bovine fat cells surrounding an artery. (D) Intramuscular bovine fat cells arranged longitudinally between muscle bundles. Courtesy of William G. Moody, University of Kentucky.

typically 2 to 8, are isolated in small spaces called *lacunae* (Figure 2.12, black arrows) throughout the matrix. A thin membrane covers all non-joint associated cartilage and is called the *perichondrium. Interstitial growth* or that growth occurring from within the tissue occurs when chondrocytes divide and form new lacunae. In contrast, *appositional growth* or that growth resulting from addition of new layers is mainly accomplished at the interface between the cartilage and perichondrium. Curiously, there are no blood vessels found within cartilage. Therefore, nutrients must diffuse through the matrix to nourish chondrocytes and, as a result, most cartilage is quite thin. There are several types of cartilage depending on the type and abundance of fibers located in the matrix.

Hyaline cartilage (Figure 2.13) consists of numerous collagen fibrils that are densely packed in the ground substance. Because collagen fibrils

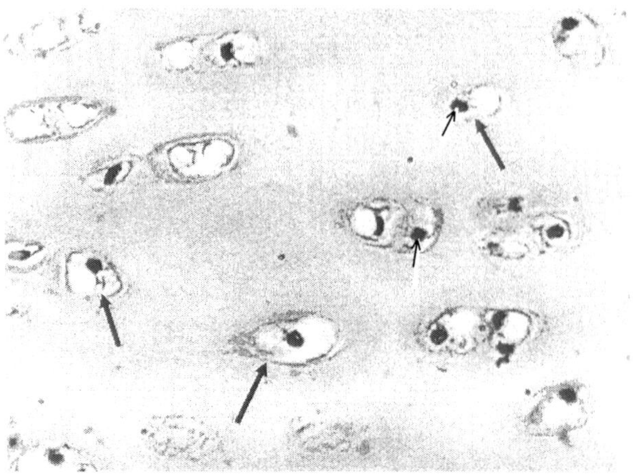

Figure 2.12 Micrograph of cartilage showing the location of chondrocytes (thin arrows) in lacunae (heavy arrows). Courtesy of D. R. Swartz, Department of Anatomy and Cell Biology, Indiana University.

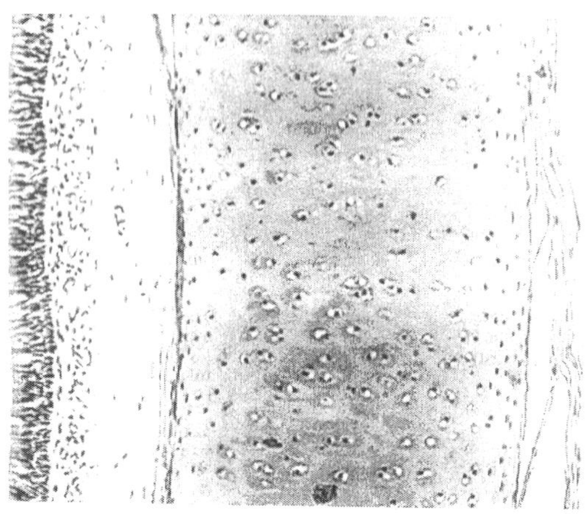

Figure 2.13 Micrograph of hyaline cartilage. Courtesy of D. R. Swartz, Department of Anatomy and Cell Biology, Indiana University.

account for nearly 40% of the total weight of hyaline cartilage, it is extremely tough and somewhat flexible. The functional characteristics make this type of cartilage extremely resistant to constant pressure and friction. Hyaline cartilage is a translucent, bluish-white substance that is most notably found on joint surfaces and acts to form the rigid ring structures of the trachea and other airways that require rigid, yet, somewhat flexible features.

Elastic cartilage as the name implies, contains numerous elastin fibers and some collagen fibers that allows for maximum flexibility with a far less rigid structure. Elastic cartilage is found in the external aspects of the ears, the Eustachian tubes and epiglottis.

Fibrocartilage (Figure 2.14) represents the extreme of hyaline cartilage and contains the bulk of its mass in collagen fibers with very little ground substance. As a result, this type of connective tissue is extremely tough. Because of its functional properties, fibrocartilage is located between the pelvic bones at the pubis, between the vertebrae of the spinal column and around selected joint capsules. One distinguishing feature of this type of cartilage is that the chondrocytes are arranged in rows.

Bone

Even though bone is a structural connective tissue as is cartilage, the functional properties of bone result in an extremely rigid structure that is quite inflexible. Because of this rigid characteristic, bone is somewhat brittle and will break if forces are exerted that surpass its capacity to resist a load. There are essentially three types of bones found in most animals:

- *Long Bones*
- *Irregular Bones*
- *Flat Bones.*

Long bones, in general, consist of enlarged ends, or structures known as *epiphyses* and a shaft region called the *diaphysis* (Figure 2.15). Located in the center of the diaphysis is a hollow void referred to as the *medullary cavity.* The medullary cavity is often filled with *bone marrow,* a substance rich in blood cell precursors, especially in young animals. In flat and irregular bones, these features are less conspicuous.

The two types of bone tissue are:

- *Compact Bone*
- *Spongy Bone* (Figure 2.16).

Compact bone, as the name implies, is very dense and is generally found at the periphery of most bones. Spongy or *cancellous* bone, on the other hand is compact bone that is arranged in small structures known as *spicules* and *trabeculae.*

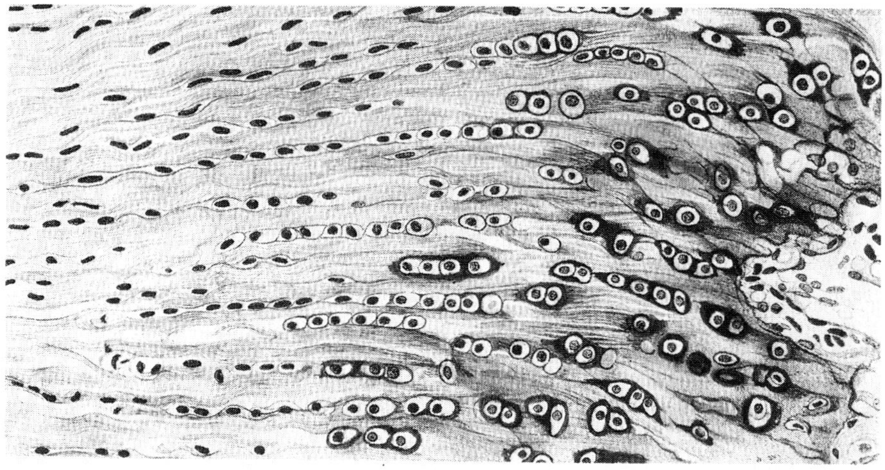

Figure 2.14 Micrograph of fibrocartilage at the insertion of tendon into bone. From A *Textbook of Histology, Ninth Edition* by William Bloom & Don Fawcett. Copyright © 1968 by Don Fawcett. Reprinted by permission.

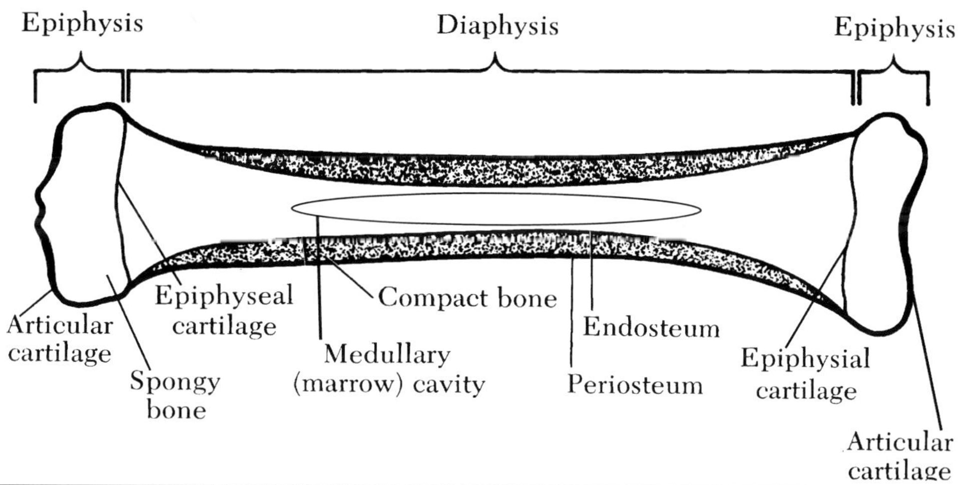

Figure 2.15 Longitudinal structure of the long bone. Courtesy of W. H. Freeman and Company.

Spicules are elongated, cylindrical structures, whereas trabeculae are long, flattened reinforcements located internal to compact bone. These structures are rather dynamic, meaning that they are continually changed or "turned-over" during the bone *remodeling* process. In fact, spicules and trabeculae are strategically located within the spongy bone matrix to resist stress. Bone remodeling occurs continually during the growth process and strengthens bones against constant and ever changing external forces that may be imposed.

Organization of bone tissue is similar to that of cartilage. Bone consists of cells that are isolated in a matrix of collagen fibers impregnated with vast amounts of calcium salts. This specialized connective tissue consists roughly of one-third collagen

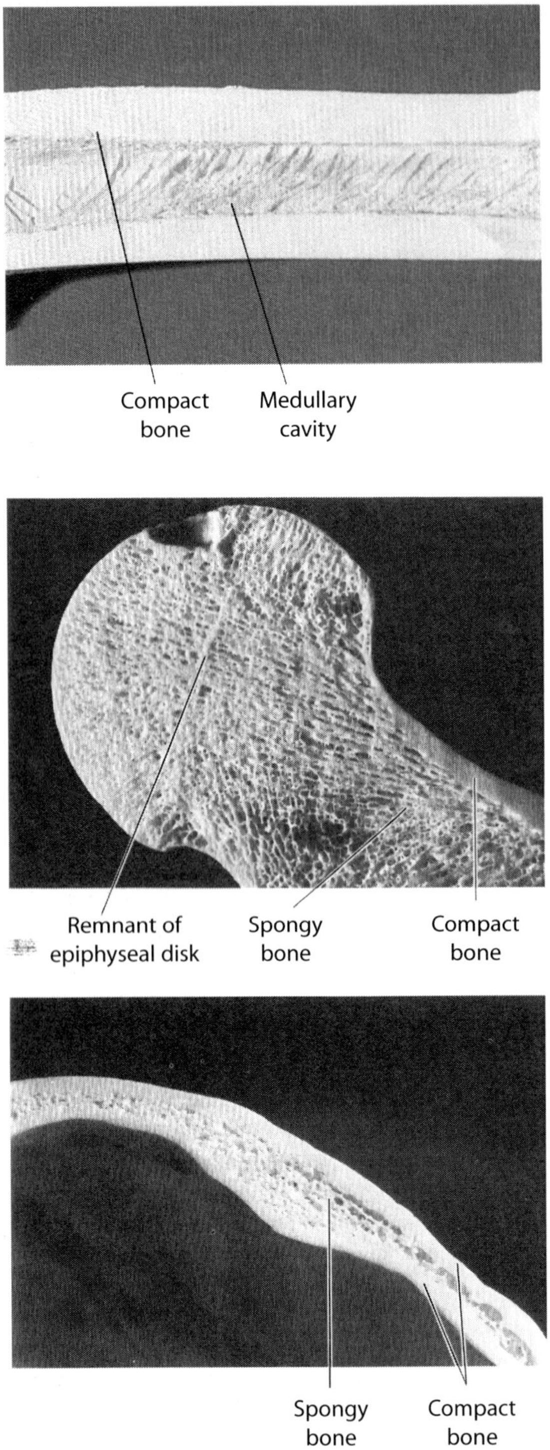

Figure 2.16 Photographs showing the different types of bone. From J. Hole, Human Anatomy and Physiology, 5th ed., Wm.C. Brown Publishers, 1990.

fibers and the remainder in calcium salts. The primary calcium salt in bone is *calcium phosphate* or *hydroxyapatite.* In addition, there are substantial amounts of *calcium carbonate* and other minerals located throughout the bone matrix. Bone cells or *osteocytes* reside in small lacunae and communicate with each other through an extensive system of canals within the bony matrix called *canaliculi.* The entire bone is covered by the *periosteum,* similar to the perichondrium of cartilage. This thin membranous structure consists of an outer connective tissue component and an inner layer of *osteoblasts, osteocyte precursers,* it acts as an avenue for blood vessels and nerves to enter the bone, and it is the primary site of appositional bone growth and bone repair. Lacunae are not merely randomly positioned throughout bone. Rather, lacunae are longitudinally and radially aligned between sheets or *lamellae* of bone. Lamellae range from 4 to 12 μm in thickness and literally "sandwich" osteocytes between each layer.

The basic functional unit in compact bone is the *osteon* or *Haversian system* (Figure 2.17). Osteons appear circular in structure when bone is viewed cross-sectionally. These structures are cylindrical in nature and their long axis is parallel to the long axis of the bone. The osteons are not randomly dispersed throughout compact bone but are highly organized to maximize their presence in the bone tissue. A venule and arteriole are located at the center of an osteon and provide circulatory exchange to cells imbedded in the structure. Located around these blood vessels are concentric circles of lamellae and osteocytes, which are locked within lacunae between the layers of bone. Because osteons are cylindrical in nature, there are regions where osteonic bone is lacking. In these areas, *interstitial lamellae* exist. Diffusion of nutrients to osteocytes occurs through the canaliculi that pass through the lamellae. Because interstitial lamellae are not associated with an osteon, per se, osteocytes located in this estranged structure derive nutrients via the closest osteon.

Interstitial growth in bone does not occur as described for cartilage. Bones grow in length at specialized structures known as *epiphyseal plates.* Epiphyseal plates are located at each end of long bones between the diaphysis and the epiphyses. These structures consist of cartilage and facilitate bone lengthening. This process will be discussed in more detail in Chapter 8. Lining the bone marrow cavity is another thin membrane known as the *endosteum.* Just beneath this membrane is a very active site of bone degradation performed by very large, multinucleated cells known as *osteoclasts.*

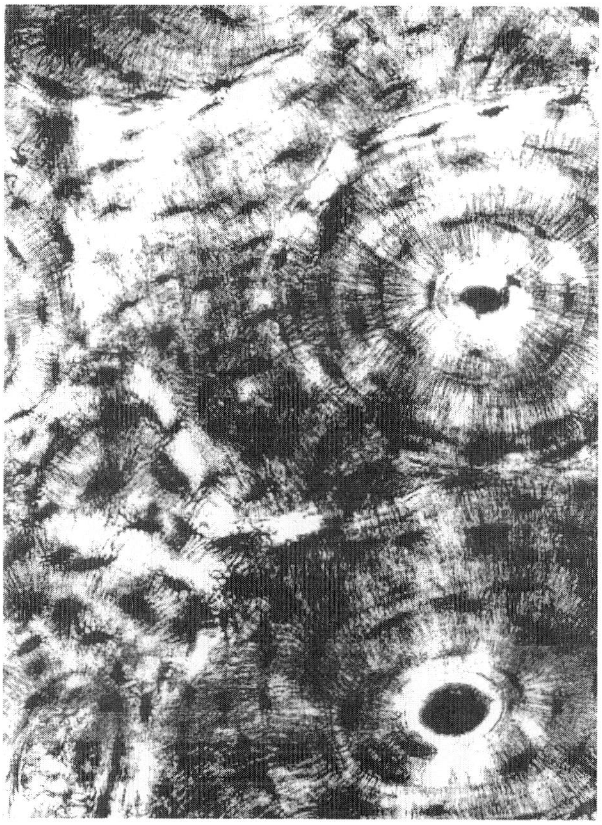

Figure 2.17 Micrograph showing the structure of an osteon. Courtesy of D. R. Swartz, Department of Anatomy and Cell Biology, Indiana University.

Nervous Tissue

Nervous tissue plays a vital role in coordinating various functions of the body and consists of *neurons* and *neuroglia* or *glial cells*. Neurons or *nerve cells* are the specialized cells that act to convey messages throughout the body via a unique and highly efficient process. The membranes of these cells possess the ability to transmit information in the form of electrical pulses. Interaction among and between nerve cells of the body form the basis for the *nervous system,* which helps organisms react to various stimuli. The *peripheral nervous system* consists of *sensory neurons,* which are responsible for detecting various stimuli, and *motor neurons,* which deliver a response message to the affected tissues. The *central nervous system,* which consists of the brain and spinal cord, is responsible for integration and coordination of body functions.

Nerve cells are elongated and possess an *axon* (Figure 2.18) that is capable of penetrating deep into various tissues. This axon is a projection of the cell and contains cytoplasm and various *cytoskeletal* proteins which help maintain the architecture of the cell. Depending on the nerve, the axon can be very long or quite short. Neurons contain a cell body or *soma,* which is an enlarged segment of the cell containing the nucleus and other organelles typically found in cells. Radiating from the cell body are additional cytoplasmic protrusions called *dendrites*. Similar to the axon of a neuron, dendrites are quite variable in length. These dendrites often form *synapatic junctions* that enable nerves to communicate with each other and other cell types.

Glial cells account for roughly half the total mass of nervous tissue in an animal. These *astrocytes, oligodendrocytes* (Figure 2.19), *miroglia* and *ependymal cells* help insulate and provide structural support to nerve cells. One such oligodendrocytic cell, the *Schwann cell,* plays a particularly important role in the function of nerves in the peripheral nervous system. These cells reside in close proximity to the axon of nerve cells and literally wrap themselves around the axon to form what is referred to as a *myelin sheath*. This myelination is interrupted along the axon at *nodes of Ranvier,* exposed portions of the nerve cell membrane. This architecture of interrupted myelination greatly increases the speed at which an action potential is propagated along the axon.

An additional characteristic of peripherial nerves is their existence as bundles, especially those emanating from the spinal cord. These bundles are organized in various layers of connective tissue. The outermost layer of connective tissue is known as the *epineurium* and encapsulates the entire bundle of nerves. The *perineurium* surrounds smaller bundles and contains numerous arteries and veins. Finally, the *endoneurium* (Figure 2.20) surrounds each individual neuron and associated glial cells.

Muscle Tissue

Muscle serves many purposes within an organism. First and foremost, muscle is the basis for all movement. It provides a means of locomotion, digestion, breathing, vision, circulation and other essential biological activities. This is accomplished by the unique ability of muscle to contract and shorten;

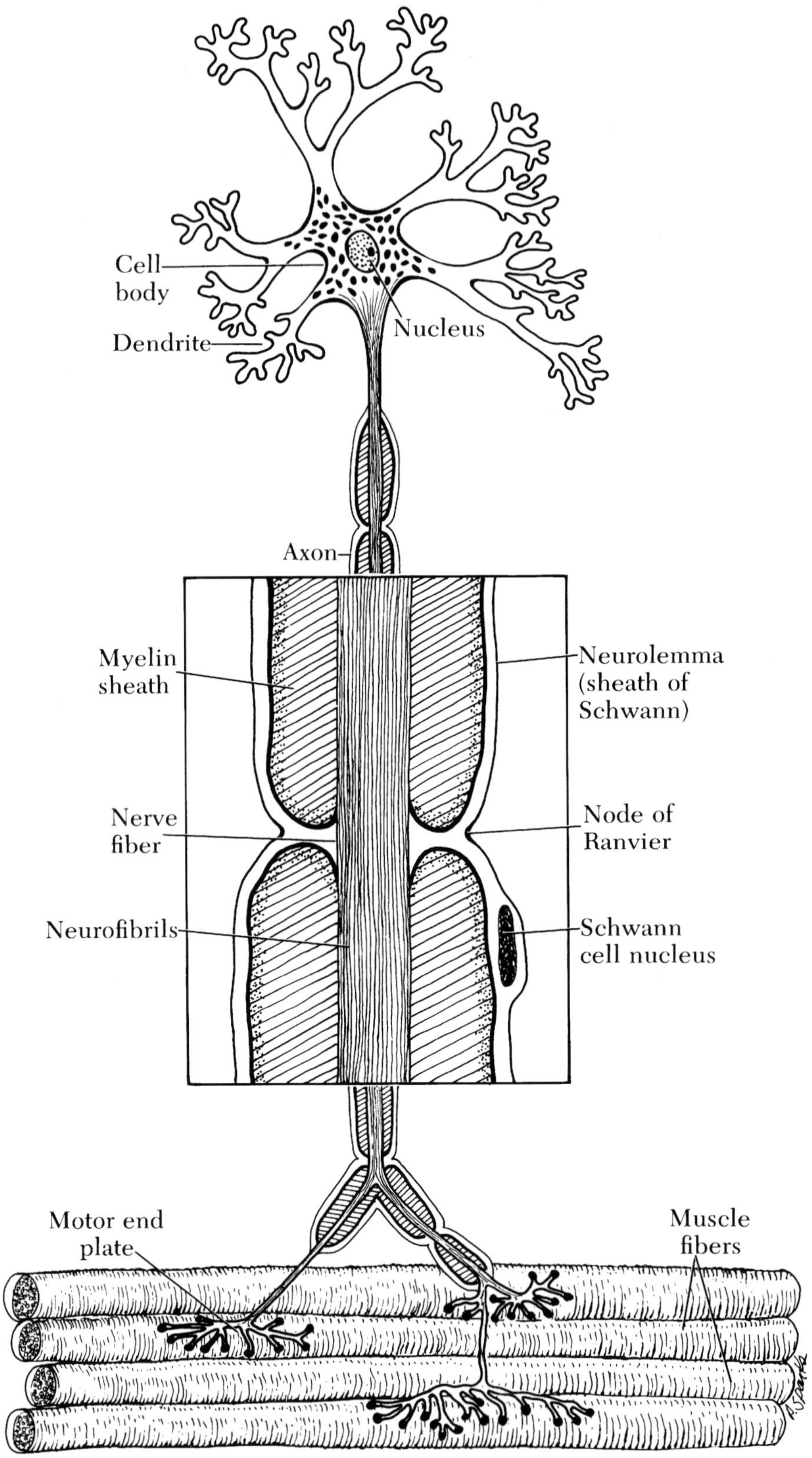

Figure 2.18 Diagrammatic illustration of a neuron and the motor end plates associated with it. The magnification of the axon (center) shows details of the nerve fiber, neurofibrils, myelin sheath, Schwann cell nucleus, and node of Ranvier. Modified from J. E. Crouch, *Functional Human Anatomy,* 2nd ed, Lea & Febiger, Philadelphia, p. 480, 1985.

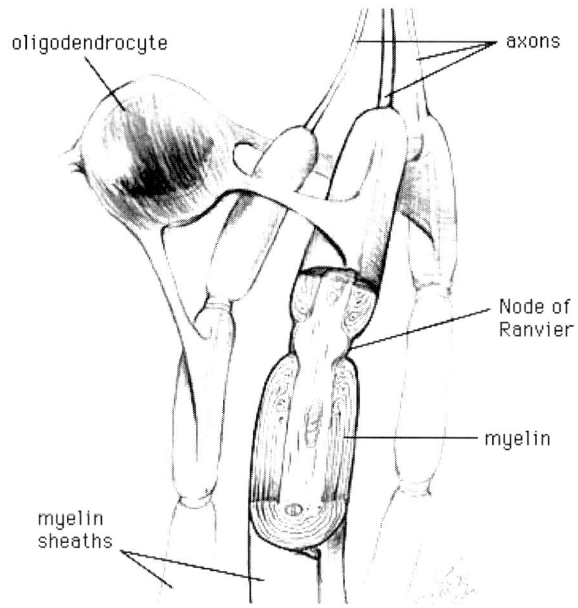

Figure 2.19 Location of oligodendrocytes.

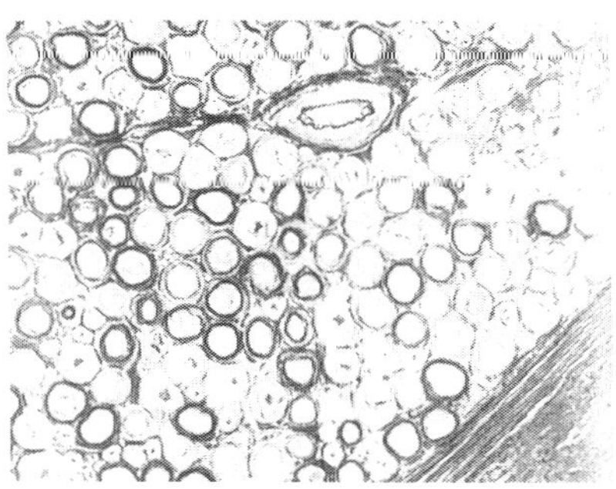

Figure 2.20 Micrograph of a nerve bundle cross section. Courtesy of D. R. Swartz, Department of Anatomy and Cell Biology, Indiana University.

therefore it is capable of moving various structures in the body. Of course, this ability to convert chemical energy to mechanical energy is only 20 to 25% efficient and much energy is lost as heat. Hence, the second most vital function of muscle is to maintain body temperature in warm-blooded animals.

Muscle is used as food and is an excellent source of high quality protein. Meat animals in the developed countries of the world contribute significantly to the quality of the human diet.

Muscles are used in a number of ways to communicate. In addition to direct communication via various voice related structures, the muscles are involved with indirect communication accomplished by body movements, which express attitudes and subtle meanings.

Optimizing muscle growth for meat production is a major goal of the meat animal industry. Although many other animal tissues are used for human consumption, skeletal muscle represents the most significant component of animal production. From an agricultural standpoint, the production of muscle is an important goal. Muscle makes up 30 to 40% of the total body mass. Consequently, it is the most important energy user of the tissues of the body and is the major drain on energy efficiency.

Muscle tissue, as other tissues of the body, performs several specialized functions. There are three types of muscle tissue in the body:

- *Skeletal Muscle*
- *Cardiac Muscle*
- *Smooth Muscle.*

Classification of these tissue types is based mainly on their microstructure and the means by which they are stimulated to contract. Muscle tissue contains a very unique set of proteins arranged in filaments that are capable of interacting by sliding past each other to provide contractile force. Muscle tissue containing highly organized filaments is *striated* tissue, whereas muscle tissue lacking this organization is *non-striated.* Striations (Figure 2.21) may be observed only via light microscopy which also provides insight into the highly organized nature of organelles within muscle cells.

Skeletal Muscle

Skeletal Structure

Muscles exist in many sizes and shapes.

- The bovine *semimembranosus* muscle, for example, weighs ten-thousand times more than the smallest flexor muscles of the lower leg of a mature animal.
- The *semitendinosus* muscle is a long cylindrical muscle and is characteristic of common concepts of a muscle architecture.

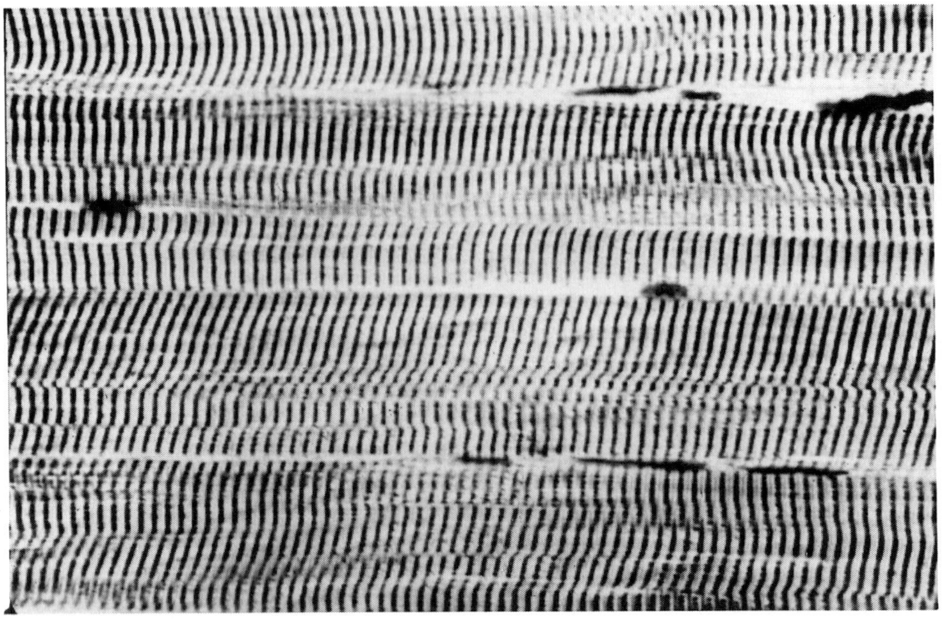

Figure 2.21 Photomicrograph of skeletal muscle fibers (× 630). From *Principles of Meat Science, Fourth Edition* by Aberle et al. Copyright © 2001 by Kendall/Hunt Publishing Company. Reprinted by permission.

- In contrast, the *masseter* muscle of the jaw is circular in shape.

Regardless of their gross anatomical structure, all muscles originate (*origin*) on one side of a joint and terminate (*insertion*) on the other side of the articulation. Sometimes the attachment to either the origin or insertion occurs via a *tendon,* which is a dense connective tissue extension of the structural framework of the muscle. This structure is commonly referred to as the *myotendinal junction.* On occasion, the origin and (or) insertion are somewhat obscure because the muscle may terminate on a thin sheet of connective tissue called a *fascia* or an *aponeurosis.* These broad, diffuse attachments are especially evident with muscles that are flat and thin, and help compress or diffuse force over a broad area.

Knowing the names of some six hundred muscles in the body of domestic animals seems a daunting task; however, much information regarding muscles is contained within the nomenclature of muscles. The muscle name can imply the function, location, size, attachment, movement or structure of the muscle. For example, one may appreciate that the *biceps femoris* contains two functional entities or "two heads." Furthermore, the *sternomandibularis* originates in the area of the mandible and inserts on the sternum. The extensor muscles of the lower leg act to extend the leg. These clues in nomenclature of muscle can help locate muscles in the body.

Organization

Compared with other tissues of the body, muscle tissue possesses relatively large cells that are harnessed in a vast network of connective tissue (Figure 2.22), similar to that of a nerve bundle (Figure 2.20). The connective tissue encapsulating these large cells merges with additional connective tissues that act as a harness to transmit contractile force to bones or other tissues to provide movement. An individual muscle is constructed in a manner that holds many individual units together to facilitate the overall function of the organ. The basis for structural integrity of muscles is connective tissue. The outer-most layer of connective tissue is called the *epimysium* and serves to separate individual muscles into distinct units. Epimysial connective tissue provides avenues for nerves and blood vessels to enter and exit muscles. Each muscle is supplied with at least one artery, vein and bundle of nerves. Within this connective tissue sheath, are a number of smaller *fasciculi* (bundles of muscle fibers) that

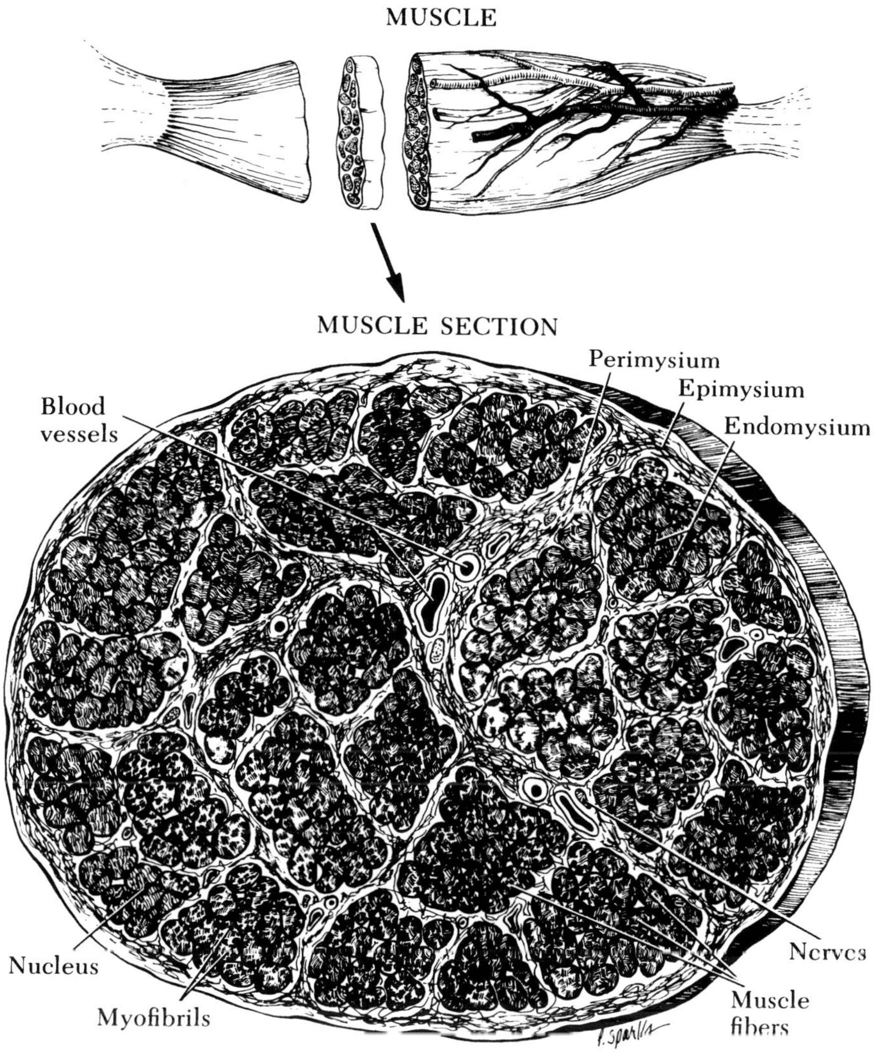

Figure 2.22 Drawing of a skeletal muscle in cross section showing muscle fibers, bundle arrangement, supporting connective tissues, nerves, and blood vessels. From J. C. Forrest, E. D. Aberle, H. B. Hedrick, M. D. Judge, and R. A. Merkel, W. H. Freeman and Co., New York.

are surrounded by connective tissue which forms the *perimysium.* There are roughly 20 to 40 muscle fibers per *primary bundle* and a group of primary bundles form a *secondary bundle,* which is encased in perimysial connective tissue as well. Perimysial connective tissue differs from epimysial connective tissue in the types of proteins present. Futhermore, *intramuscular fat* or *marbling* is fat deposited within the perimysial connective tissues around and between bundles. The number and size of the fasciculi (or fascicles) vary with muscle location and function. Muscles commonly used to maintain posture tend to have a more extensive fascicular network that provides elasticity to the muscle. Connective tissue amount and structure may differ within a muscle. For example, muscles of the shoulder tend to have higher concentrations of connective tissue in the center of the muscle that form tendinous attachments to the skeleton.

Within each fascicle are muscle fibers surrounded by yet another connective tissue known as the *endomysium.* This connective tissue matrix is considered the major connective tissue component responsible for meat tenderness and changes with age. The endomysium lies immediately adjacent to the muscle cell membrane.

Muscle Fiber

Immediately adjacent to the endomysium is the *basement membrane* or *basal lamina,* which consists mainly of interwoven reticular fibers and serves to attach the endomysium to the cell membrane of a *muscle cell, myocyte, muscle fiber,* or *myofiber* (synonyms). The multinucleated muscle fiber is the basic cellular structure of muscle or meat (Figure 2.23). Muscle fibers may extend the entire length of a muscle, but normally extend only a partial length (Figure 2.24). Fibers terminating within the muscle are called *intrafascicularly terminating fibers.* Their growth in length increases the number of fibers in the muscle cross section thereby increasing the girth of the muscle.

The cell membrane or *sarcolemma* is similar to other classical bilipid cell membranes of other cell types. Unique to it are small invaginations evenly spaced along the entire surface of the cell which are continuations of the cell membrane and act as communication channels that extend into deep portions of the cell.

Muscle Cytoskeleton

The term *cytoskeleton* literally means the framework supporting the internal structure of a cell. Cellular cytoskeletons consist mainly of three types of filaments made of protein:

- *Microfilaments*
- *Intermediate filaments*
- *Microtubules*

These structural elements maintain the shape of cells, interconnect organelles and form a network to attach all cellular components to the membrane (Figure 2.25). In muscle cells, the microfilaments, those ~7 nm diameter, containing actin, comprise the majority of the filamentous cytoskeleton. The largest class of filaments, microtubules containing tubulin and measuring ~23 nm in diameter, are not present in appreciable amounts in muscle cells and therefore, will not be discussed further. The remaining intermediate filaments, termed such because their diameter is intermediate (~10 nm) to those

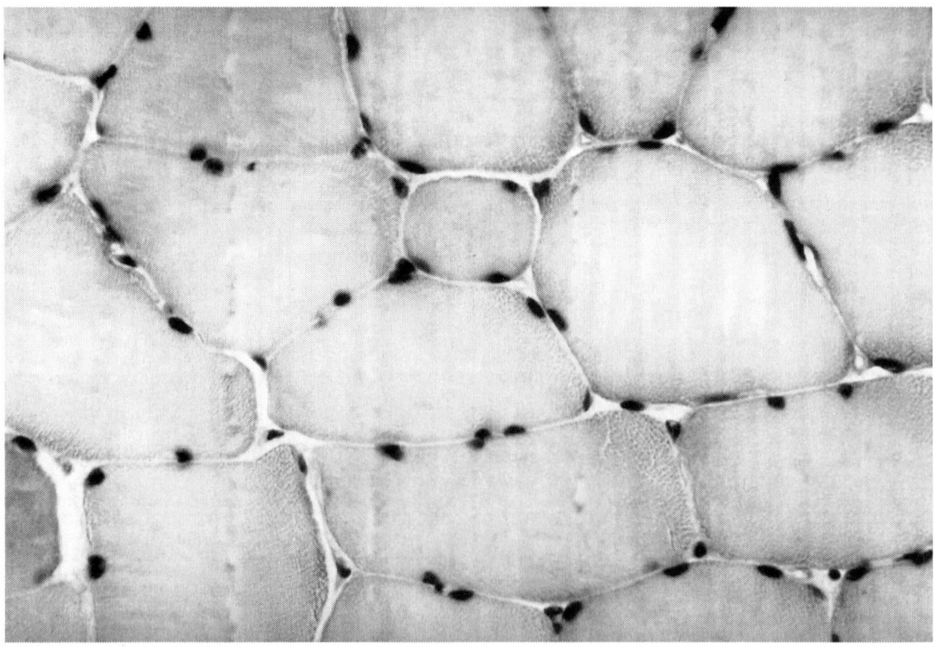

Figure 2.23 Photomicrograph of a cross-section of skeletal muscle fibers of the pig, showing their polygonal shape and the peripheral position of nuclei (× 630).

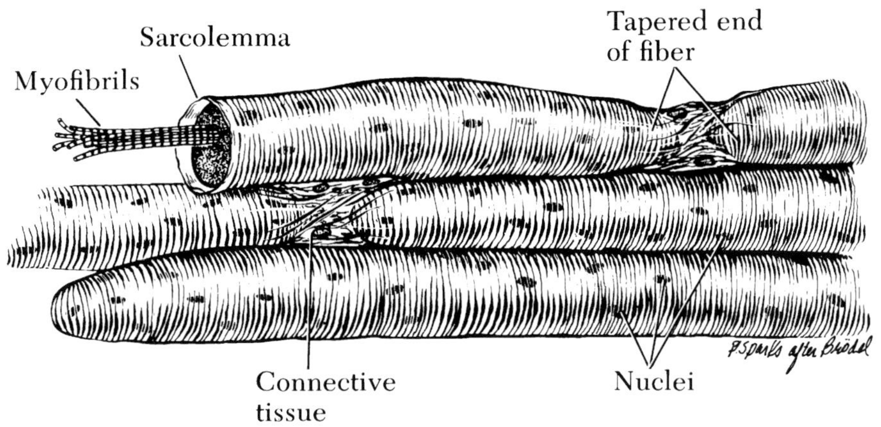

Figure 2.24 Drawing of skeletal muscle fibers showing structural features and their longitudinal orientation. From Brodel, M., *Bulletin of the Johns Hopkins Hospital,* 61 (1937), 295. © John Hopkins University Press. Reprinted with permission of Johns Hopkins University Press.

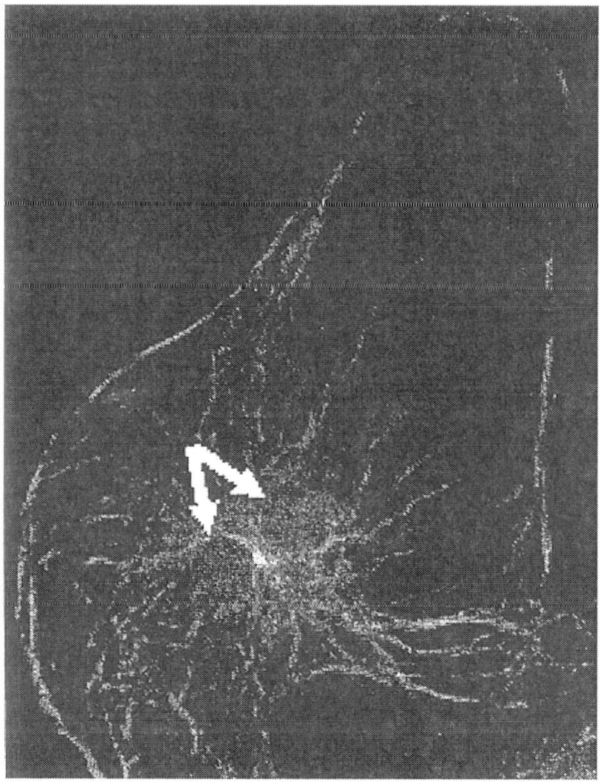

Figure 2.25 Micrograph of a cardiomyocte showing the elaborate structure of cytoskeleton. Arrows identify the position of the nuclei. Courtesy of D. R. Swartz, Department of Anatomy and Cell Biology, Indiana University.

larger and smaller filaments of the cytoskeleton, are prevalent in muscle cells and serve to hold the microfilaments (grouped into myofibrils) in concert with the muscle cell membrane.

Myofibrils

Myofibrils Figures 2.26 and 2.27 show the "inner makings" of a skeletal muscle cell. *Myofibrils* are unique microfilamentous organelles of muscle fibers. Although, the number of myofibrils in a cell varies greatly, on the average, adult muscle cells contain 1,000 to 2,000 myofibrils. Cross-sectional analyses of a myofibril reveal that they are highly organized, cylindrical structures that extend the entire length of the muscle cell and measure about 1 μm in diameter. These organelles, like all other organelles in the cell, are bathed in cytoplasm, which is rich in protein.

The smallest contractile unit of muscle is the *sarcomere* (meaning "muscle unit"). The average length of a sarcomere is 2.5 μm in the resting (uncontracted) state but can range from 0.9 μm in super-contracted muscle to 4.0 μm in severely stretched muscle. This repeating structure represents the organization of myofibrils. Thin filaments radiate in opposite directions from the *Z-line,* the outer demarcation of the sarcomere. The Z-line helps organize the sarcomeres across a given muscle fiber and is composed primarily of the protein *alpha-actinin.* Thick filaments are positioned in the center of the sarcomere and overlap with thin filaments. Because these sarcomeres are in register and aligned

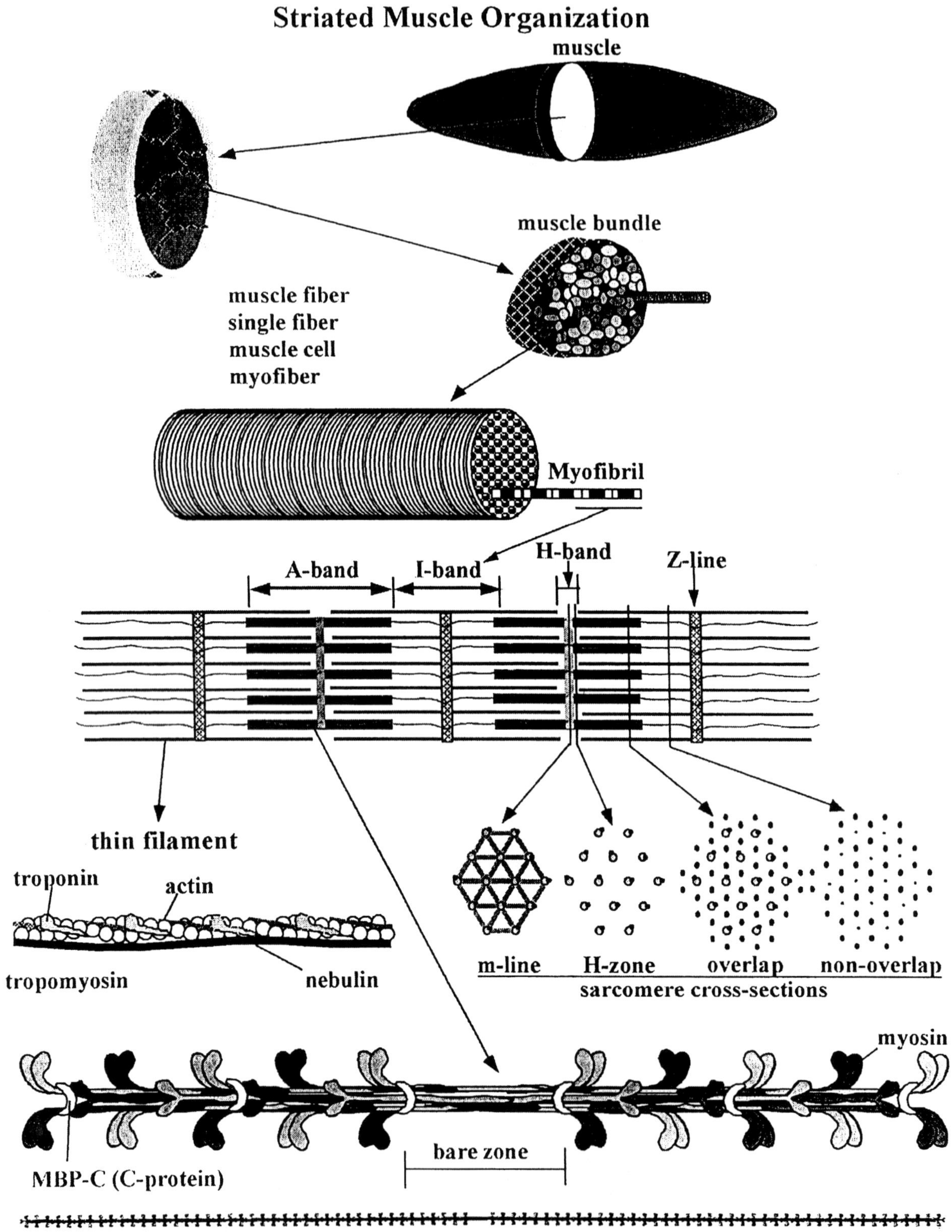

Figure 2.26 Skeletal muscle organization. Courtesy of D. R. Swartz, Department of Anatomy and Cell Biology, Indiana University.

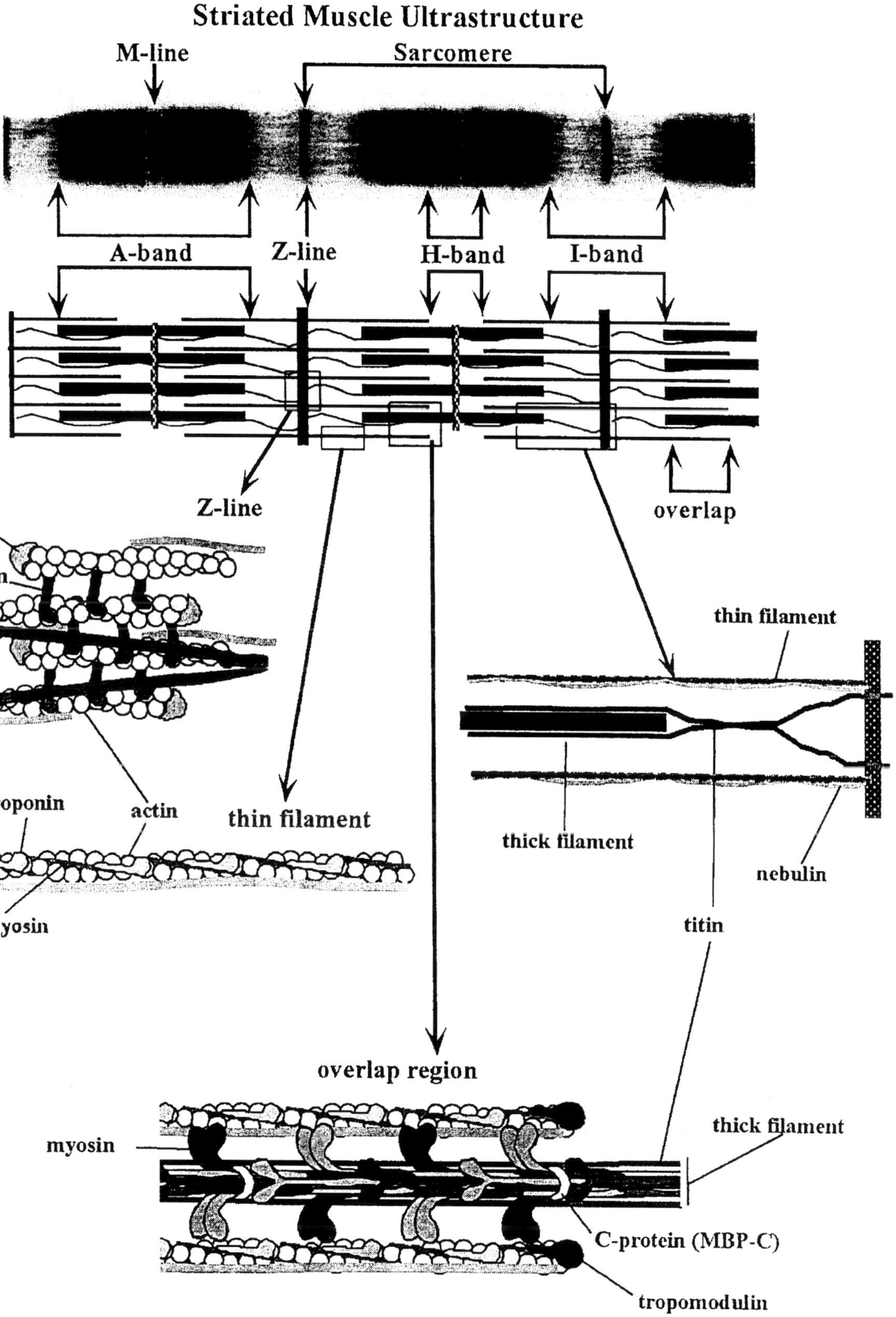

Figure 2.27 Ultrastructure of the sarcomere. From *Proceedings of the American Society of Animal Science,* 1998. Copyright © 1998 by American Society for Animal Science. Reprinted by permission.

across the entire muscle fiber, alternating light and dark banding patterns are readily apparent microscopically. This striated appearance of myofibrils is the hallmark of skeletal and cardiac muscle.

Understanding the banding pattern in myofibrils is important for several reasons. For example, the length of sarcomeres is related to meat tenderness. A shorter sarcomere length results in greater overlap of thick and thin filaments. This leads to greater physical resistance during mastication (chewing). Although understanding the ramifications of sarcomere length with regard to meat tenderness is beyond the scope of this text, it is import to realize that economically important traits often have biological bases.

The dense area located in the middle of a sarcomere where the thick filaments are located is known as the *A-band* (Figures 2.26 and 2.27). The small area within the A-band where thin filaments terminate from each half of the sacromere, and only thick filaments are present, is called the *H-zone.* The highly dense area in the center of the sarcomere is the *pseudo H zone* and is probably due to the presence of additional proteins that serve to align and stabilize the thick filaments in a three-dimensional space. During contraction, the *I-band,* consisting of the Z-line and thin filament regions from adjacent sarcomeres, shortens as a result of the sliding of filaments. In contrast, the width of the A-band remains constant regardless of the state of contraction because filaments do not change in length with contraction.

Myosin Myofilaments occupy 80 to 87% of the total volume of muscle fibers. The major myofilaments in skeletal muscle are the thick or *myosin* filaments and the thin or *actin* filaments. The myosin filament is composed mainly of *myosin heavy chain* and *myosin light chain* molecules. The most abundant contractile protein in muscle is myosin which accounts for only 5% of the wet muscle tissue but 55 to 60% of the total myofibrillar protein. Recall that the myosin filament is located in the center of the sarcomere and overlaps the actin filaments. This filament is the real power component of muscle contraction and sarcomere shortening because its unique structure allows it to "rachet" or pull actin filaments together. As a result of this ability to pull actin filaments, Z-line movement is toward the center of a sarcomere.

Myosin is a 470 kilodalton protein consisting of two myosin heavy chains (~220 kilodaltons each) and two myosin light chains (~20 kilodaltons each). The myosin molecule contains a tail region which interacts (coils) with the tail region of another myosin molecule (Figure 2.28). At one end of the myosin molecule is a globular head that possesses ATPase activity and has the ability to bind actin. During contraction the myosin head binds to actin and cleaves ATP. After ATP is cleaved and ADP and inorganic phosphate is released, the head undergoes a conformational change and swivels to pull the

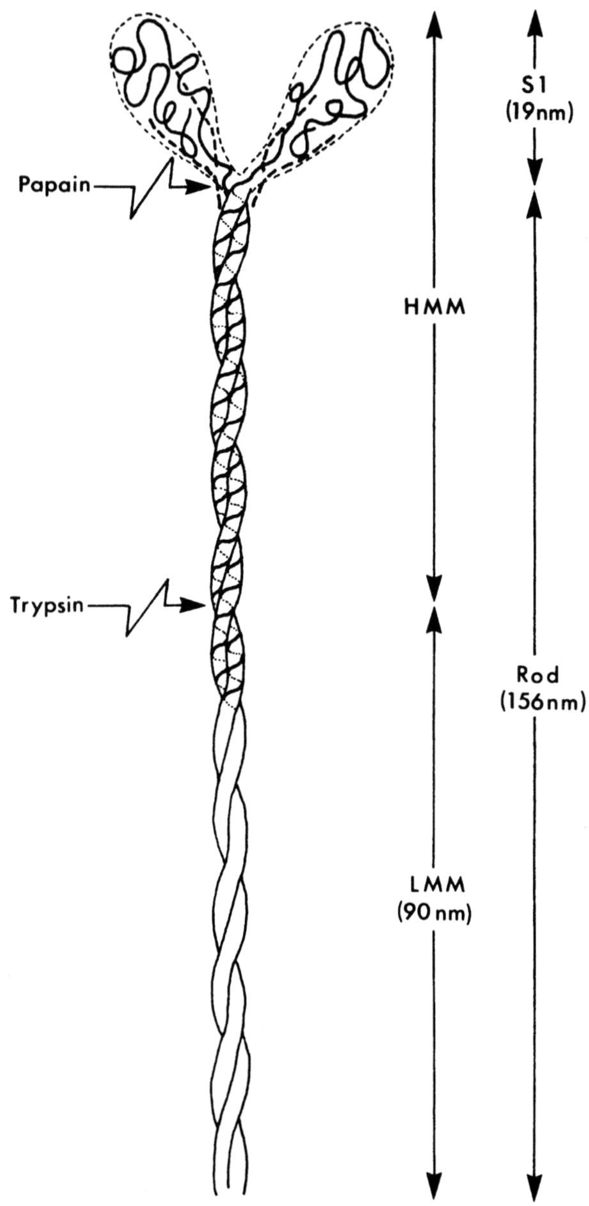

Figure 2.28 Diagram of the myosin molecule and its subunits. From A. G. Engel and C. Franzini-Armstrong, *Myology, Basic and Clinical,* 2nd ed, McGraw-Hill, Inc., New York.

actin filament toward the center of the sarcomere. Approximately 360 myosin molecules are aggregated on a core protein and are held in place with C-protein (MBP-C, Figures 2.26 and 2.27) to form *filamentous myosin* or a *myosin filament.* This structure of about 14 to 16 nm in diameter and filament formation is energetically favorable and myosin will spontaneously aggregate in solution.

Electrophoretic separation of isolated myosin treated with the proteinase trypsin causes the formation of two distinct, large fragments i.e. *light meromyosin* and *heavy meromyosin* that are approximately 70 and 170 kilodalton, respectively (Figure 2.28). Addition of another proteinase, *papain,* which is often used as a meat tenderizer, will cleave heavy meromyosin into smaller fragments. The heavy meromyosin subfragment 1 (S1) possesses the globular head and all of the ATPase activity. The light meromyosin portion is mainly responsible for formation of myosin filaments.

There are two myosin light chains associated with each myosin heavy chain molecule, i.e., *proximal* and *distal light chains.* These proteins are substantially smaller molecules bound to the globular region of myosin near the papain cleavage site. Neither light chain is essential for myosin ATPase activity but they can modify the speed at which myosin will hydrolyze ATP and therefore, are responsible for fine-tuning muscle function.

Understanding the way in which myosin molecules are organized and assembled into the thick filaments is difficult. Because the thick filament is positioned toward the center of the sarcomere, it can exert a force in both directions. This is accomplished by the arrangement of myosin molecules in opposite directions or *bi-directionally* on a core protein. Because myosin possesses a substantial tail region, there is a small region in the center of the thick filament that does not contain globular heads and therefore, the possibility for generating force in this region does not exist. Of course, this is not necessary, as thin filaments normally do not exist in this region. Recall that myofibrils are cylindrically shaped structures, therefore, there must be a mechanism by which myosin molecules can be held in register or alignment across a myofibril (Figure 2.29). The protein primarily responsible for this is *m-protein* or *skelemin,* which is located at the center of a myosin molecule and attaches adjacent myosin filaments to each other. As a result of protein being complexed at the center of the sarcomere, an *m-line* is observed in the banding pattern (Figure 2.26).

Myosin molecules, as well as other contractile proteins in muscle, can exist as *isoforms,* meaning there are several types of myosins depending on functional requirements. In fact, subtle differences in the protein sequence of the globular head, will dramatically alter the ability of muscle to contract. This is the basis for different muscle fiber types and will be discussed in more detail in Chapter 6.

Actin Actin is the next most abundant contractile protein in muscle fibers consisting of nearly 20% of total myofibrillar protein and 2% of wet muscle tissue. The structure of *filamentous actin (F-actin)* is complex (Figure 2.30). The subunit of the actin myofilament is *G-actin* or *globular actin.* This 40 kilodalton protein polymerizes with other G-actin proteins to develop a helical F-actin that is roughly 6 to 8 nm in diameter. Because this process occurs spontaneously in solution, a protein called *Cap Z* is added to the end of F-actin to control thin

Figure 2.29 Electron photomicrograph of a cross-section of a myofibril, showing the thick and thin filaments and their hexagonal arrangement. Adapted from D. S. Smith, Muscle, Academic Press, Inc., New York, p. 14, 1972.

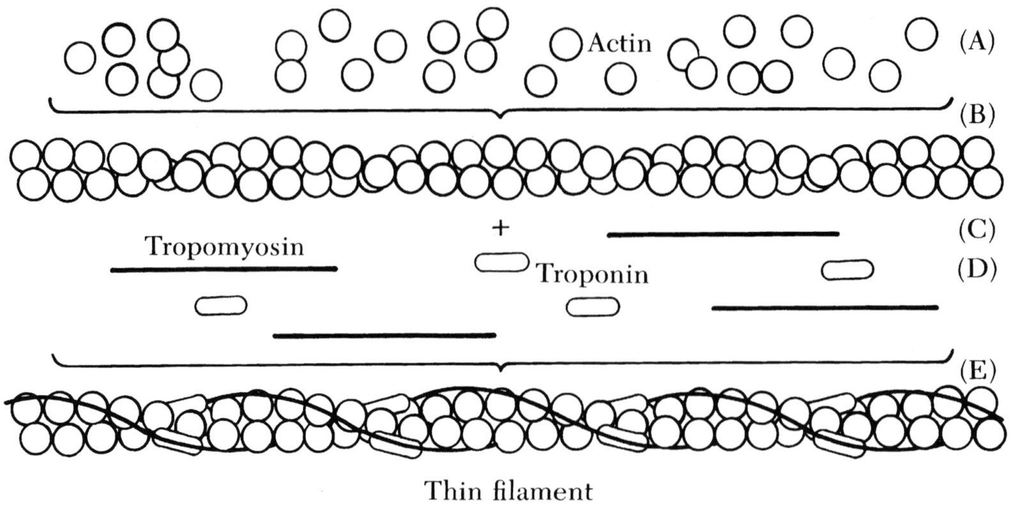

Figure 2.30 Diagrammatic representation of the construction and fine structure of an actin filament, showing the relation of tropomyosin and troponin to its (A) G-actin molecules, (B) two chains of polymerized actin (F-actin) coiled about each other, (C) long thin molecules of tropomyosin, (D) pellet-shaped troponin molecules, and (E) an actin filament, showing how two coiled chains of F-actin and polymerized tropomyosin molecules form two long strands of tropomyosin, one strand being associated with each of the F-actin chains. The periodic location of troponin along the actin filament every seven G-actin molecules, and its association with the tropomyosin strand, are also illustrated. Modified from J. M. Murray and A. Weber, "The Cooperative Action of Muscle Proteins." George V. Kelvin/Scientific American, Inc.

filament length (Figure 2.27). There are roughly 400 G-actin molecules in one F-actin. Because F-actin is a double helix, there is a prominent groove running the entire length of the F-actin filament (Figure 2.30). In striated muscle, additional proteins bind to F-actin and this structure is collectively referred to as the thin filament.

Regulatory Proteins

Tropomyosin, another filamentous protein of the thin filament, occupies this F-actin groove (Figure 2.31). A simple analogy to visualize this structure is to compare F-actin to two strands of pearls. The two strands of pearls are wrapped around each other and a groove forms between the strands. Adding tropomyosin with F-actin would be analogous to attaching small cords to the strands, as the cord would likely favor alignment in the groove generated by the two strands of pearls. Located at every seventh G-actin molecule, a trimeric protein called *troponin* is bound. Troponin binds to the F-actin and tropomyosin. Interaction between these proteins and the thin filament is key to regulation of cross-bridge development during muscle contraction.

Trimeric troponin consists of *troponin T, I* and *C.* Troponin T as the name implies, is associated with tropomyosin. During contraction, calcium is released into the cytoplasm and immediately binds to calcium-sensing domains of troponin C. Calcium binding to troponin C during initiation of contraction creates a conformational change in trimeric structure, which changes the location of the tropomyosin in the actin groove and moves the inhibitor (I) troponin away from the myosin-binding site on the actin filament, thereby facilitating contraction.

Intrasarcomeric Proteins

Additional proteins exist within the myofibril that help maintain and organize the structure. These proteins are termed *intrasarcomeric proteins* or proteins within the sarcomere.

Titin or *connectin* is the largest protein ever discovered and represents nearly 10% of the myofibrillar protein in muscle, making it the third most abundant protein in muscle next to actin and myosin. For years, titin mystified researchers because it was so large (2,800 kilodalton, or over 10 times the size of a single myosin molecule). There is much known about this gigantic protein, yet the exact function remains under investigation. The end of the molecule is located at the Z-line, while the bulk of the molecule spans the entire length of half a sarcomere and terminates at the M-line (Figure 2.27). The structure of titin has provided clues to the function of the protein in skeletal muscle. As one may gather, the portion of the molecule located at the Z-line serves as means to attach the protein to the Z-line. The rest of the molecule, especially that located in the I-band, is "spring-like." The portion of the molecule located in the A-band contains numerous repeating amino acid sequences that facilitate interaction with the repeating structure associated with thick filaments. Therefore, titin helps maintain the structure of a half sarcomere and is responsible for the *resting tension* of muscle or that tension existing in a muscle when it is not stretched. If structures like titin were not present in muscle and the sarcomeres were stretched beyond the overlap regions of actin and myosin, it would be virtually impossible to re-align the sarcomeres.

Another rather large muscle protein (600 kilodalton), *nebulin,* localizes in the sarcomere at the same location as actin, radiating out of the Z-line parallel to the long axis of the myofibril through the I-band and entering the A-band. The exact role of nebulin in maintaining muscle structure is not known but it may bind along the actin filament and provide a template for building and maintaining filamentous actin (Figure 2.27).

Intersarcomeric Proteins

Those proteins positioned outside the myofibril are known as *intersarcomeric proteins* and serve to align the cylindrical myofibrils in a mucle fiber. To visualize this, one could use the analogy of a raft made of logs (Figure 2.31). If the logs are not tied together with cord, they will independently float in different directions. Much like this scenario, if there is no structure responsible for holding the logs together when one log is moved, contracting myofibrils will not work together unless there are structural proteins holding myofibrils in register. The protein primarily responsible for "tying" the myofibrils together is called *desmin.* Desmin is a 53 kilodalton protein that encircles the myofibrils at the Z-line and radiates out perpendicular to the long axis of the myofibril and attaches adjacent myofibrils to one another. Two additional proteins, *paranemin* and *synemin* bind in the same location and help stabilize myofibril-to-myofibril interactions.

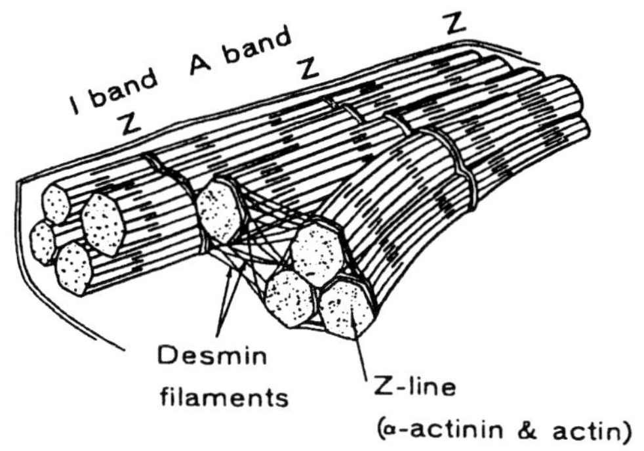

Figure 2.31 Illustration of the location of desmin around myofibrils. From A. M. Pearson, *Muscle and Meat Biochemistry,* Academic Press, Inc, San Diego.

Myofibrils are of little use during contraction if they are not attached to the sarcolemma, as the contractile force generated must be ultimately attached to the connective tissue elements of a muscle and subsequently to tendinous attachments. *Costameres* essentially act to anchor myofibrils to the sarcolemma and helps transfer the force generated from contracting myofibrils to the cell membrane (Figure 2.32). The structure responsible for this task in a muscle fiber is the costamere, which is a discrete collection of several aggregated proteins regularly spaced along the sarcolemma (Figure 2.33).

- One such protein, *filamin,* is located at the most peripheral limits of the Z-line as well as just beneath the sarcolemma.
- *Vinculin,* another costameric protein, is a lipid binding protein and anchors the costamere to the bilipid membrane of the muscle cell.
- Another protein, *talin,* is associated with the costamere but its exact function is not known.

A complete understanding of all the proteins in muscle and how they interact to facilitate the proper function of muscle is still lacking. However, Table 2.4 summarizes the size and abundance of each known muscle protein and where they are located within the muscle cell.

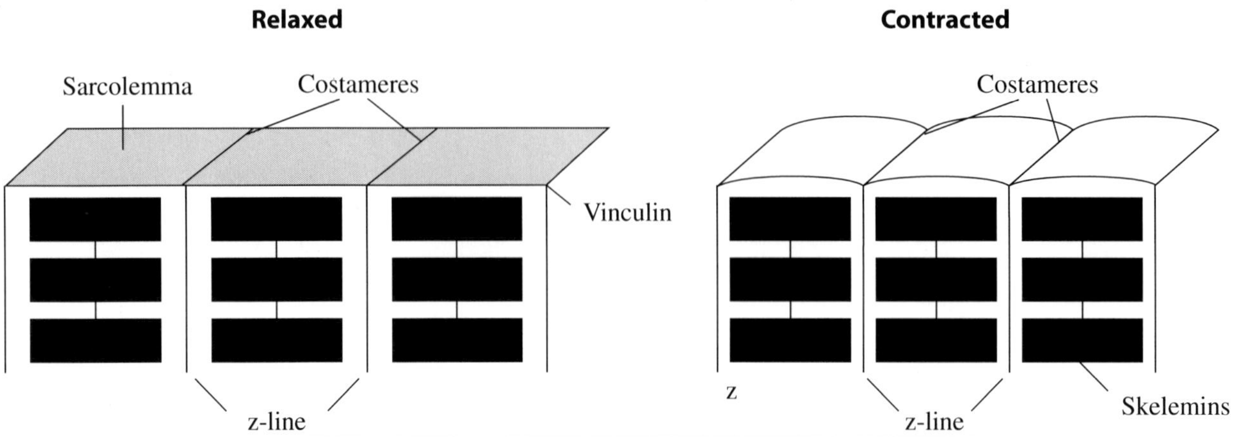

Figure 2.32 Illustration showing of the importance of costameric attachments in facilitating muscle shortening. From M. Greaser, Proceeding of the 44th Annual Reciprocal Meat Conference, National Live Stock and Meat Board, Chicago, as modified from Pardo et al. (1983), Journal of Cell Biology, vol. 97, pg. 1081.

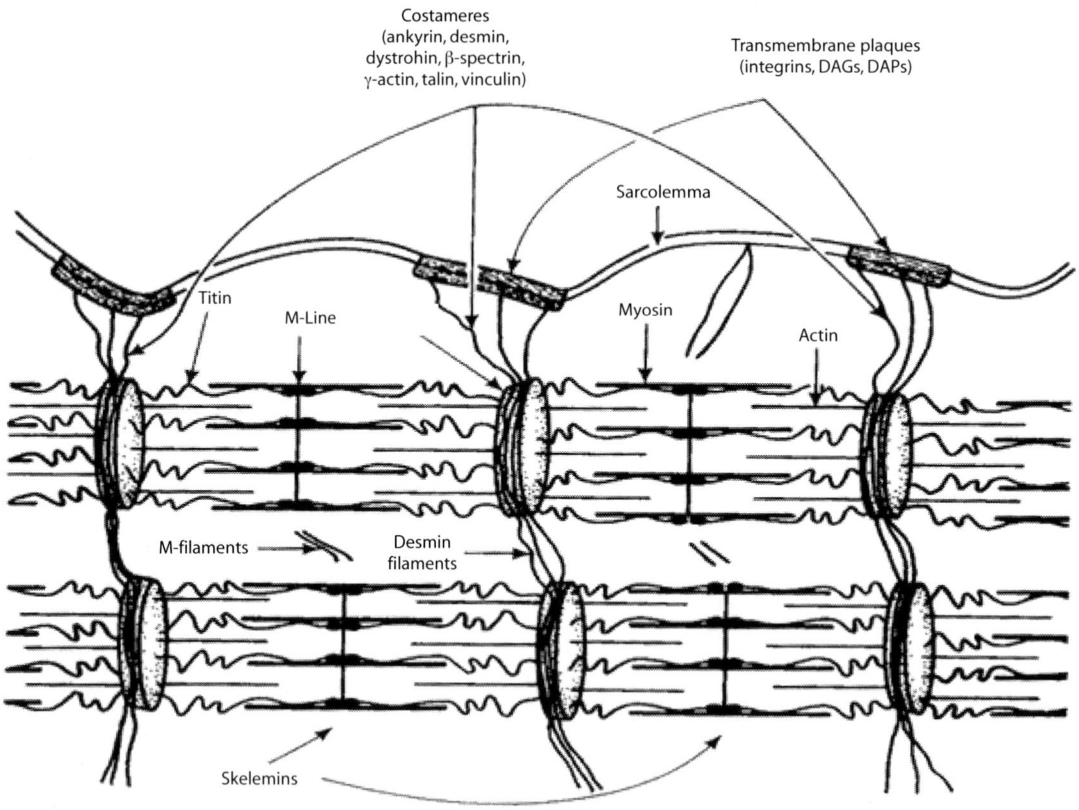

Figure 2.33 Diagram showing the structures participating in the organization of the costameres in skeletal muscle in relation to the Z-line. From R. G. Taylor, G. H. Geesink, V. F. Thompson, M. Koohmarie and D. E. Goll, Is Z-disk degradation responsible for postmortem tenderization? (1995), Journal of Animal Science, vol 73, pg. 1352.

TABLE 2.4 SELECTED MYOFIBRILLAR AND OTHER CYTOSKELETAL PROTEINS LOCATED IN MUSCLE FIBERS

Protein	Location	Approximate % of Cytoskeleton	Approx. MW	(# of Subunits)
Myosin	Thick filament	45	520,000	(6)
C-Protein (MyBP-C)	Thick filament	2	130,000	(1)
H-Protein (MyBP-H)	Thick filament	1	74,000	(1)
Myomesin	M-line	2	185,000	(1)
M-Protein	M-line	1	165,000	(1)
Creatine Kinase	M-line	<1	80,000	(2)
Skelemin	M-line (peripheral)	<1	195,000	(1)
Actin	Thin filament	20	42,000	(1)
Tropomyosin	Thin filament	5	66,000	(2)
Troponin	Thin filament	5	69,000	(3)
Tropomodulin	Thin filament-free end	<1	41,000	(1)
Titin	Longitudinal sarcomeric filaments (Z- to M-line)	10	3,700,000	(1)
Nebulin	Parallels (part of) thin filaments to the Z-line	3	773,000	(1)
α-Actinin	Z-line (integral)	2	204,000	(2)
Cap Z	Z-line (integral)	<1	66,000	(2)
Desmin	Z-line (peripheral)	<1	212,000	(4)
Paranemin	Z-line (peripheral)	<1	356,000	(2)
Synemin	Z-line (peripheral)	<1	372,000	(2)
Filamin	Z-line (peripheral) Costamere-Membrane	<1	560,000	(2)
Dystrophin	Costamere-Membrane	<1	854,000	(2)
Talin	Costamere-Membrane	<1	536,000	(2)
Vinculin	Costamere-Membrane	<1	116,000	(1)

From R.M. Robson, E. Huff-Lonegran, F.C. Parrish, C.-Y. Ho, M.H. Stromer, T.W. Huiatt, R.M. Bellin, and S.W. Sernett. Proceedings of the 50th Annual Recprocal Meat Conference, National Live Stock and Meat Board, Chicago.

Additional Components of Muscle Fibers

Sarcoplasmic reticulum and T-tubules Small invaginations of the sarcolemma exist at very discrete intervals all along the muscle fiber. These channels serve as the outer portion of the communication system of a muscle fiber. These *transverse-* or *T-tubules* are bilipid membranes that surround myofibrils at the A-I band junction (Figure 2.34). Extending between two adjacent T-tubules is an elaborate structure developed from the endoplasmic reticular system of the cell known as the *sarcoplasmic reticulum.* The sarcoplasmic reticulum consists of the two structures. The *terminal cisternae* or "ending reservoir" lies adjacent to the T-tubules. This portion of the sarcoplasmic reticulum further thins out to form the *fenestrated collar* directly over the A-band. The lumen of terminal cisternae and the fenestrated collar are calcium storage locations during muscle relaxation. The anatomical structure formed from two terminal cisternae, located on each side of a T-tubule, and one T-tubule is a *triad.* As there is no physical connection between the T-tubule system and the lumen of the terminal cisternae, communication between these two structures occurs

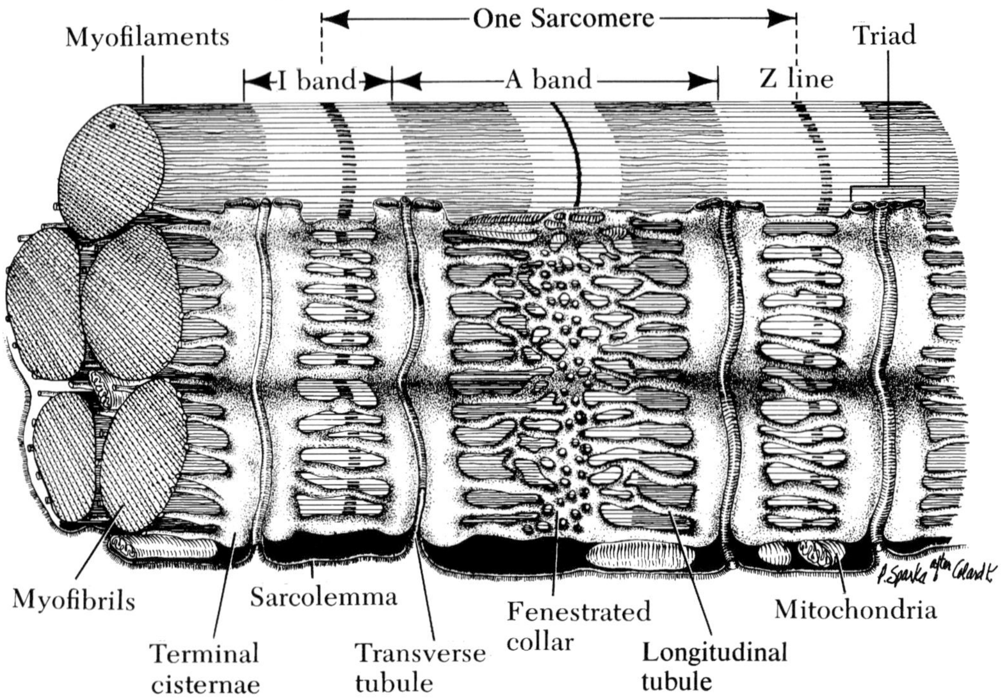

Figure 2.34 Diagrammatic representation of the sarcoplasmic reticulum and T tubules, and their relation to the myofibrils of mammalian skeletal muscle. From *A Textbook of Histology, Ninth Edition* by William Bloom & Don Fawcett. Copyright © 1968 by Don Fawcett. Reprinted by permission.

via a voltage sensing protein, the *dihydropyridine receptor* and the *ryanodine receptor* or the *sarcoplasmic calcium channel* (Figure 2.35). As soon as an electrical impulse reaches the dihydropyridine receptor of the T-tubule, the conformation of the receptor changes. This, in turn, alters the conformation of the calcium channel of the terminal cisternae and allows calcium into the cytoplasm. Because there are more calcium channels than voltage sensors (dihydropyridine), there must be a means whereby additional calcium channels can be opened without coming into close proximity to a voltage sensor. This is accomplished because the calcium channels possess calcium-binding domains that are capable of acting independently of direct dihydropyridine receptor mediation in order to release calcium into the cytoplasm after stimulation. Other proteins imbedded in the membrane of the terminal cisternae continually pump calcium ions back into the sarcoplasmic reticulum where it is bound to a protein that has high capacity for binding calcium but at reduced strength. Proteins in the membrane that facilitate this partitioning of calcium are known as *calcium pumps,* while the protein responsible for binding calcium in the sarcoplasmic reticulum is known as *calsequestrin.*

Nuclei Muscle fibers are multinucleated cells and, therefore contain numerous nuclei located just beneath the sarcolemma. These elliptical nuclei are located at the periphery of muscle fibers (Figure 2.24). The long axis of the nucleus runs parallel to the long axis of the muscle fiber. Nuclei are randomly located along the long axis of the muscle fiber and there are roughly 33 nuclei per millimeter of length.

Mitochondria Mitochondria are oblong cellular organelles that are responsible for generating energy substrates for the cell, traditionally known as the "powerhouse of the cell." Mitochondria are usually clustered around the periphery of the cell, probably forced there by the contractile apparatus (myofibrils) in the center of the muscle fiber. Mitochondria are about 1 μm in diameter and 3 to 4 μm in length and

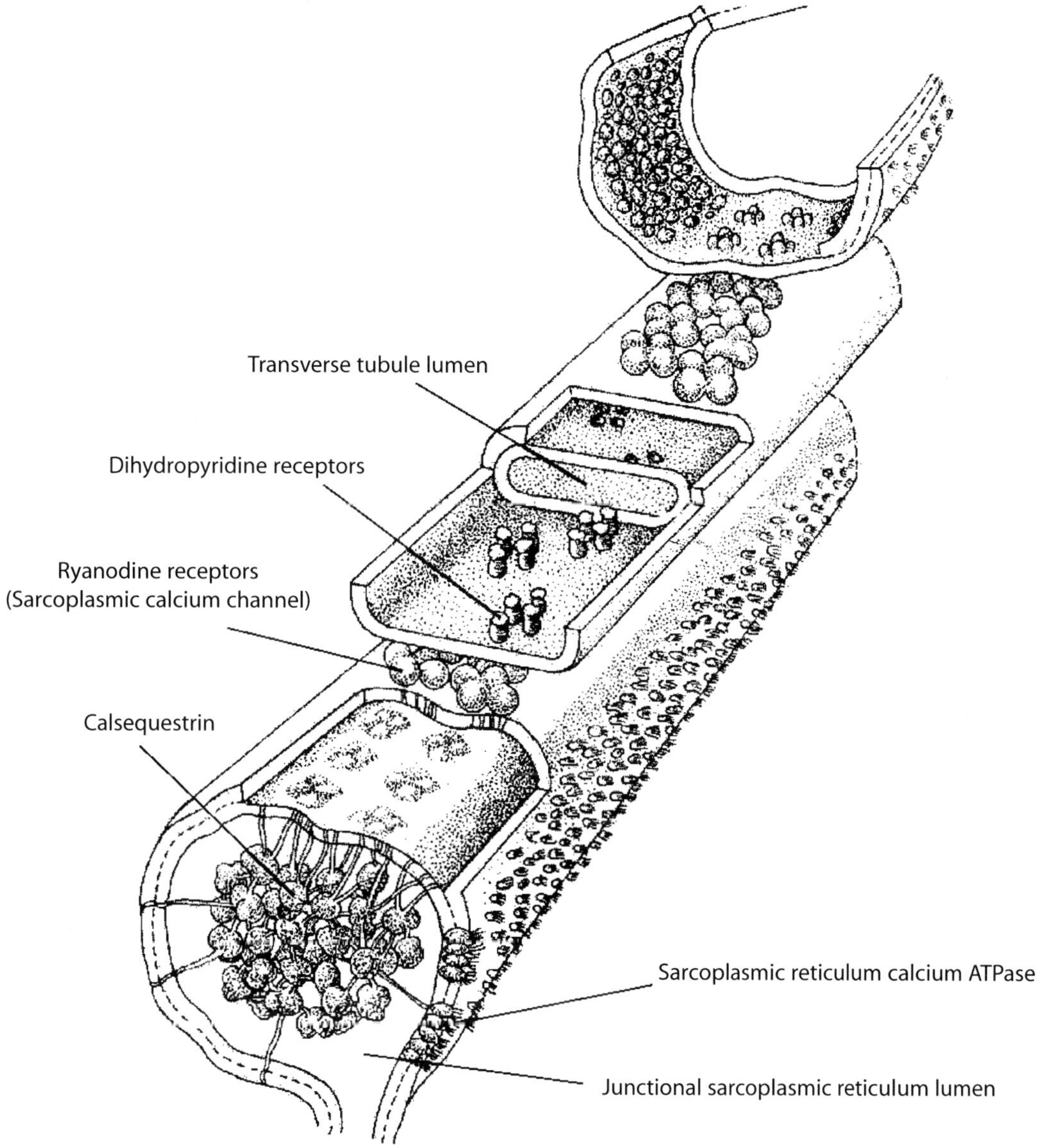

Figure 2.35 Organization of the sarcoplasmic reticulum, transverse tubules and associated structures. Reproduced from *The Journal of Cell Biology,* 1988, 107, pp. 2587–2600, by copyright permission of The Rockefeller University Press.

contain numerous enzymes necessary to conduct oxidative metabolism. Mitochondria account for 1.1 to 5.2% of the volume of the muscle fiber.

Sarcoplasm The cellular cytoplasm of a muscle cell is referred to as *sarcoplasm.* Of course, this component represents only a small fraction of the total volume of the muscle cell. Like that of most other cell types, the cytoplasm contains many soluble substrates needed for general metabolism such as *lysosomes, glycogen granules,* and *ribosomes* plus a number of other housekeeping structures similar to those found in other cells.

Cardiac Muscle

Cardiac muscle, like skeletal muscle is striated and operates essentially the same, except the control of contraction is involuntary rather than voluntary (Figure 2.36). *Cardiocytes* or cardiac muscle fibers

(A)

(B)

Figure 2.36 Photomicrographs of cardiac muscle. A longitudinal section (A), showing fiber arrangements and nuclei locations and, (B) an electron photomicrograph showing the banding pattern, intercalated disks, mitochondria, and glycogen granules. From *A Textbook of Histology, Ninth Edition* by William Bloom & Don Fawcett. Copyright © 1968 by Don Fawcett. Reprinted by permission.

are relatively small, only 5 to 10 μm in diameter and 50 to 120 μm in length, and each cell has centrally located nuclei, usually 1 to 2 nuclei per cell. Cardiac muscle cells have an elaborate communication network that allows the heart to coordinate repeated, uniform contractions. Communication of this nature occurs at small connections between cells known as *intercalated discs*. These discs appear as dark irregular demarcations that run perpendicular to the long axis of the muscle fiber (Figure 2.36). The structures allow an electrical impulse to migrate through each cell. Uniform contraction is guaranteed because cardiac myofibrils are attached directly to the intercalated discs and allow the heart to act in *syncytium* or as a group.

Another feature of cardiac muscle that makes it unique to skeletal muscle is that the thin filaments (actin) are quite variable in length. In skeletal muscle, the thin filaments are uniform in length and if muscle is stretched to a point where minimal overlap between actin and myosin occurs, little to no force can be generated because sufficient cross-bridge formation is hampered. Unlike skeletal muscle, cardiac muscle must contract even under these extreme conditions. Thus, even in cardiac muscle that has experienced extreme stretching, there will be sufficient overlap of the thick and thin filaments to generate a contraction and continued circulation.

Another unique characteristic of cardiac muscle fibers is that they are branched cells. Cardiac muscle cells also have more mitochondria than skeletal or smooth muscle and may contain variable amounts of lipid droplets.

Smooth Muscle

Smooth muscle is found as sheet-like muscle masses surrounding various organs. Smooth muscle plays a critical role in regulating fluid movement through blood vessels and digesta through the intestinal tract. Special features of smooth muscle are its extreme elasticity and pliability. A prime example of this plasticity is found in the smooth muscle of the uterus *(myometrium)*. In a gravid uterus, the muscle layer is quite thick, whereas the uterus at full term is a few millimeters in thickness. Such elasticity allows smooth muscle cells to stretch to incredible lengths, yet retain their capability of performing contraction at such an insulted state. Extension of skeletal muscle to such limits would render it useless because the normal overlap of the thick and thin filaments would be abolished.

Morphologically, smooth muscle differs substantially from skeletal muscle (Figure 2.37). Microscopic

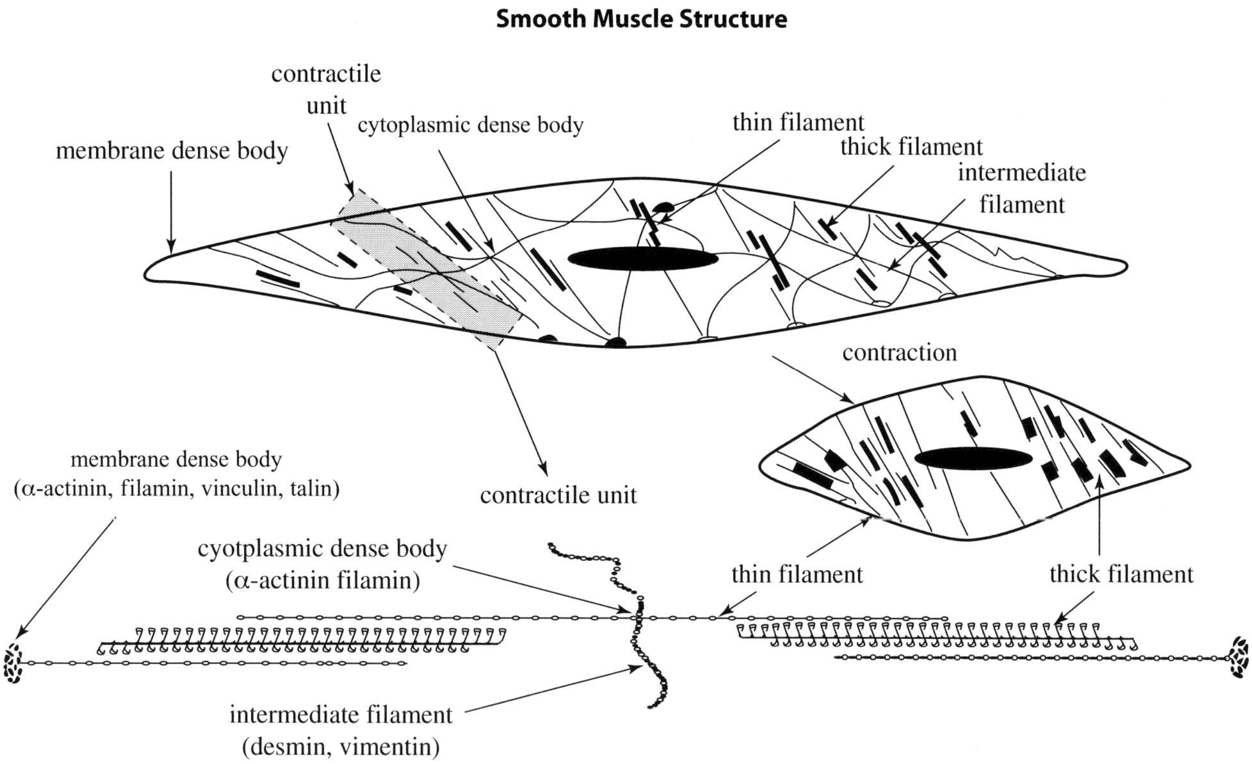

Figure 2.37 Organization of smooth muscle elements. Courtesy of D. R. Swartz, Department of Anatomy and Cell Biology, Indiana University. From *Proceedings of the American Society of Animal Science,* 1998. Copyright © 1998 by American Society for Animal Science. Reprinted by permission.

observation of smooth muscle reveals that each muscle cell contains only one centrally located nucleus in contrast to peripherally located nuclei in skeletal muscle fibers. Furthermore, smooth muscle is more triangular in cross sectional appearance compared to the typical polygonal shape of skeletal muscle cells.

Smooth muscle cells range from 5 to 10 μm in diameter and from 30 to 260 μm in length. The functional difference between striated muscles and the non-striated smooth muscle is based on the organization of myofilaments. Smooth muscle cells, like skeletal muscle cells, contain actin and myosin throughout the entire cell. However, the thick and thin filaments are not systematically arranged into repeating sarcomere structures and, therefore, smooth muscle does not have a striated appearance (Figure 2.37). In smooth muscle cells, there are *dense zones* or *dense bodies* along the cell membrane and other locations that are analogous to Z-line structures of striated muscle. Just as actin molecules radiate out from the Z-lines in striated muscle, actin molecules are attached to dense bodies in smooth muscle cells. Thick filaments (myosin filaments), likewise are positioned strategically throughout the cell, overlapping with actin filaments. As a result, smooth muscle can contract and generate a force in multiple directions. This type of contraction is especially valuable to organs that require uniform pressure exerted from many directions such as the uterus and bladder, and tubular structures that regulate flow of body functions. Like cardiac muscle, smooth muscle is involuntary and therefore, requires different stimuli to cause contraction.

Chemical Composition

Understanding overall composition of muscle is essential for understanding growth because some elements of its composition may change with physiological modifications associated with growth (Table 2.5). The principal constituent of muscle is water. Water accounts for 65 to 80% of the total volume of muscle. Water is located between the cells in

Table 2.5 Approximate Composition of Mammalian Skeletal Muscle (Percent Fresh Weight Basis)

Component (range)	Percent	Component	Percent
Water (65 to 80)	75.0	**Non-protein nitrogenous substance**	1.5
Protein (16 to 22)	18.5	Creatine and creatine phosphate	
Major contractile proteins		Nucleotides	
myosin		(adenosine triphosphate (ATP),	
actin		adenosine diphosphate (ADP),	
Regulatory proteins		(etc.)	
tropomyosin		Free amino acids	
troponin (complex)		Peptides	
tropomodulin		(anserine, carnosine, etc.)	
Cytoskeletal proteins		Other nonprotein substances	
intrasarcomeric proteins		(creatinine, urea, inosine	
titin		monophosphate (IMP), nicotinamide	
nebulin		adenine dinucleotide (NAD),	
C-protein		nicotinamide adenine dinucleotide	
myomesin		phosphate (NADP)	
creatine kinase		**Carbohydrates and non-nitrogenous**	
α-actinin		**substances** (0.5 to 1.5)	1.0
H-protein		Glycogen	
M-protein		Glucose	
intersarcomeric proteins		Intermediates and products of	
desmin		cell metabolism	
filamin		(hexose and triose phosphates,	
vinculin		lactic acid, citric acid, fumaric	
synemin		acid, succinic acid, acetoacetic	
skelemin		acid, etc.)	
Cap Z		**Inorganic constituents**	1.0
paranemin		Potassium	
dystrophin		Total phoshorous	
talin		(phosphates and	
Sarcoplasmic proteins		inorganic phosphorous	
soluble sarcoplasmic and		Sulfur (including sulfate)	
mitochondrial enzymes		Chlorine	
myoglobin		Sodium	
hemoglobin		Others	
cytochromes and flavo-		(including magnesium, calcium,	
proteins		iron, cobalt, copper, zinc,	
Stromal proteins		nickel, manganese, etc.)	
collagen			
elastin			
other insoluble proteins			
Lipids (range 1.5 to 13.0)	3.0		
Neutral lipids			
Phospholipids			
Cerebrosides			
Cholesterol			

the extracellular spaces and forms an avenue for chemicals to diffuse in and out of cells from the vascular bed. Furthermore, the ions associated with this fluid are crucial for establishing *membrane potentials* or charges that exist across cell membranes.

Protein, accounts for 16 to 22 % of the mass of muscle. Proteins may be assigned to three distinct classes:

- *Myofibrillar Proteins*
- *Sarcoplasmic Proteins*
- *Stromal Proteins*

Myofibrillar protein represents the bulk of proteins in muscle and includes any protein associated with the myofibrils. Three subclasses of myofibrillar proteins are:

- The contractile proteins, myosin and actin
- The regulatory proteins which are generally restricted to tropomyosin and the troponin complex (T, I and C), and
- The cytoskeletal proteins involved in the scaffolding structure of the myofibrils.

Myofibrillar proteins are soluble in solutions with high ionic strength (such as high salt solutions).

Sarcoplasmic proteins are those associated with the cytoplasm and are chiefly enzymes and other factors that facilitate cellular activities, such as energy metabolism. The sarcoplasmic proteins are readily soluble in low salt solutions.

Stromal proteins are those proteins associated with the structural framework or connective tissues of muscle. In contrast to sarcoplasmic proteins, the stromal proteins are quite insoluble. Collagen, reticulin and elastin are examples of stromal muscle proteins.

Proteins contain nitrogen but all compounds containing nitrogen are not necessarily proteins. This constituent of muscle tissue is often referred to as *non-protein nitrogen* and consists of compounds such as ATP, vitamins, nucleic acids and small peptides. This constituent represents a fairly small proportion of the total mass of muscle (~1.5%), yet is critical to the overall function of the tissue.

Lipid is the most variable constituent of muscle, ranging from 1.5% to 13% or more of wet weight. Lipid is primarily present as neutral lipids or triglycerides, however a portion of the total arises from the phospholipid that is normally found in all bilipid membranes. The majority of the neutral lipid arises from adipose tissue that is deposited within the perimysium of muscle as marbling.

As one might expect, muscle tissue does not contain an appreciable amount of carbohydrate. In fact, only about 0.5 to 1.3 % of the total mass of muscle tissue is carbohydrate. Most of the compounds comprising the carbohydrate fraction of muscle are associated with glycolytic metabolism. In particular, glucose, glycogen, lactate and other intermediates of energy metabolism represent the bulk of muscle carbohydrate. One exception to this, however, is highly glycosylated proteins that significantly contribute to the overall abundance of carbohydrate in muscle.

The remaining constituents of muscle are inorganic compounds and represent the least of the total mass. Various cations and anions are distributed throughout the intra- and extracellular spaces and are vital to the function the tissue. Specifically, sodium, magnesium, potassium, calcium, iron, phosphate, selenium and chloride ions are abundant in muscle tissue.

Chapter 3

Embryonic Growth

Introduction

The complete process of reproduction is a vast field of interest. A detailed description and understanding of this essential process is well beyond the scope of this text. However, an understanding of prenatal development is critical to understanding postnatal growth, as some events occurring during this period impact later growth.

The process of creating offspring is a very complex one that requires the orchestration of a myriad of events that must occur at the appropriate time with great precision (Figure 3.1). In fact, any deviation from the normal developmental program will likely result in death of the forming organism and termination of pregnancy. Furthermore, the time spent *in utero* or in the uterus represents a significant proportion of the entire lifetime of an animal. In particular, sheep, cattle and pigs, raised solely for the purpose of meat production, spend nearly 85, 60 and 70%, respectively, of their entire existence in the prenatal stage. In poultry, the time spent *in ovo* or in the egg relative to the postnatal period of growth is considerably less, but nonetheless represents a significant formative period in the life cycle of an animal. In fact, factors that impact prenatal growth and development can severely impact postnatal growth. Therefore, the old adage "that you have what you have developed" is truc, with some minor but important exceptions.

Development of literally all tissues of the body starts at fertilization of the egg (Figure 3.2). Through a myriad of *cellular replications,* various *cell lineages* or cell families develop, which later develop into different tissues of the body. From a single fertilized cell, there is a massive amplification of cells that requires many cell *divisions* or replications. During each division, cells become more *restricted,* in terms of the type of cells that they are capable of becoming. Another way of stating this is that a fertilized cell is *multipotent* or *totipotent* and has the potential of developing into multiple cell types or tissues (Figure 3.1), but as development continues, each generation of new cell types loses ability to differentiate into different cell types. Therefore, after each division, cells may become more and more restricted to a specific cell lineage, or fate.

This developmental process requires a significant amount of time to reach completion and, depending on the size of the animal at birth, the time needed to allow for increases in cell numbers may differ with species. In fact, the rate of prenatal growth in any species is related to its postnatal growth rate (Figure 3.3). Animals that grow slower at the cellular or tissue levels likely do so because they require additional time to develop more cells, which ultimately translates into additional weight and mass at birth. As a result, however, these animals havc grcatly cxpandcd ccll numbcrs, and consequently, have exaggerated growth rates and larger adult body sizes. This is probably a reflection of all cells growing at the same time.

This same sort of phenomenon can be observed within species as well. For example, "Continental" breeds of cattle, such as Charolais, Simmental and Limousin have much larger mature body sizes and longer gestation periods than their smaller "British" breed counterparts (~287 versus 282 days, respectively). Therefore, even within species, larger animals require more time than smaller animals during

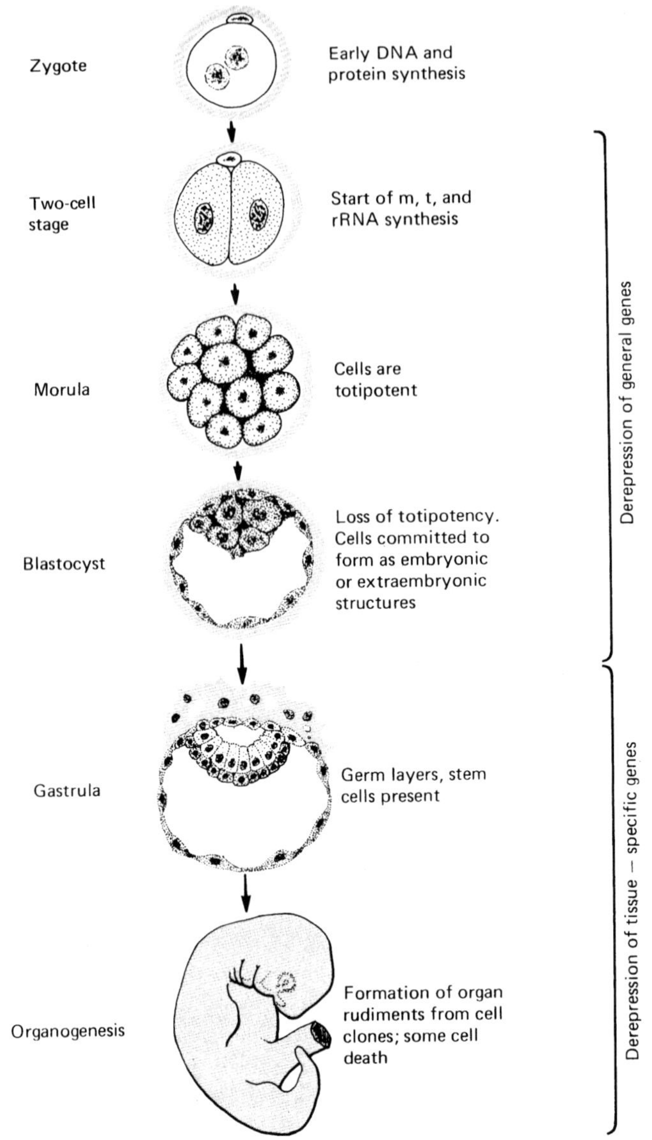

Figure 3.1 Overview of mammalian development. From B. M. Carlson, *Patten's Foundations of Embryology,* McGraw-Hill Co., New York.

Table 3.1 Implantation in a Variety of Domestic Farm Animals

Species	Maternal recognition (days)	Attachment begins (days after ovulation)	Types of Placenta
Cattle	16–17	28–32	Cotyledonary
Sheep	12–13	14–16	Cotyledonary
Swine	10–12	12–13	Diffuse
Horse	14–16	35–40	Diffuse

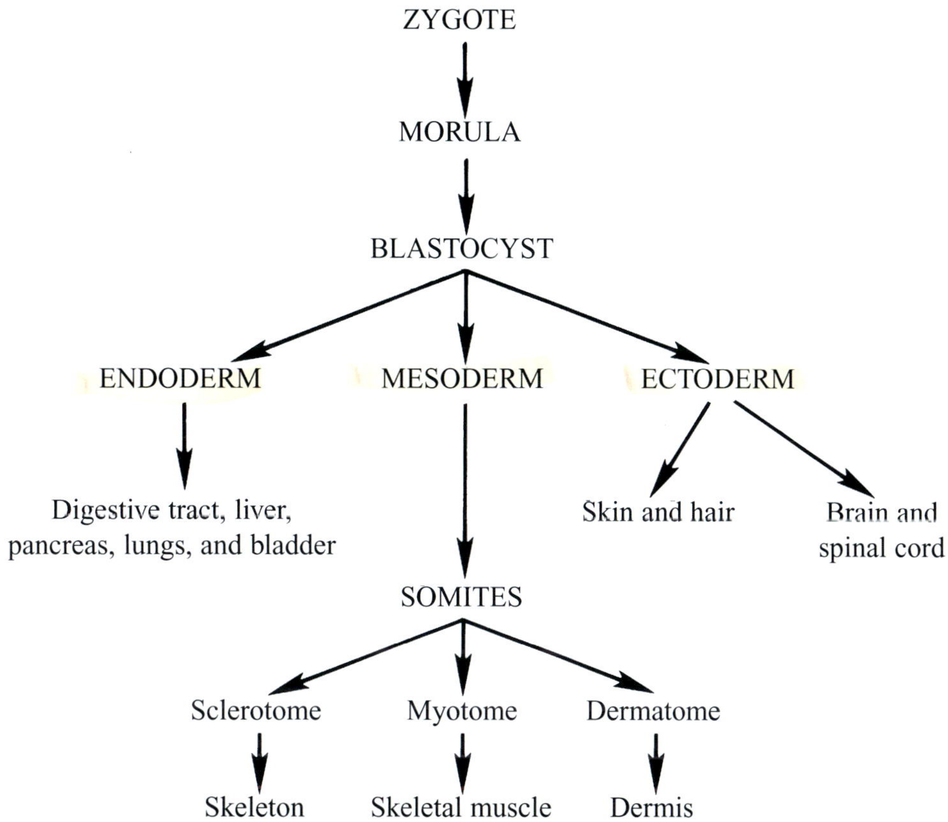

Figure 3.2 Differentiation of tissues during prenatal growth and development.

prenatal development mainly because they are generating more cells. As a result of this increased number of cells developed, postnatal growth rate is greater in larger animals because there are more cells, which collectively, result in exaggerated growth postnatally.

Prenatal Development

In utero growth and development of mammals can be divided into:

- *Ovum Phase*
- *Embryonic Phase*
- *Fetal Phase*

Although these terms are often used interchangeably, the phases of development need to be defined, but one should realize that this type of classification relies heavily on subjective criteria. By most definitions the ovum phase of development is the period of time from fertilization of the ovum to implantation. The embryonic phase, is completed when the developing organism begins to reflect its mature form (e.g., a developing pig begins to take on the form and appearance of a pig). From the point where species identification can be established onward, a developing organism is considered a *fetus.* Obviously, the fetal phase of development, at least in most animals, is the longest phase. The phase shortly after birth is considered the *neonatal phase* of development and will be discussed in later chapters.

Ovum Phase

As background for this discussion, recall that *ova, oocytes* or unfertilized eggs are released from female *gonads* or *ovaries* (Figure 3.4). A single egg (*ovum*) is surrounded by a tough, elastic coat known

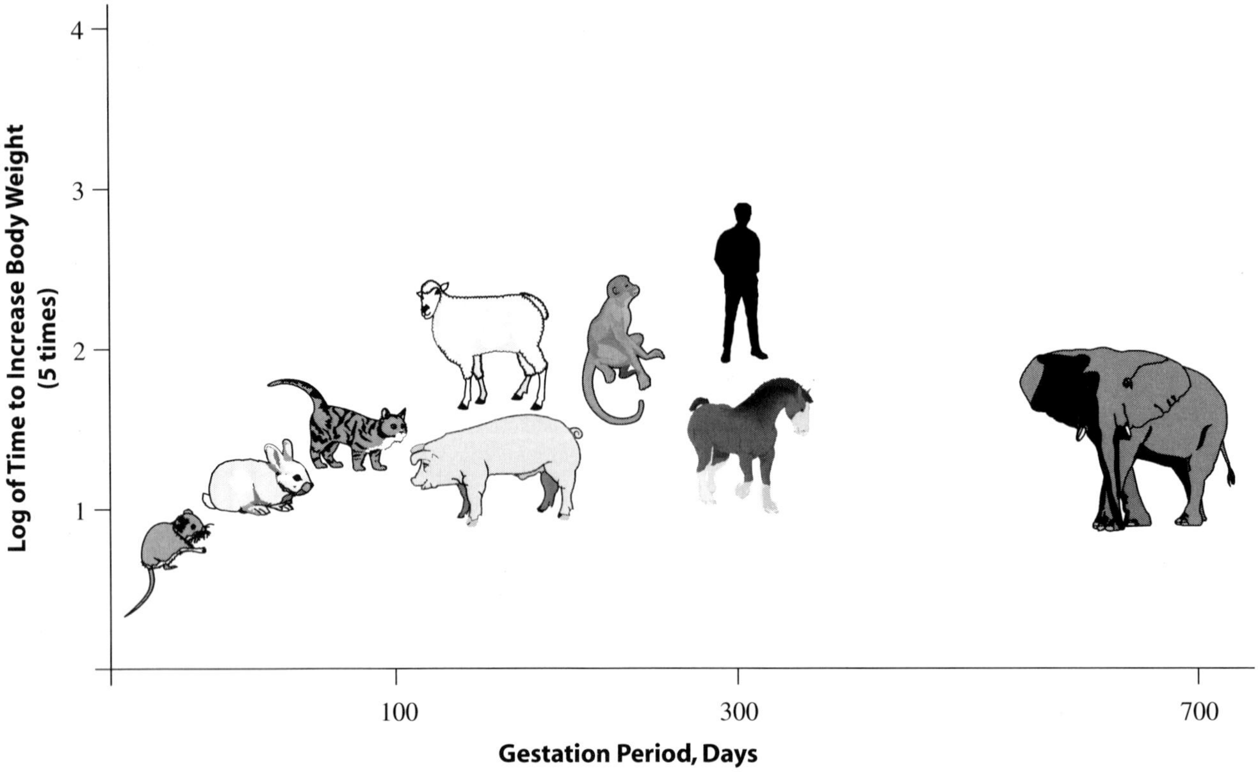

Figure 3.3 Relationship between gestation length and body weight gain.

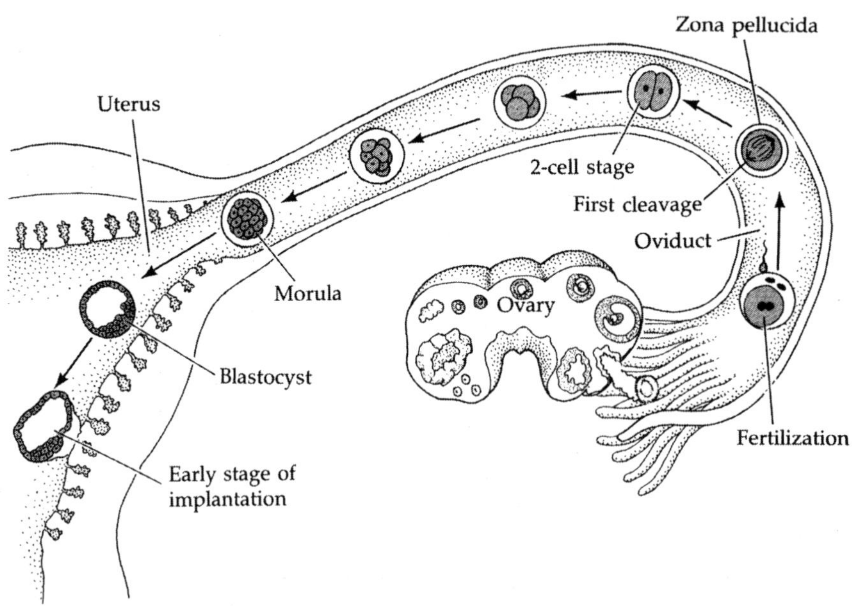

Figure 3.4 Movement and development of human embryos during early embryo development. From S. Gilbert, *Developmental Biology, Third Edition,* Sinauer Associates, Inc., Sutherland.

as a *zona pellucida.* After the ovum reaches the fallopian tubes the male gamete, *sperm,* must penetrate the zona pellucida and fertilize the ovum, depending on species, at various locations in the female reproductive tract. This event results in the formation of a *zygote* (fertilized egg).

The entire process whereby sperm penetrate the zona pellucida and ultimately the ovum, sets in motion a cascade of events known as *oocyte activation* and, in due course, development of viable offspring.

Immediately after penetration by the male gamete, the rate of metabolism in the oocyte increases dramatically, which increases the vigor of the developing organism.

Meiosis, which essentially ensures proper chromosome number, occurs to expel the second *polar body* or nonfunctioning DNA complements. At this point, the only DNA present in the oocyte is contained within the two haploid (1N) male and female pronuclei. A *pronucleus* (1N) is simply an enlarged area of concentrated DNA in a gamete that is positioning itself for imminent segregation events.

Once the pronuclei from both gametes migrate to the center of the cell, fertilization is complete and the appropriate number of chromosomes (2N) is restored. At this point, *cytokinesis* events or development of daughter cells ensues.

The ovum phase of development is relatively short (11 to 14 days) and is characterized by rapid cell division known as the period of *cleavage.* The process of cleavage is completed when the ovum comes in contact with the uterine wall (Table 3.1). Figure 3.5 shows that during the cleavage phase of development, there is a massive increase in cell number. This process occurs in the absence of protein synthesis and ultimately results in an increase in the DNA to protein ratio. It cannot be overemphasized at this point that the primary impact of cleavage on the developing embryo results in an increased cell number and a decrease in cell size, the latter being more typical of cells in the adult animal. This process of increasing numbers of cells without change in overall size is illustrated in Figure 3.6.

Morphologically, the once single-celled structure becomes transformed into many uniformly sized, loosely associated embryonic cells or *blastomeres.* At approximately the 8-cell stage, the structure goes through a stage of *compaction* (Figure 3.7) during which the inner cells become closely associated. Additional divisions, giving rise to a morula (20 to 30 cells, depending on the species) signal the end of the ovum and cleavage phases of development.

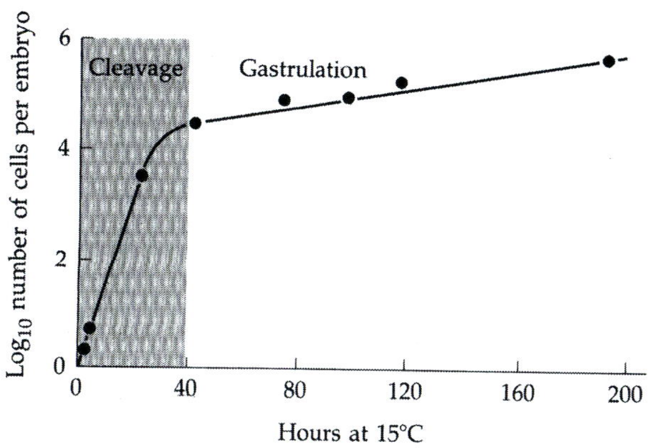

Figure 3.5 Change in embryo cell number during early development. From S. Gilbert, Developmental Biology, 3rd ed., Sinauer Associates, Inc., Sutherland.

After the fertilized egg reaches the uterine *lumen* or cavity, the zona pellucida is lost (Figure 3.4). This process, in some cases resembles, in part, the liberation of a chick from a shelled egg and is often referred to as the process of *hatching.* Although physical cracking of the zona pellucida occurs in cattle, in most other species the zona pellucida is merely digested or degraded away from the egg.

Blastomeres are characterized as having great developmental *plasticity* or flexibility. Embryos that contain damaged cells still retain the capacity of undergoing normal development. This demonstrates that each cell in the early embryo is not restricted to a particular assignment during development, or destined to develop a particular tissue. In fact, many blastomeres can be removed from a developing embryo without severe detriment to embryo survival. To illustrate this plasticity even further, *embryo splitting* has been performed successfully in a number of species including cattle. During this process, embryos are physically divided into two separate clusters of cells. These clusters or split embryos then may be transferred into a female using standard embryo transfer techniques and development of identical twins will result. Even though viability and pregnancy rates diminish, successive embryo splitting may be accomplished. This amazing capability likely ensures development by

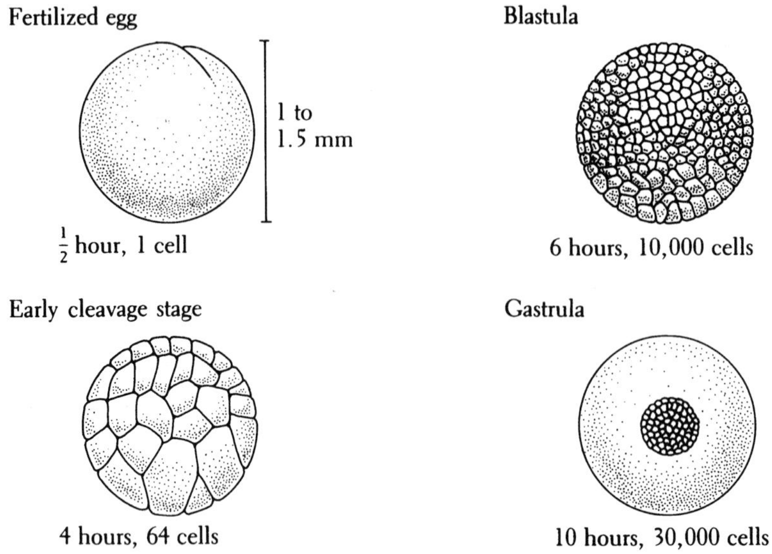

Figure 3.6 Relative change in cell size and number during early frog development.

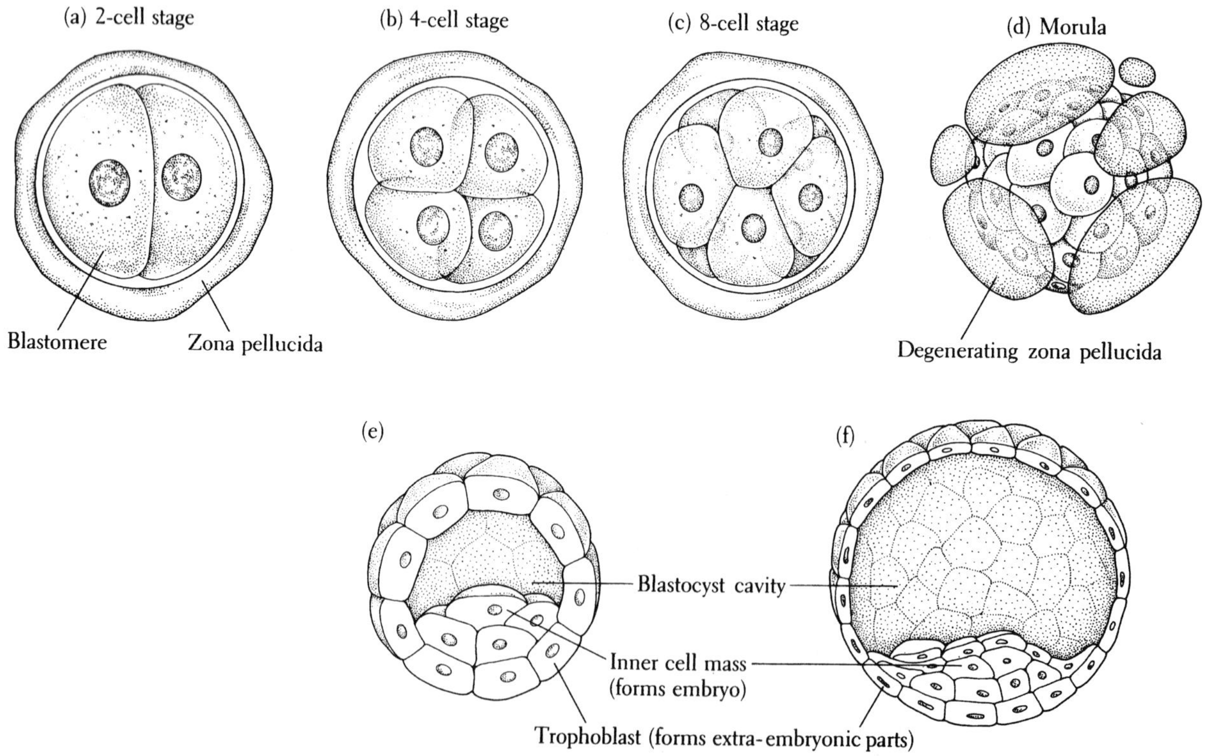

Figure 3.7 Blastocyst formation in the human embryo. From *Developmental Biology* by N. Holder and V. Walbot. Copyright © 1987 by McGraw-Hill Companies, Inc. Reprinted by permission of McGraw-Hill Companies, Inc.

allowing each cell the ability to control developmental programs independently.

Morphogenesis

Morphogenesis or the "creation of shape" is a term used to refer to the collective changes in the embryo as directed by changes in the internal and external structures. There are several processes that enable an embryo to take on a vast number of morphological changes. Of particular interest is the development of multiple cell layers. The process of a single cell layer giving rise to multiple layers is particularly important during morphogenesis and can occur by different means (Figure 3.8).

- *Invagination* yields multiple cell layers by creating a depression in the surface of the embryo. As the depression becomes deeper and deeper, the sides of the depression come together, forming multiple layers.
- *Involution* occurs when new layers are formed by the inward rolling of cells along an existing membrane.
- *Ingression* occurs when cells begin to replicate on the internal aspects of an existing cellular layer and form another layer.
- *Epiboly* is the increase in cell numbers on the outer surface of an embryonic layer.
- *Delamination,* as the name implies, develops new structures by dividing an existing layer into two separate structures.

Embryonic Phase

Progression from the cleavage phase to the blastula phase or *blastulation* occurs very rapidly and is characterized by two main events. First, by increasing the number of blastomeres, a hollow ball of cells forms known as the *blastocyst.* The fluid-filled cavity within this structure is referred to as the *blastocoele* (Figure 3.7). Second, there is the formation of a flattened layer of cells, or trophoblasts, that surround the blastocoele. Furthermore, a more compact mass of cells (*inner cell mass*) forms. This collection of cells pushes into the blastocoele as a consequence of rapid cell proliferation on one side of the blastocyst at a point known as *Koller's Sickle* or the *posterior marginal zone* (Figure 3.9). As the number of cells at this location increases dramatically, some begin to migrate across the blastocyst

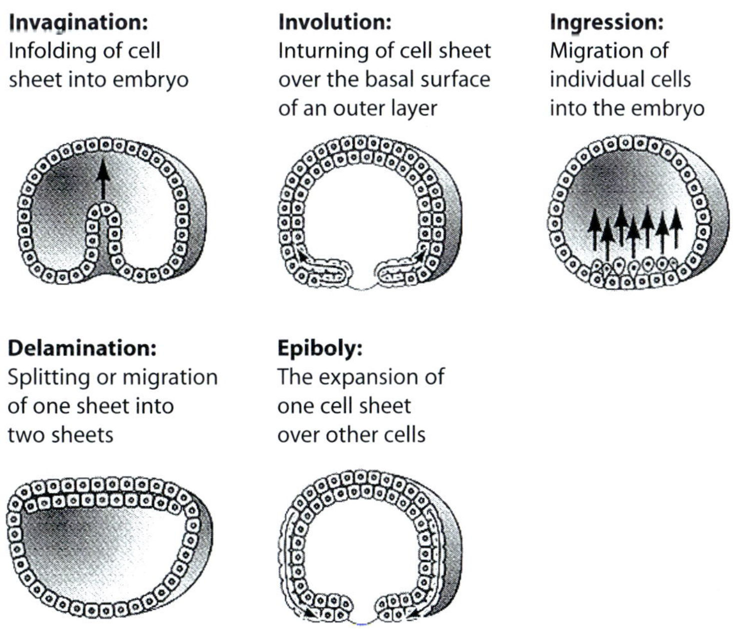

Figure 3.8 Types of cell movements during gastrulation. From *Developmental Biology, Third Edition* by Scott Gilbert. Copyright © 1991 by Sinauer Associates, Inc. Reprinted by permission.

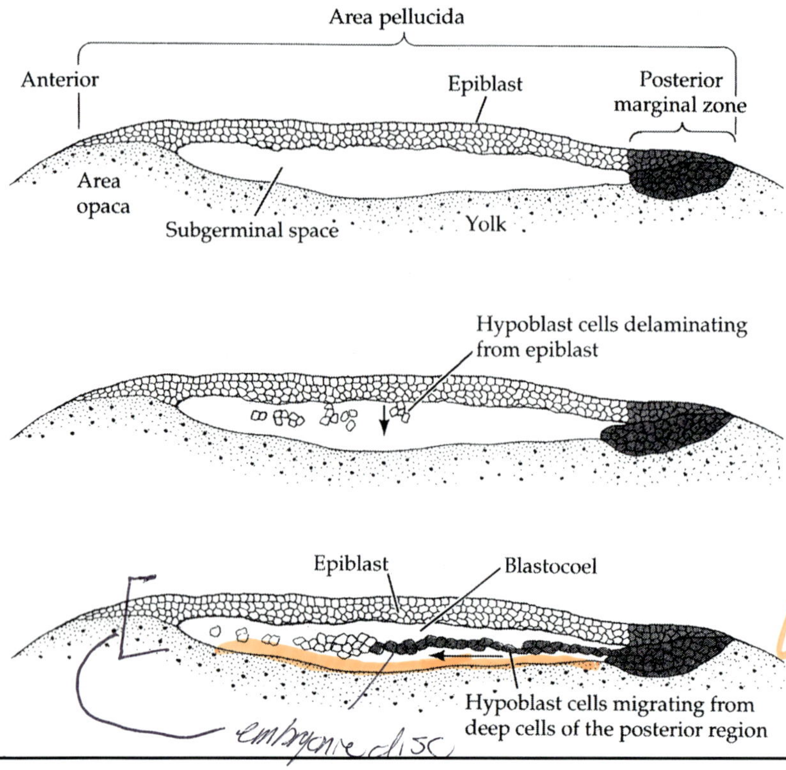

Figure 3.9 Illustration of hypoblast formation during avian development.

and form the *embryonic disc* in the area known as the *area pellucida.* At this point, the embryonic disc now consists of two layers, the *epiblast* that resides toward the *animal pole* and is the origin of all tissues of the animal and the *hypoblast* that directly underlies the epiblast layer (Figure 3.9). The hypoblast cells further divide and participate in the formation of the remainder of the extraembryonic tissues.

Gastrulation

After formation of the blastula, the embryo enters the *gastrulation* phase of development. Gastrulation, derived from the Greek word, *gaster,* which means stomach, begins a period when regions begin to exert more control over development rather than individual cells controlling their own fate i.e., *cellular autonomy.* Furthermore, a great deal of cellular reorganization and movement occurs during this phase of prenatal development.

At this point, cells of the epiblast layer begin to replicate in a region known as the *Hensen's Node* that is located in the more dorsal and caudal (towards the posterior end) end of the developing embryo (Figure 3.10). As the cells continue to replicate and increase in number in this region, a thickened structure begins to form parallel to the long axis of the embryo known as the *primitive streak* (Figure 3.10). This structure covers the most caudal two-thirds of the embryonic disc. The anterior portion of the embryonic disc, lacking the primitive streak, ultimately participates in the formation of the head of the animal. Shortly after the primitive streak is formed, a small depression develops within the dorsal most aspects of the streak creating the *primitive groove.* At this point, cells begin to migrate medially into this groove and ventrally through a structure known as the *primitive knot* (Figures 3.10 and 3.11a). As more and more epiblasts migrate beneath the epiblast layer, just above the hypoblast layer, a three-layered embryo is developed (Figure 3.11a). The resulting three layers comprise the three primary *germ layers* or rudimentary tissues of the developing embryo:

1. The *ectoderm,* the outer most cell layer of the embryo forms cells that ultimately form the skin and neuronal tissues.

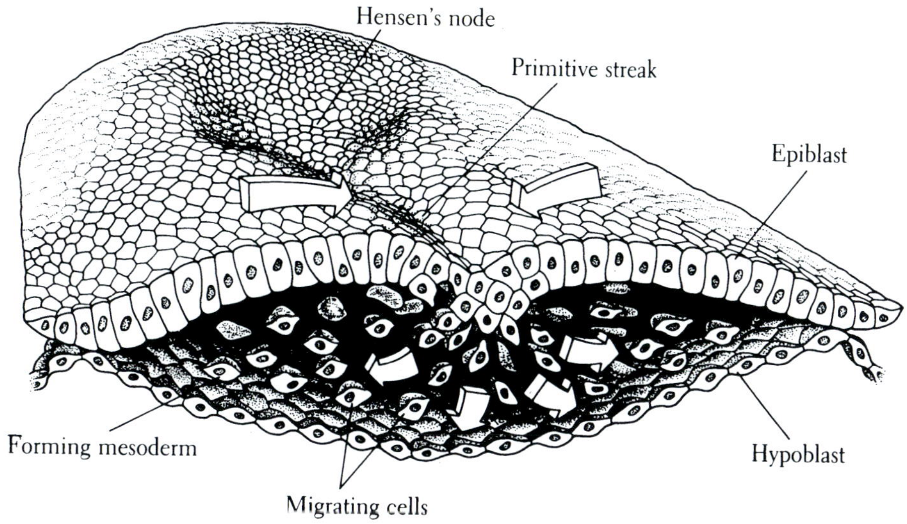

Figure 3.10 Migration of cells during early avian development. From *Developmental Biology* by N. Holder and V. Walbot. Copyright © 1987 by McGraw-Hill Companies, Inc. Reprinted by permission of McGraw-Hill Companies, Inc.

2. The *mesoderm* or middle germ layer participates in the development of muscle, fat and bone.
3. The *endoderm* develops into tissues of the respiratory and digestive tracts.

Neuralation

Neuralation is an event during development that results in the formation of the spinal cord precursors and occurs simultaneous with events of gastrulation (Figures 3.11b,c and 3.12). As elements of the primitive groove continue to develop, the ridges forming the outer limits of the groove elevate and become more pronounced. The outermost layer of cells representing the primitive groove are now known as the *neural ectoderm* and are much different than those cells of non-neuronal ectoderm. As development continues, these ridges begin to converge and ultimately fuse together. As a result, a *neural tube* is formed. This neural tube, as well as the smaller structure, located just beneath or ventral to the neural tube, the *notochord,* form the earliest identifiable structures of the spinal cord found in a developing embryo. The apical or upper most aspects of the grooves form the *neural crest.* Once the neural tube is closed, cells from the neural crest migrate to peripheral aspects of the embryo to form the peripheral nerves.

Somitogenesis

As mesodermal cells become more and more prominent, they form morphologically distinct ridges of cells along the primitive groove. These two ridges, adjacent to the neural tube, are known as *paraxial mesoderm* or *lateral plate mesoderm.* Upon further embryonic maturation, these elongated collections of cells begin to organize into cuboidal masses known as *somites* (Figures 3.13 and 3.14). This highly ordered event during development occurs in a *rostral* (toward the head) to caudal (toward the tail) fashion and each somite is paired to a corresponding somite on the other side of the neural tube (Figure 3.14). Soon after somite formation, these cuboidal cell masses are clustered together in a fairly random nature. However, shortly after formation, these somitic cells transform into cells possessing a more columnar, epithelial morphology. The most apical portion of the cell is oriented toward a hollow center known as the *somitocoele* (Figure 3.15). The function of this "hollow" center is not yet known but is likely filled with a hormone-rich fluid that is important in regulating development of this structure.

Shortly after somitocoele formation, two populations of somitic cells are evident, the *dermomyotome* and the *sclerotome* (Figure 3.15). Collectively,

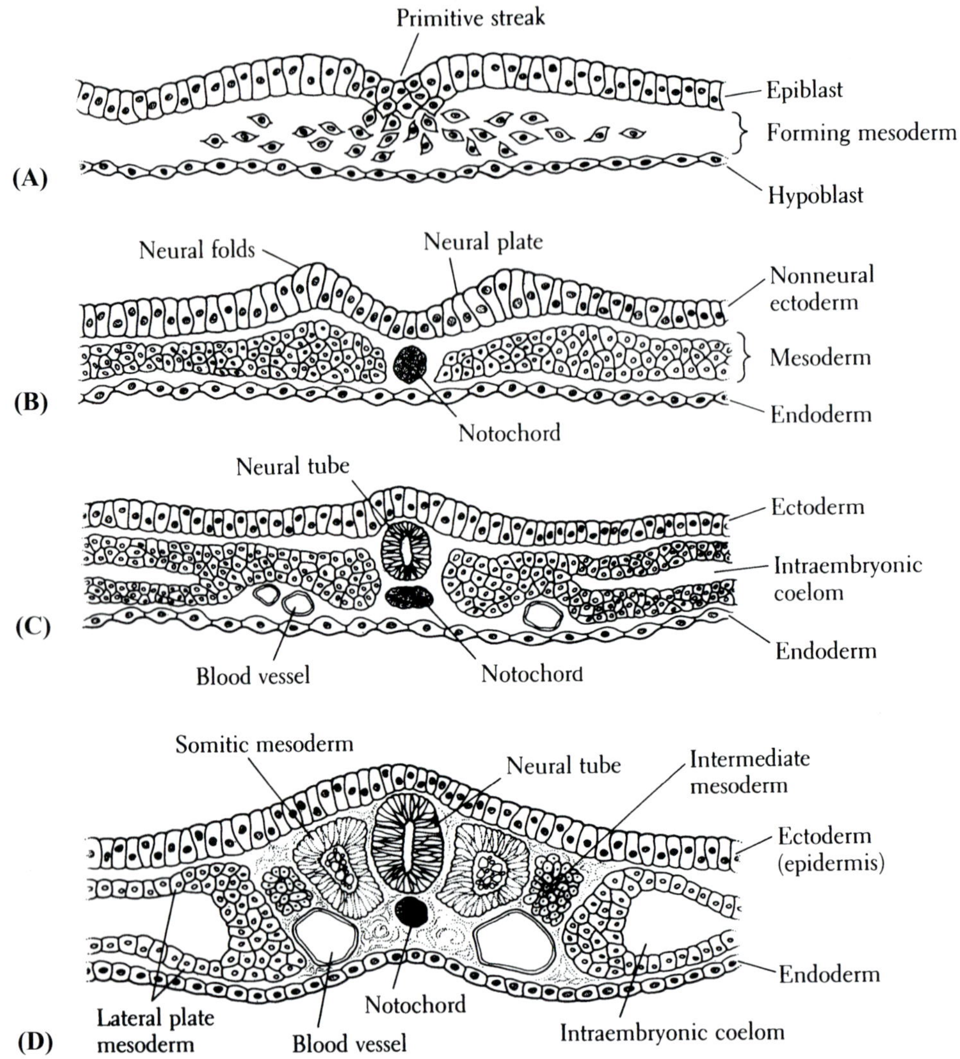

Figure 3.11 Formation of various structures during avian neurulation. From *Developmental Biology* by N. Holder and V. Walbot. Copyright © 1987 by McGraw-Hill Companies, Inc. Reprinted by permission of McGraw-Hill Companies, Inc.

the dermomyotome contains two types of *progenitor* or precursor cells. The sclerotomal cells, which are closest to the neural tube, *delaminate* or separate from the rest of the somite and migrate around the neural tube and form the vertebral column of the developing organism (Figure 3.14, 3.15, and 3.16). Cells of the dermomyotome, delaminate and form two layers as well. The cells located closest to the outer ectoderm form the *dermatome* while the cells just beneath and ventral to the structure form the *myotome* (Figure 3.16). The cells of the myotome are essentially responsible for developing all the muscles of the body. Those myotomal cells located in the myotome are located at the most lateral aspects of the structure, closest to developing limbs, migrate to form the muscles in the limb. Cells of the remaining dermatome form cells of the dermis. Remaining myotomal cells migrate to other locations in the body and form muscle tissue of the axial skeleton. These migrations involve *chemotaxis,* which means that myoblasts follow a "chemical trail" to their intended destiny in the body. Alternatively, other factors or *morphogens,* are produced in one location of a developing embryo but exert an effect on distant

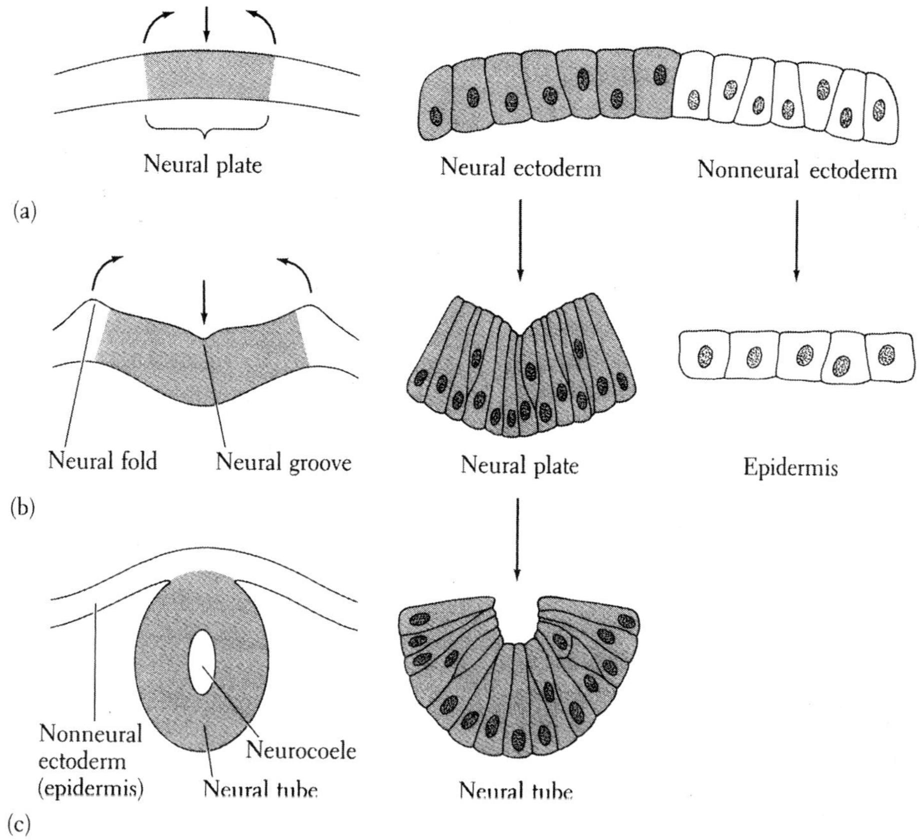

Figure 3.12 Changes in cell morphology during the formation of the neural tube. From *Developmental Biology* by N. Holder and V. Walbot. Copyright © 1987 by McGraw-Hill Companies, Inc. Reprinted by permission of McGraw-Hill Companies, Inc.

cells. Defining the signals that are responsible for developing different structures and the migration of cells to their various locations of propagation is a very active area in biological research. Understanding those mechanisms may allow for development of strategies to control muscle development in domestic animals.

Limb Bud Formation

Although somewhat peripheral to a text of this nature, limb bud development is one of the most well-studied events of embryonic development. For reasons yet unknown, cells from somites closest to the point of limb formation migrate to a point just under the surface ectoderm to form the limb bud *primordial* or mass of precursor cells (Figures 3.17 and 3.18). Of particular importance to the development of limbs is the presence of various *organizing centers* (Figure 3.19), which act to coordinate the development of tissues in a given area. The *zone of polarizing activity* is located at the most posterior aspect of the developing wing bud and organizes the developing limb in the anterior to posterior axis. In contrast, the *apical ectodermal ridge* is responsible for maintaining the outgrowth of the limb by keeping the underlying cells in the *polarizing zone* in an undifferentiated state. The remaining dorsal and ventral ectoderm is responsible for maintaining position in the final dimension.

Fetal Phase

The bulk of organ and tissue development is completed during the earliest stages of prenatal development. Although much simplified, once the animal

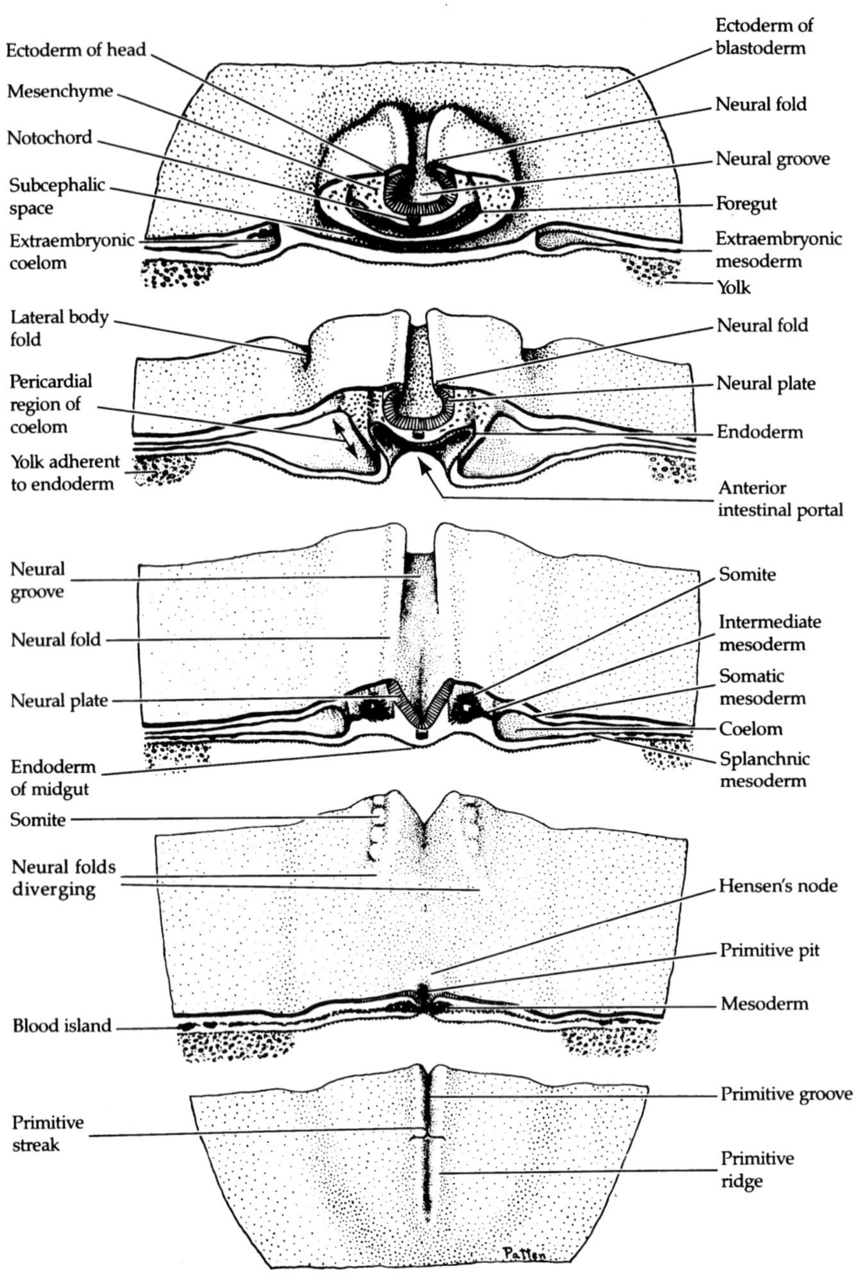

Figure 3.13 Illustration showing the changes in somite morphology during chick embryo development. From *Developmental Biology* by N. Holder and V. Walbot. Copyright © 1987 by McGraw-Hill Companies, Inc. Reprinted by permission of McGraw-Hill Companies, Inc.

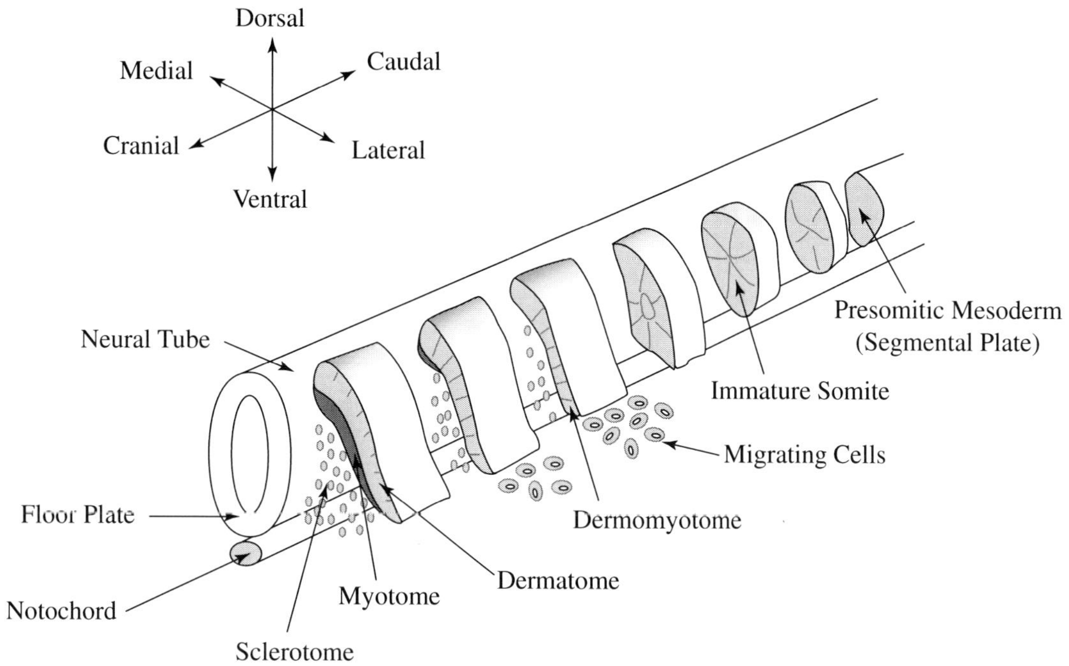

Figure 3.14 Formation and organization of newly formed somites adjacent to the neural tube. From M. Buckingham, *Molecular Biology of Muscle Development* (1994), Cell, vol 78, pg. 15.

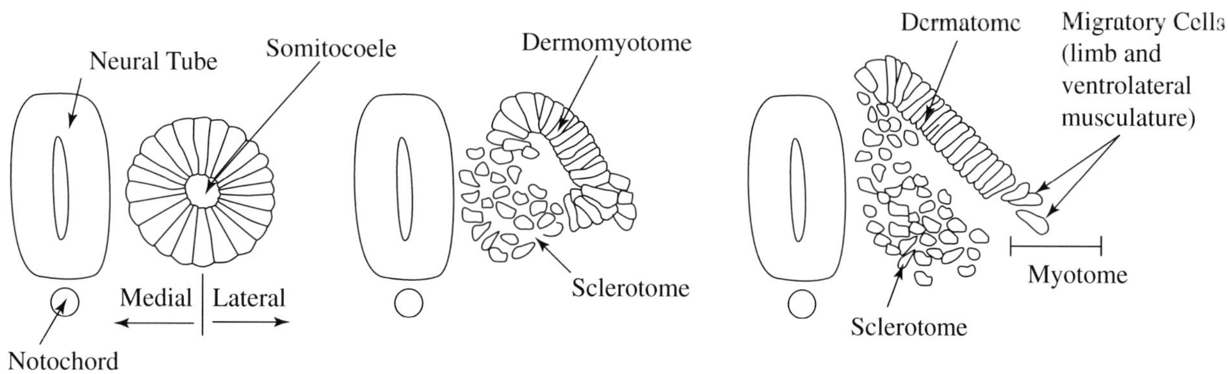

Figure 3.15 Maturation of a somite into the dermatome, myotome, and sclerotome. From *Developmental Biology, Third Edition* by Scott Gilbert. Copyright © 1991 by Sinauer Associates, Inc. Reprinted by permission.

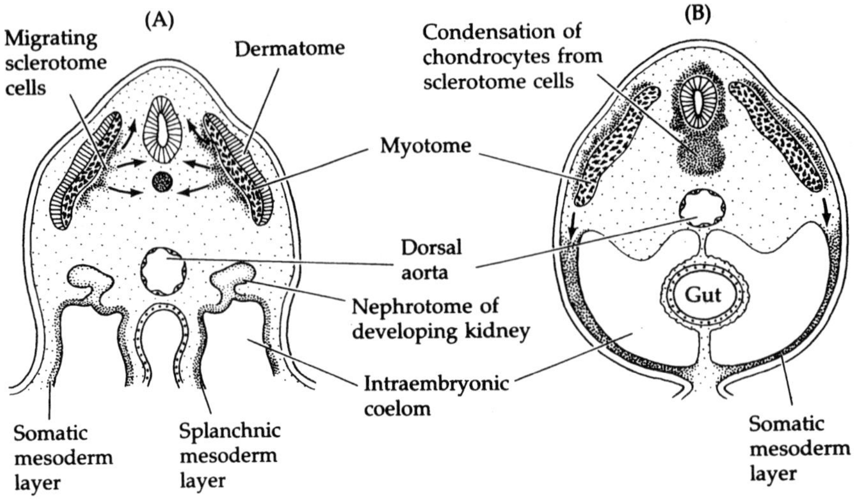

Figure 3.16 Migration of sclerotomal cells in preparation for formation of the vertebral column. From *Developmental Biology, Third Edition* by Scott Gilbert. Copyright © 1991 by Sinauer Associates, Inc. Reprinted by permission.

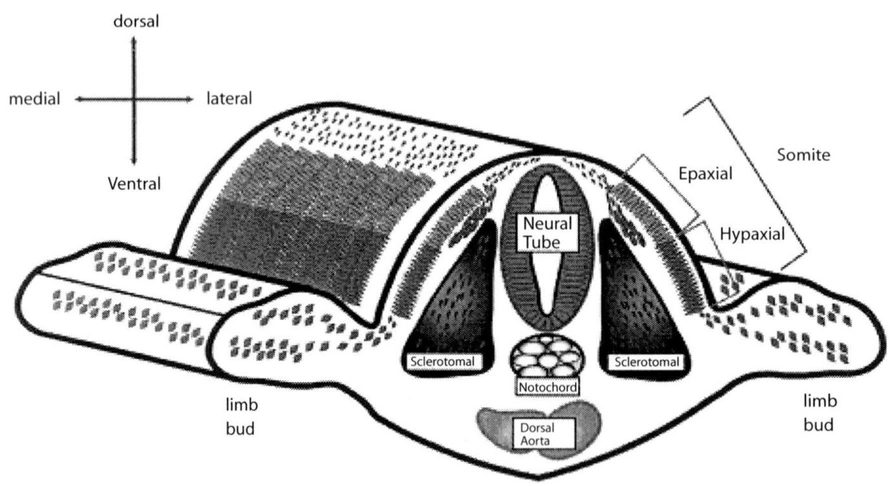

Figure 3.17 Migration of mesodermal cells and formation of a limb bud. From T. Hawke and D. J. Garry, *Myogenic Satellite Cells: Physiology to Molecular Biology* (2001), *Journal and Applied Biology,* vol. 91, pg. 534.

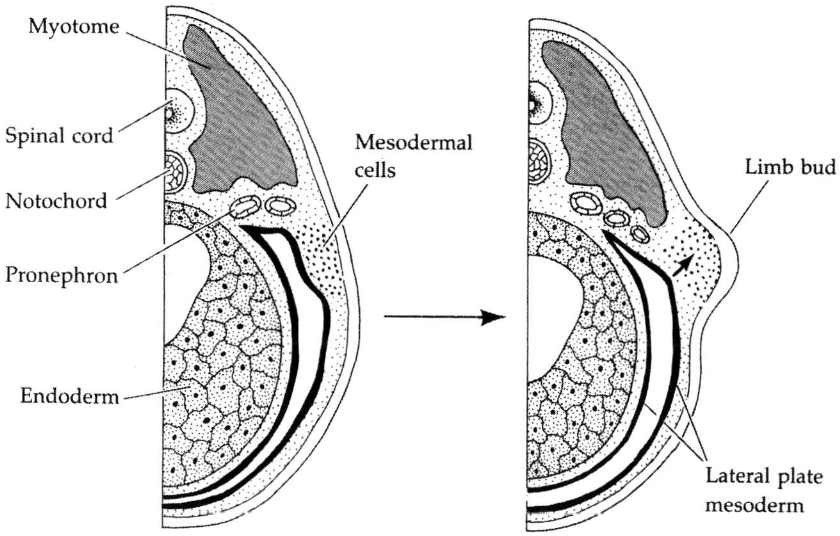

Figure 3.18 Organizing centers during limb bud formation. From J. Capdevila and J. C. I. Belmonte, Patterning mechanisms controlling vertebrae limb development in *Annual Review Cell Development and Biology,* 2001 vol. 17, pg. 87.

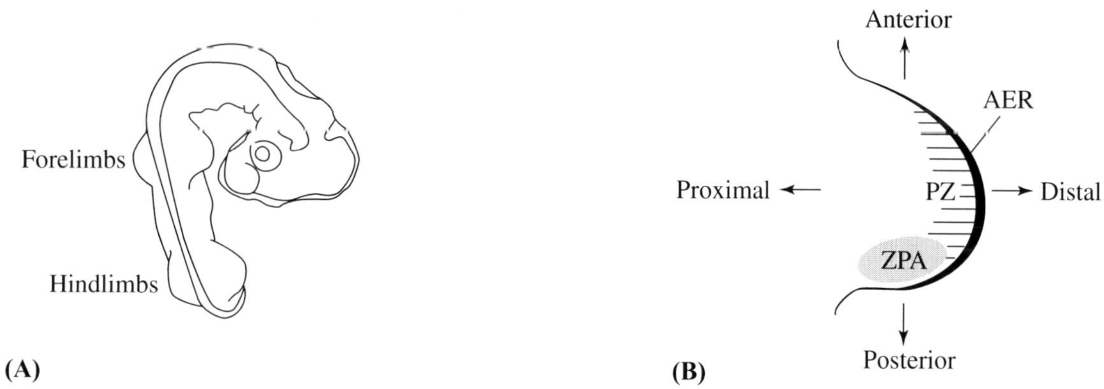

Figure 3.19 The relative location of limb buds development during early development (A) and the location of the apical ectoderm ridge (B) (AER), the progress zone (PZ) and the zone of polarizing activity (ZPA). From J. Capdevila and J. C. I. Belmonte, "Patterning Mechanisms Controlling Vertebrae Limb Development" in *Annual Review Cell Development and Biology,* 2001 vol 17, pg. 87.

has taken on the morphology and distinct features characteristic of a species, the main activity left to complete is to increase in size. Therefore, the fetal phase of prenatal development entails some further maturation and specialization of tissues, but basically is characterized by a dramatic increase in size of existing organs and tissues.

Maternal Membrane Development

Not all cells arising from a fertilized egg become cells of the embryo. In fact, a large portion of the cells of a blastocyst (Figure 3.7) participates indirectly to the viability of the developing animal by forming a specialized set of tissues to facilitate nutrient uptake from the uterus. These cells, in particular, are located immediately beneath the zona pellucida and are known as *trophoblasts*. The name of these cells is derived from the Greek root meaning to feed or nourish. The trophoblast layer of cells plays an important role in development of *extraembryonic* or *placental tissues,* which is a collection of tissues responsible for transporting nutrients and waste to and from the developing fetus, respectively. Membranes developed from the trophoblast include the *amnion, yolk sac, chorion* and *allantois,* and originate solely from the embryo. The fetal mem-

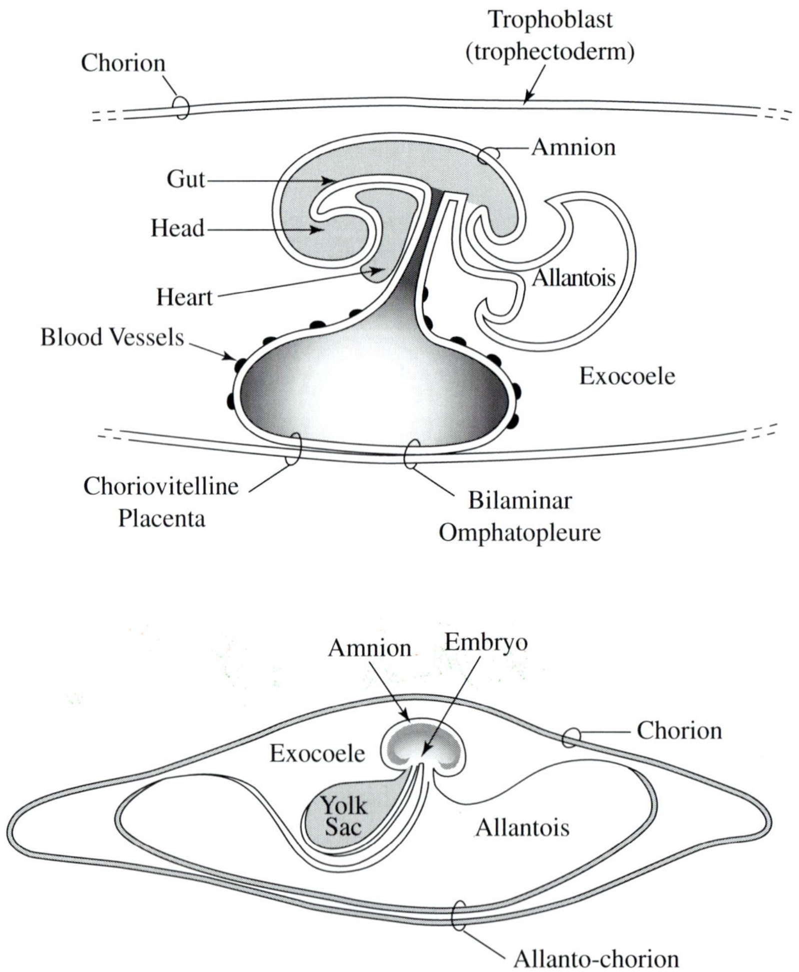

Figure 3.20 Pig fetal membrane organization. From *Journal of Reproduction & Fertility,* (1981), 62, 321–335,"The Mammalian Fetal Membranes" by J. S. Perry. Copyright © 1981 by Society for Reproduction & Fertility. Reprinted by permission.

branes and the developing fetus are often referred to collectively as the *conceptus* (Figure 3.20).

As soon as the ovum comes in close proximity or touches the maternal epithelium of the uterus, the membranes participate in the process of *implantation.* Implantation is an essential process which facilitates fetal membranes coming in close apposition to maternal tissues in order for nutrient flow to occur. The movement of gases, such as oxygen and carbon dioxide, and nutrients, such as amino acids and glucose, occurs over a larger and larger surface area as the animal continues to develop. The ability to move these materials from the maternal circulation to the fetal circulation is related, in part, to the relative proximity of each to the other's circulatory system (Figure 3.21). The extent of implantation and association with the maternal tissues varies with species. In some cases, the developing egg literally penetrates the uterine epithelium and is known as *interstitial implantation.* In contrast, fertilized eggs of other species merely come in close contact with the maternal epithelium and fetal tissues attach by a *superficial implantation* process.

The type of physical connection between the fetal and maternal tissues varies greatly with species. The *epitheliochorial* condition is best characterized by attachment of the placenta with the maternal epithelium, as is the case in pigs (Figure 3.22). This connection, therefore, results in two membranes, placental epithelium and maternal epithelium, separating the maternal and fetal blood supplies. This could be considered the most conservative of maternal-fetal connections, or the least extensive connection. The *endotheliochorial* connection, like that of most carnivores, is more extensive because the placental epithelium invades the maternal epithelium and literally positions itself immediately adjacent to the maternal blood vessels. It is obvious that this type of interface

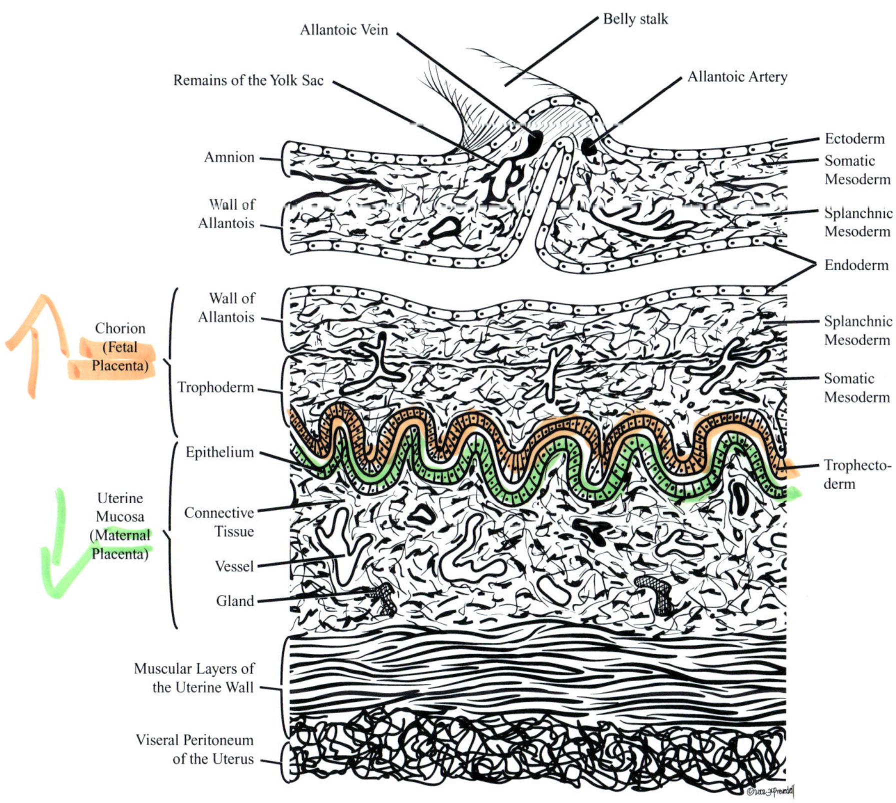

Figure 3.21 Illustration showing the association between maternal and fetal tissues. Courtesy of Faye C. Prevedell.

is more intimate than that previously discussed. In yet other animals, there is a *syndesonchorial* type of placentation. This type of interface is similar to the endotheliochorial condition but there is an added structure, a connective tissue layer, which resides between the placental epithelium and the maternal vasculature. This type of interface is commonly observed in most ruminant species. An even more profound fetal-maternal interface, *hemochorial,* exists when conceptus blood is literally capable of escaping the maternal capillary bed and allowed to circulate unimpeded against the placenta. This interface is the type occurring in humans and rodents.

The amount and type of interaction between the maternal and fetal tissues is, likewise, species dependent. In a *diffuse* attachment, as occurs in pigs, the entire placenta attaches to the maternal epithelium (Figure 3.22). The means by which this occurs is important because it does so in a manner that nearly doubles the surface area of both membranes and therefore, increases nutrient uptake potential by the conceptus. *Villi* or small protrusions, not to be confused with cellular villi projects, form from both epithelia and become *interdigitated* at the interface.

In ruminants, interlocking structures are limited to discrete islands of attachments known as *cotelydons* or *placentomes* (Figure 3.23). The corresponding structures on the maternal side are the *carnucles,* which are predestined to form during pregnancy. There are some slight variations in the structures depending on ruminant species, but for the most part the placenta forms a structure cupping discrete areas of maternal epithelium. In the horse, smaller versions of this type of attachment are found and termed *microcotyledonary* (Figure 3.24).

Measurements of Prenatal Development

The most effective means of defining the stage of development for any organism is the time post-conception. Little variation will be detected if actual days post-conception are used to evaluate animal development. Unfortunately, breeding dates are not available on every female and reasonable methods for determining the age of a developing animal or *conceptus* are needed.

Somite Formation

During early embryonic development, many embryologists rely heavily on the number of developed paired somites for determining age. Two species used extensively to study development are mice and poultry. The number of somites present has been an accurate index of embryo age, almost to the nearest hour.

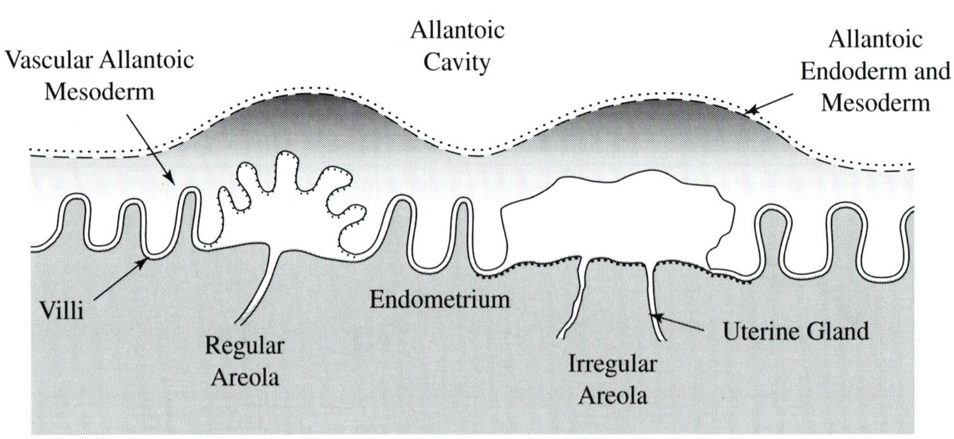

Figure 3.22 The unique interlocking maternal and fetal membranes in the pig. From *Journal of Reproduction & Fertility,* (1981), 62, 321–335, "The Mammalian Fetal Membranes" by J. S. Perry. Copyright © 1981 by Society for Reproduction & Fertility. Reprinted by permission.

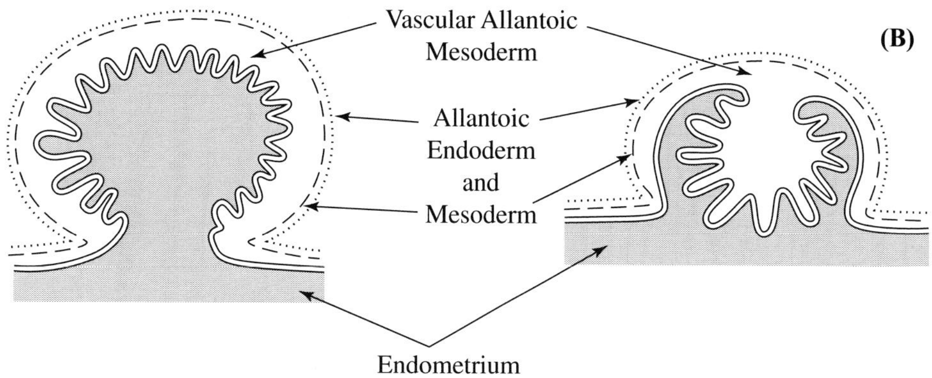

Figure 3.23 Cotyledon structure in cattle and sheep. From *Journal of Reproduction & Fertility,* (1981), 62, 321–335, "The Mammalian Fetal Membranes" by J. S. Perry. Copyright © 1981 by Society for Reproduction & Fertility. Reprinted by permission.

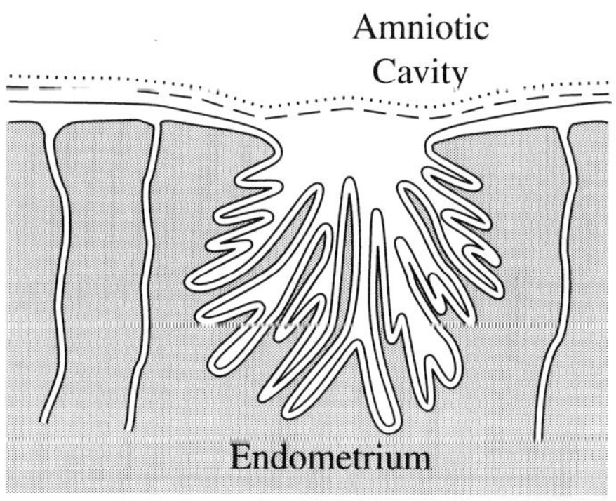

Figure 3.24 Microcotyledon structure in the horse. From *Journal of Reproduction & Fertility,* (1981), 62, 321–335, "The Mammalian Fetal Membranes" by J. S. Perry. Copyright © 1981 by Society for Reproduction & Fertility. Reprinted by permission.

Weight

Weight is a good estimator of conceptus age but significant variation is associated with this measurement. It is interesting to note the phenomenal rates at which these organisms grow. In fact, during early embryonic development, it is not unusual for an embryo to double its weight in just three days.

Length

Length is very useful for determining conceptus age. Like weight, however, some variation exists in the accuracy of this indicator. *Crown-rump length* (CRL) is commonly used to determine the age of bovine fetuses. The equation is:

$$\text{age (months)} = \sqrt{\text{CLR (inches)} \times 2}$$

Crown-rump length is measured from the most dorsal aspect of the cranium (poll) to the most caudal portion of the hips (the ishium of the pelvis). When a conceptus is developing, the physical constraints of the uterus tend to force the fetus into a ball. After fetuses are liberated from this environment, their spines are somewhat flexed and great care must be taken when evaluating crown-rump measurements. Nonetheless, this is a rather useful calculation but should not be used during the embryonic phases of development or during the later phases of fetal growth. Figure 3.25 shows that length does not increase linearly during all early development. The reason for substantial variation during late fetal development may be related to expression of fetal genetic potential and (or) variability in nutritional plane of the dam. For example,

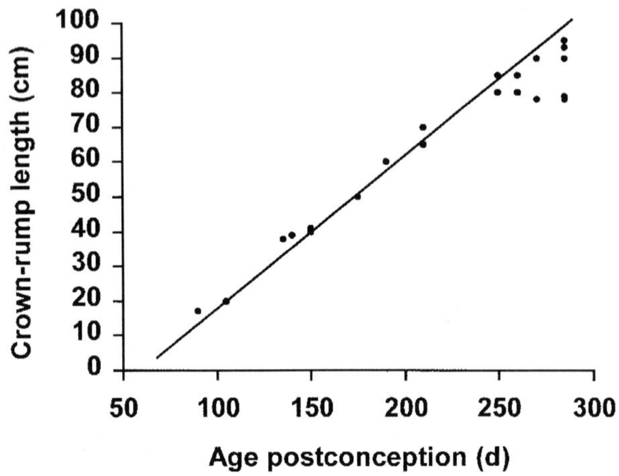

Figure 3.25 Relationship between crown-rump length (CRL) and gestational age. From Robelin, J., A. Lacourt, D. Bechet, M. Ferrara, Y. Briand and Y. Geay. Muscle differentiation in the bovine fetus: a histological and histochemical approach (1991), Growth, Development and Aging, vol. 55, pg. 151.

the average length of a large-framed Charolais calf would be greater than the average crown-rump length for a small-framed Hereford calf at constant age.

Anatomical Indications

Although reasonably accurate in some cases, visual evaluations of gross developmental changes are subjective and therefore, are not generally used for pinpointing embryonic and fetal age. Table 3.2 shows the relative time of development for different anatomical features of various species. Depending on the necessity for deriving embryonic or fetal age, this methodology may or may not be useful. For example, if an aborted conceptus is found, visual determination of various physical traits could easily provide an animal breeder sufficient detail for determining the length of gestation within a 2 to 3 week period. If, however, a more accurate determination of age is necessary, this method would not suffice.

Table 3.2 Anatomical Features, Sizes and Gestational Ages of Various Species During Prenatal Growth

Characteristic	Eyelids Forming		Hair Covering		Birth	
Species	Time(d)	Size	Time(d)	Size	Time(d)	Size
Dogs	30	20 mm	45	80 mm	60	160 mm
Cats	28	30 mm	46	85 mm	60	150 mm
Horses	55	5 cm	270	95 cm	330	110 cm
Pigs	28	28 mm	—	—	115	390 mm
Cattle	38	3 cm	230	25 cm	284	98 cm
Sheep	27	20 mm	116	375 mm	150	500 mm

Chapter 4

Myogenesis

Introduction

Even though the origin of muscle or *myogenesis* actually begins at fertilization, scientists have known for a long time that muscle develops from the mesodermal somites, more specifically the myotome. Recent scientific studies have provided understanding of molecular mechanisms underlying the process by which muscle is developed. Scientists have been able to identify portions of DNA that, when introduced into cell types other than muscle cells, are capable of reprogramming those cells to a muscle lineage. These developments are possible due to advances in molecular biology techniques. In conjunction with these advances in understanding of myogenesis, it is important to recognize that changes in physical structure or *morphology* may accompany changes in gene expression. Furthermore, the primary goal of skeletal muscle tissue is to contract; therefore, development of the unique apparatus that allows muscle cells to contract must occur before muscle tissue is considered fully developed. Finally, muscle is not inert and is continually modified during growth and development. Because so many influential factors come into play in myogenesis, we will attempt to include both genetic and morphological descriptions as we characterize myogenesis in this chapter.

Molecular Aspects of Myogenesis

Proliferation and Replication

One of the requisites for development of an organism is an increase in cell numbers via *mitosis* or *cell division* (Figure 4.1). Certainly in the context of development, the process of increasing cell numbers, *proliferation* or *replication,* is not as simple as generating two identical *daughter cells* as products of one cell division. However, for the sake of simplicity, we will make that assumption.

Scientists initially hypothesized that multinucleated muscle cells were somehow capable of nuclear replication without *cytokinesis* or cell division. This hypothesis was proven incorrect as scientists demonstrated that mature cells are *postmitotic* or incapable of division after attaining a multinucleated status. It became evident that the number of muscle fiber precursor cells or *myoblasts* increase initially, then come together through a membrane fusion process that ultimately yields a multinucleated cell. With this understanding, it is evident that a critical point in controlling muscle development must involve proliferation of muscle precursor cells. Assuming a specific number of myoblasts are needed to produce a given amount of muscle during normal development, it follows that large increases in muscle mass would result if all of those myoblasts were capable of making just one additional division. This reasoning has led many growth biologists to investigate mechanisms controlling cell replication.

The Cell Cycle

All cells exist in a theoretical *cell cycle* (Figure 4.2), which implies that replication competent cells can proceed through a well-characterized set of stages that ultimately lead to development of two daughter

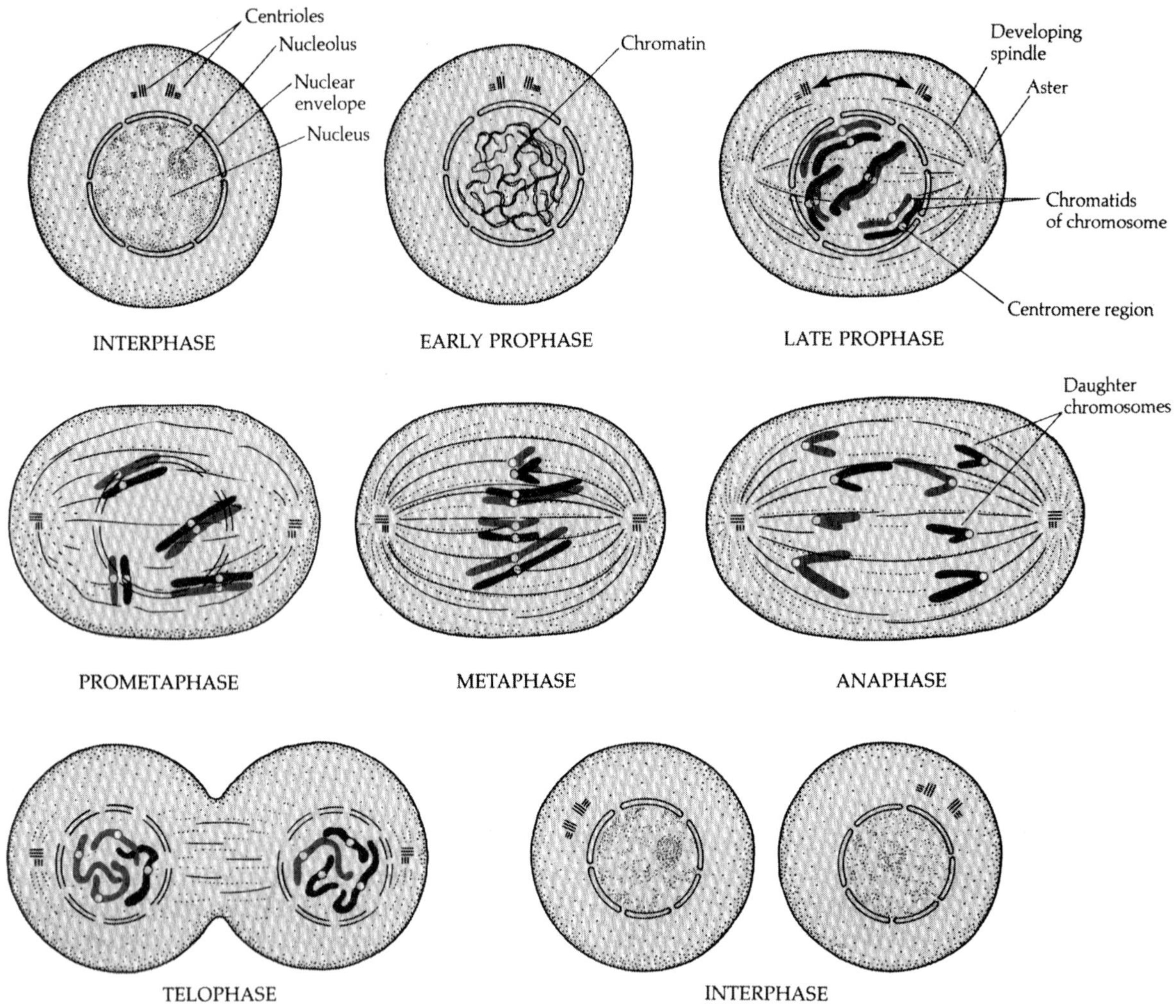

Figure 4.1 Schematic illustrating the events of mitosis in animal cells. From *Developmental Biology, Third Edition* by Scott Gilbert. Copyright © 1991 by Sinauer Associates, Inc. Reprinted by permission.

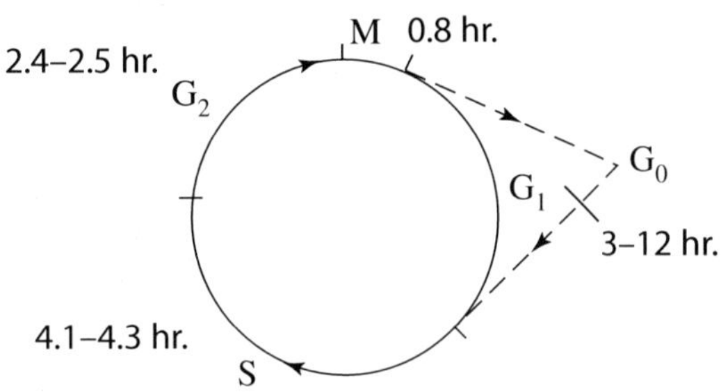

Figure 4.2 Stages of the typical cell cycle. G1 (Gap 1), S (synthetic), G2 (Gap 2), M (Mitosis) and G0 phases vary in length. From A. M. Pearson and R. B. Young, Muscle and meat biochemistry, Academic Press, Inc., New York.

cells via cell division or mitosis. Accordingly, it is possible that this process could continue indefinitely, unless a signal exists to stop the cycle. During the process of mitosis, the entire *genome* or DNA must be replicated and each duplicate of DNA must be moved to divergent cell locations so that cytokinesis can proceed yielding two identical daughter cells. Therefore, it is understandable that arresting this process at any point is virtually impossible, and may result in cell death. In regard to the myoblast cell cycle, some of these critical points are well-known.

There are four distinct phases of the cell cycle (Figure 4.2). All cells enter the cell cycle from the phase referred to as the *Gap 1 (G1)* phase. This phase is the most variable in length and can range from 3 to 12 hours for cells in culture. Cells in this

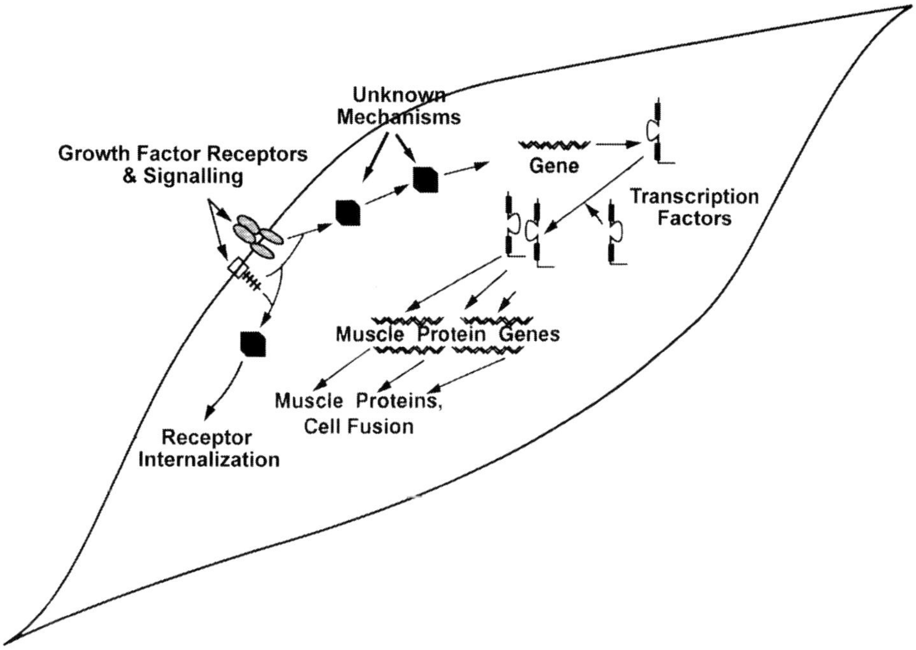

Figure 4.3 Illustration showing the action of growth factors on a cell. Modified from J. R. Florini, K. A. Magri, D. Z. Ewton, P. L. James, K. Grindstaff, and P. S. Rotwein (1991), "Spontaneous" differentiation of skeletal myoblasts is dependent upon autocrine secretion of insulin-like growth factor-II. Journal of Biological Chemistry, vol. 266, pg. 15917.

phase respond to cues in their environment to signal continuance to the next phase of the cycle or remain in the G1 phase. At some point, there is a *restriction point* contained within the G1 phase, a theoretical point at which cells are obligated to proceed through the cell cycle.

The *synthetic* or *S-phase* is the period of time the cell dedicates to DNA replication or synthesis. Understandably, this phase is rather constant and lasts between 4.1 and 4.3 hours in cultured myoblasts. The phase following the S-phase is called *Gap 2 (G2) phase.* This phase consistently ranges from about 2.4 to 2.5 hours in length for many muscle cell types and represents the stage when the intracellular architecture remodels itself to accommodate physical division of the cell mass in the *mitosis* or *M-phase.* The M-phase is the shortest, lasting for just under an hour (0.8), but is very constant. After mitosis, the cell re-enters the G1-phase of the cell cycle and readies itself for repeating the entire cycle.

During the G1-phase, cells are extremely sensitive to external cues from the environment. *Growth factors* signal the cell to remain in the cell cycle or to exit the cycle and continue in development and specialization of a tissue. Growth factors are often small proteins that bind to *cell surface receptors* and signal to the nuclei the status of the environmental milieu (Figure 4.3). Not all growth factors are proteins. Steroids are capable of binding receptors in cellular cytoplasms and migrating to the nucleus to effect a response. These factors can stimulate replication and are considered *mitogens.* Some examples of mitogenic growth factors are *fibroblast growth factors, insulin-like growth factor- I* and *-II, testosterone,* and *transforming growth factor.* Growth factors also can inhibit replication. In fact, insulin-like growth factors are capable of stimulating and inhibiting replication depending on amount available and duration of exposure. Growth factors and hormones will be covered in more depth in Chapter 10. Albeit, cells in the G1 phase of the cell cycle are highly susceptible to cues from the environment.

Another alternative for cells in the G1-phase is the opportunity to exist in a *protracted G1 phase* or *G0 phase.* This *compartment* or theoretical location in the cell cycle is a metabolic state in which cells

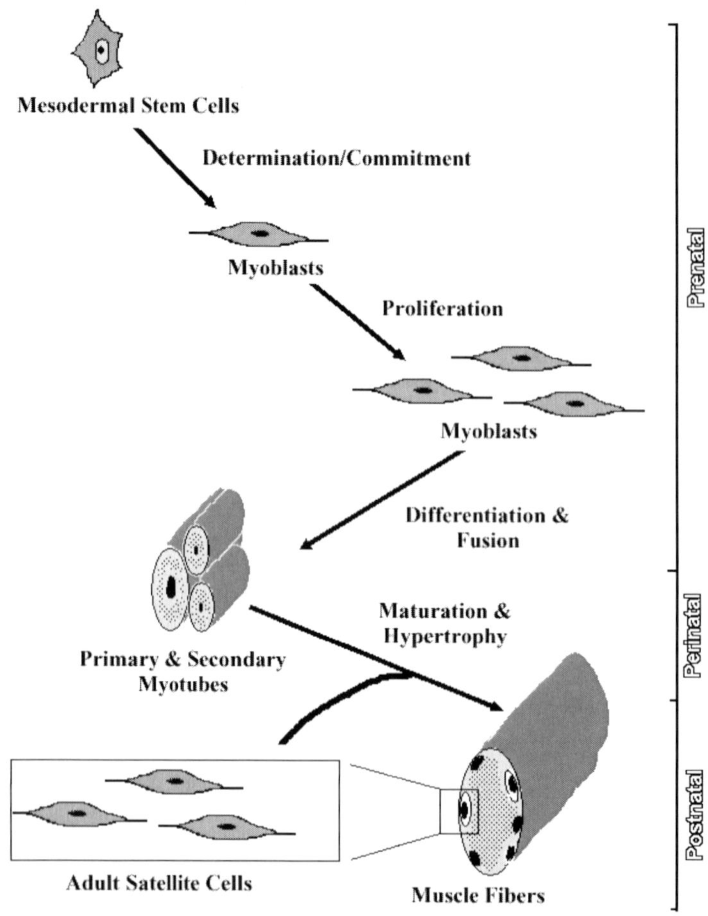

Figure 4.4 Major events occurring during myogenesis.

remain capable of re-entering the proliferative cell cycle without dividing. Cells in this stage often are referred to as being *quiescent* or dormant, continuing basal metabolism necessary to remain viable, yet not participating in replication or other more specialized events. These quiescent cells in muscle tissue, referred to as *satellite cells,* are very important in postnatal muscle growth, and will be discussed in more detail in Chapter 5.

Determination

Recall that cells must migrate from the somites to locations in the body where muscle development is needed, but details regarding the exact mechanisms of migration to various places remain obscure. However, some cells do so because they are "destined" to develop muscle fibers. Therefore, these cells are *restricted* or constrained to a *muscle cell lineage* and are considered to be *determined* or *committed* cells. Cells at this point are termed *myoblasts,* i.e., those muscle precursor cells that participate in formation of multinucleated myocytes or muscle fibers. Figure 4.4 depicts biochemical and molecular control of well-defined steps in myogenesis.

The event that triggers determination or commitment is expression of one or more genes referred to as *muscle regulatory factor genes* (*MRFs*). These factors collectively make up a family of transcription factors, or those factors responsible for "turning on" the transcription of other genes located in the nucleus and contain a *helix-loop-helix motif* or a specialized structure. When two of these same proteins combine (*homodimers*) or two different helix-loop-helix transcription factors combine (*heterodimers*), they can bind to the regulatory region of various muscle-specific genes and cause these genes to be expressed, or in some

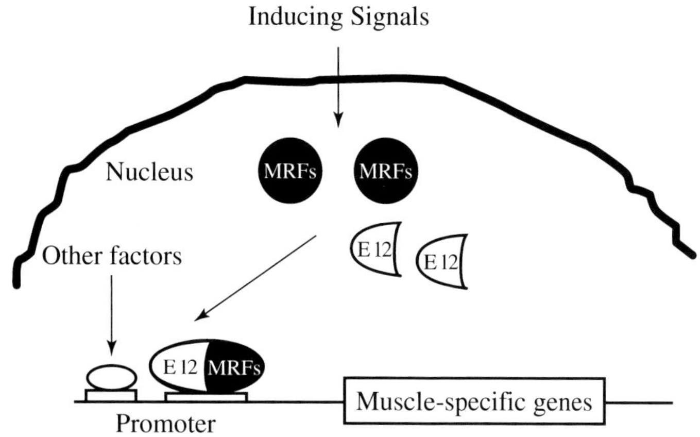

Figure 4.5 Action of myogenic regulatory factors (MRFs) in muscle cells during myogenesis.

cases, be repressed (Figure 4.5). If non-muscle cells are genetically modified to express MRFs, the cells will become myogenic. As a result, MRF expression is considered a major control point in myogenesis. Some of the well-known helix-loop-helix proteins include:

- *Myogenin*
- *MRF-4*
- *Myo-D*
- *Myf-5*

Although cells expressing helix-loop-helix genes are determined to become muscle cells, morphologically they are indistinguishable from other cell types in muscle tissue, like fibroblasts, without use of special molecular techniques.

Differentiation and Fusion

Even though determined muscle cells, or myoblasts, express muscle regulatory factor genes, these cells are *replication competent,* meaning they are capable of progressing the cell cycle and giving rise to additional myoblasts. Myoblasts, however, do not contract because they do not contain contractile proteins, or have not attained the level of specialization where contraction is possible. Before attaining such designation, these cells must receive the pertinent environmental signals and cues that induce them to *differentiate*. At this point:

- The cells stop dividing.
- The cells begin to align with one another.
- The cell membranes begin to fuse together to form an immature muscle fiber (also referred to as a *myotube*).
- There is a concomitant up-regulation of muscle-specific genes such as *muscle creatine kinase, acetylcholine receptor subunit, myosin,* and *actin.*

It is of utmost importance at this point to understand that even though cessation of replication is used often in conjunction with differentiation, the events are mutually exclusive. In other words, myoblasts can stop replicating, but this does not imply that they will differentiate. In fact, myoblasts can remain quiescent for long periods without becoming differentiated.

The exact process by which myoblasts fuse together to form multinucleated muscle cells is not well established; however, first myoblasts come in close proximity to each other (Figure 4.6). Then small attachments or tubules become apparent between juxtaposed cells. Dense structures beneath opposing membranes are generated and form tight junctions between closely associated myoblasts. Finally, the two bilipid membranes become one and subsequently dissolve in the cytoplasm of the newly

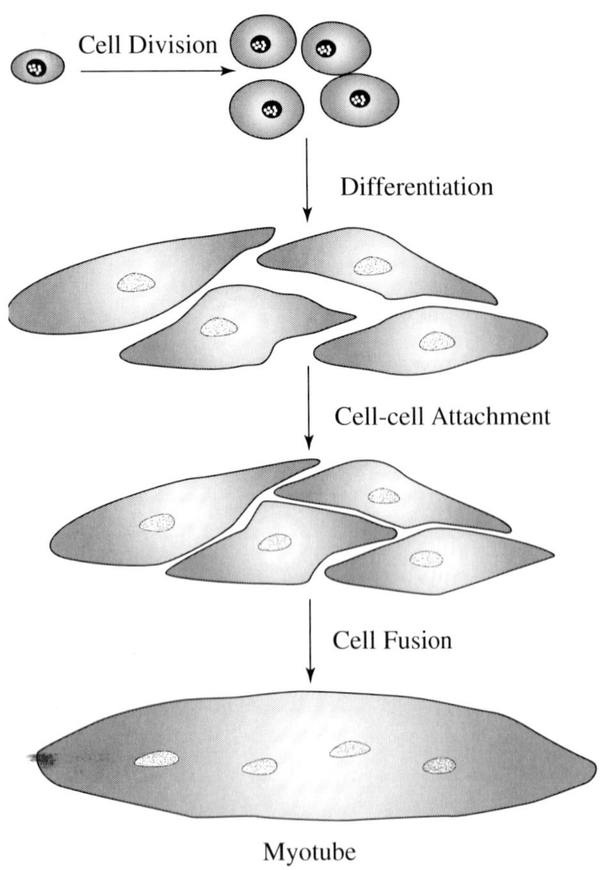

Figure 4.6 Process of myoblast fusion. Courtesy of D. R. Swartz, Department of Anatomy and Cell Biology, Indiana University.

formed multinucleated cell. Calcium ions play an important function in this fusion process.

Maturation

Once myoblasts differentiate and fuse to form myotubes, resulting cells do not continue to express a given set of genes or function the same way throughout the entire lifespan of the animal. In fact, these new muscle fibers change dramatically in the weeks and months following fusion. The process by which these cells continue their journey toward higher and higher complexity is called *maturation.* Many morphological events must occur to facilitate maximal functionality. Of course, morphological changes are preceded by changes in gene expression. In particular, those genes that are expressed in adult muscle are not necessarily expressed in fetal muscle. Therefore, a complex set of signals is sent or detected by the muscle, and ultimately deciphered for making changes in gene expression. These signals will be covered in more depth in the remainder of this and the next chapter.

Morphological Aspects of Myogenesis

In review, recall that most all skeletal muscle is derived from somites, which develop from *paraxial mesoderm* lying on either side of the neural tube and somite formation occurs in a head to tail fashion with those older or more mature somites being located toward the cranial end of the embryo (Figure 4.7). Myogenesis in the somite is rather complex, but much is known about its morphological changes. Once the somite has reached the two layered, dermomyotome stage (Figure 4.7), cells immediately begin to accumulate ventrally to form the myotome. The process by which this occurs is not completely known, but appears to involve migration of cells from the *epaxial* (dorsomedial ridge) and *hypaxial* (ventrolateral) aspects of the dermomyotome to immediately below the dermomyotome, creating the myotome (Figure 4.8). The process by which these myoblasts and myoblast precursor cells are targeted to various locations throughout the body is poorly understood. However, hypaxial cells located in the somites located adjacent to developing limb buds have been shown to migrate into developing limbs.

From a morphological standpoint, development of muscle is quite unique and intriguing. It is important to remember that muscle does not just appear during prenatal development. There are some very distinctive stages that occur resulting from the *biphasic nature* or two-stage manner of development. This is particularly evident from Figures 4.9 and 4.10, where two populations of muscle fibers are clearly evident. Just before and during myoblast differentiation, myoblasts cluster, align in a vast network of connective tissue and their membranes fuse to form an immature muscle fiber referred to as a *myotube. Primary myotubes* are the first muscle fibers to develop (Figure 4.11) and act as an architecture, or *scaffolding,* on which other myoblasts can align and begin to fuse. As the name implies, myotubes appear histologically as hollow tubes (Figure 4.10 and 4.11f). Of course, this is merely an

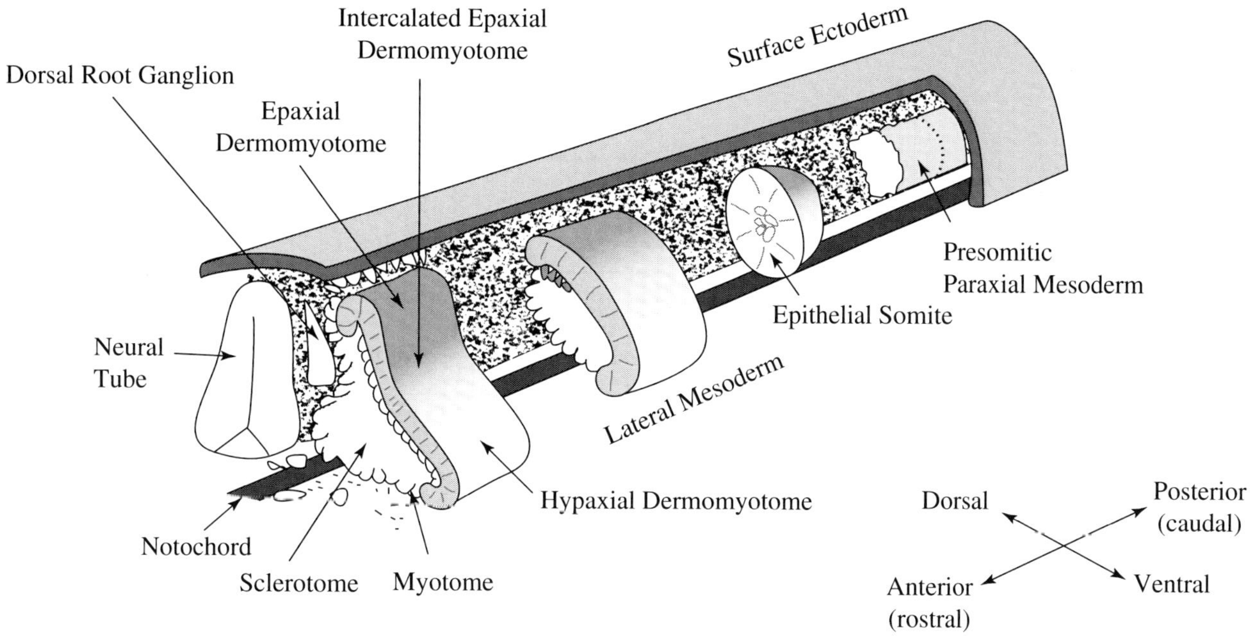

Figure 4.7 Structures of the dermomyotome. Reprinted from *Current Opinion in Genetics and Development,* Vol. 11 by Margaret Buckingham, "Skeletal muscle formation in vertebrates", copyright 2001 with permission from Elsevier Science.

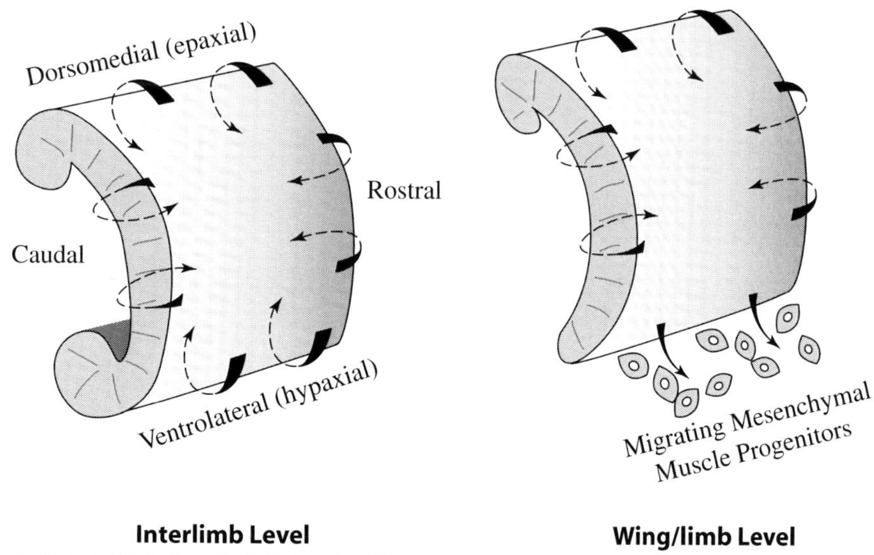

Figure 4.8 Migration of muscle precursor cells from the dermomyotome. Reprinted from *Current Opinion in Genetics and Development,* Vol. 11 by Margaret Buckingham, "Skeletal muscle formation in vertebrates", copyright 2001 with permission from Elsevier Science.

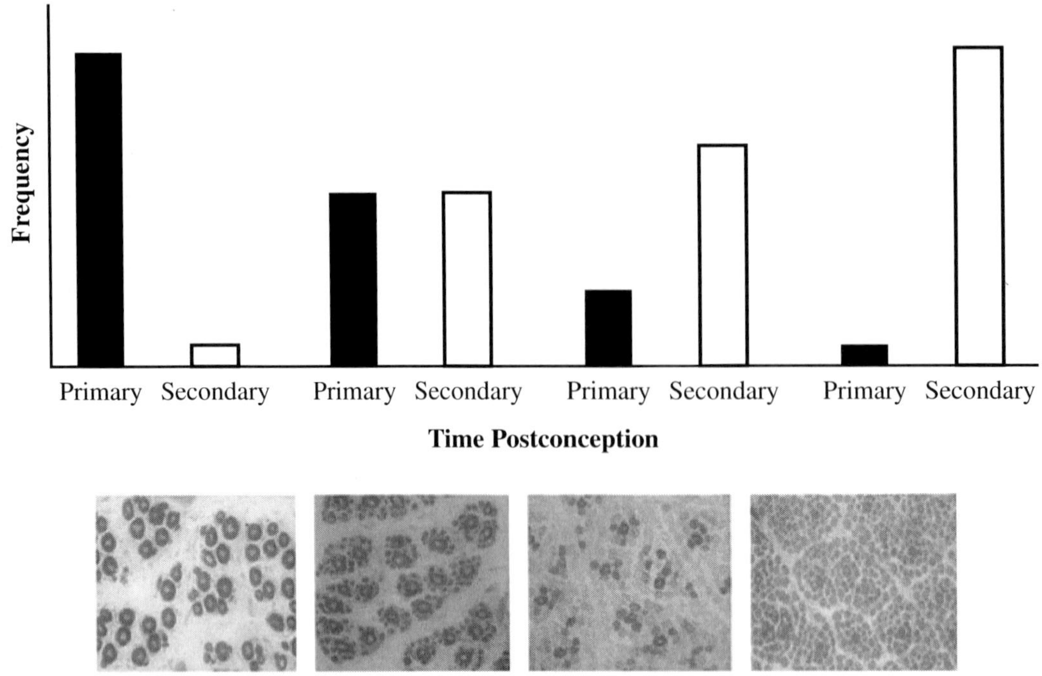

Figure 4.9 Change in the relative size and frequency of primary and secondary muscle fibers during fetal development.

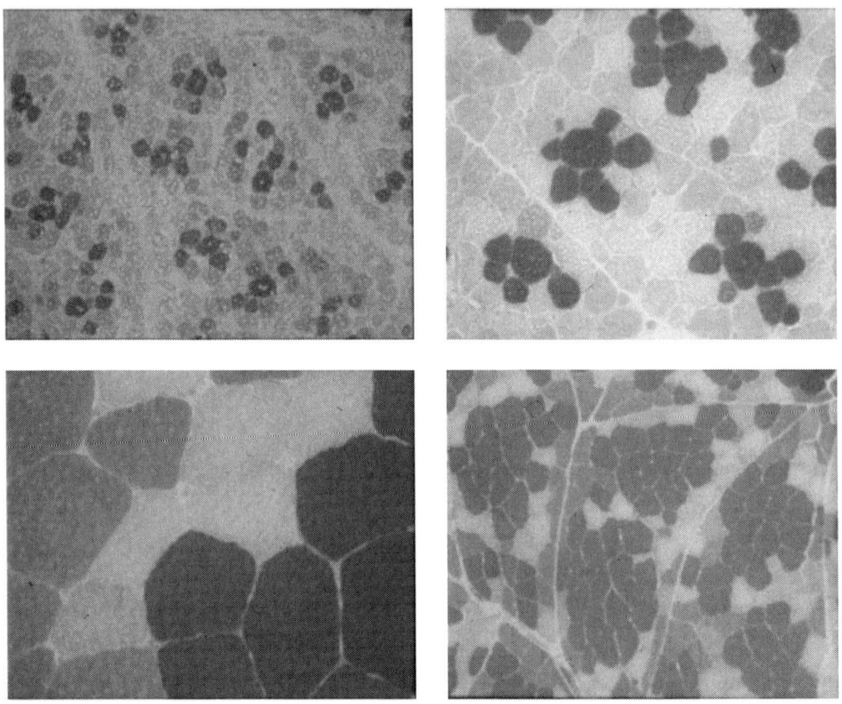

Figure 4.10 Micrographs of developing primary and secondary muscle fibers.

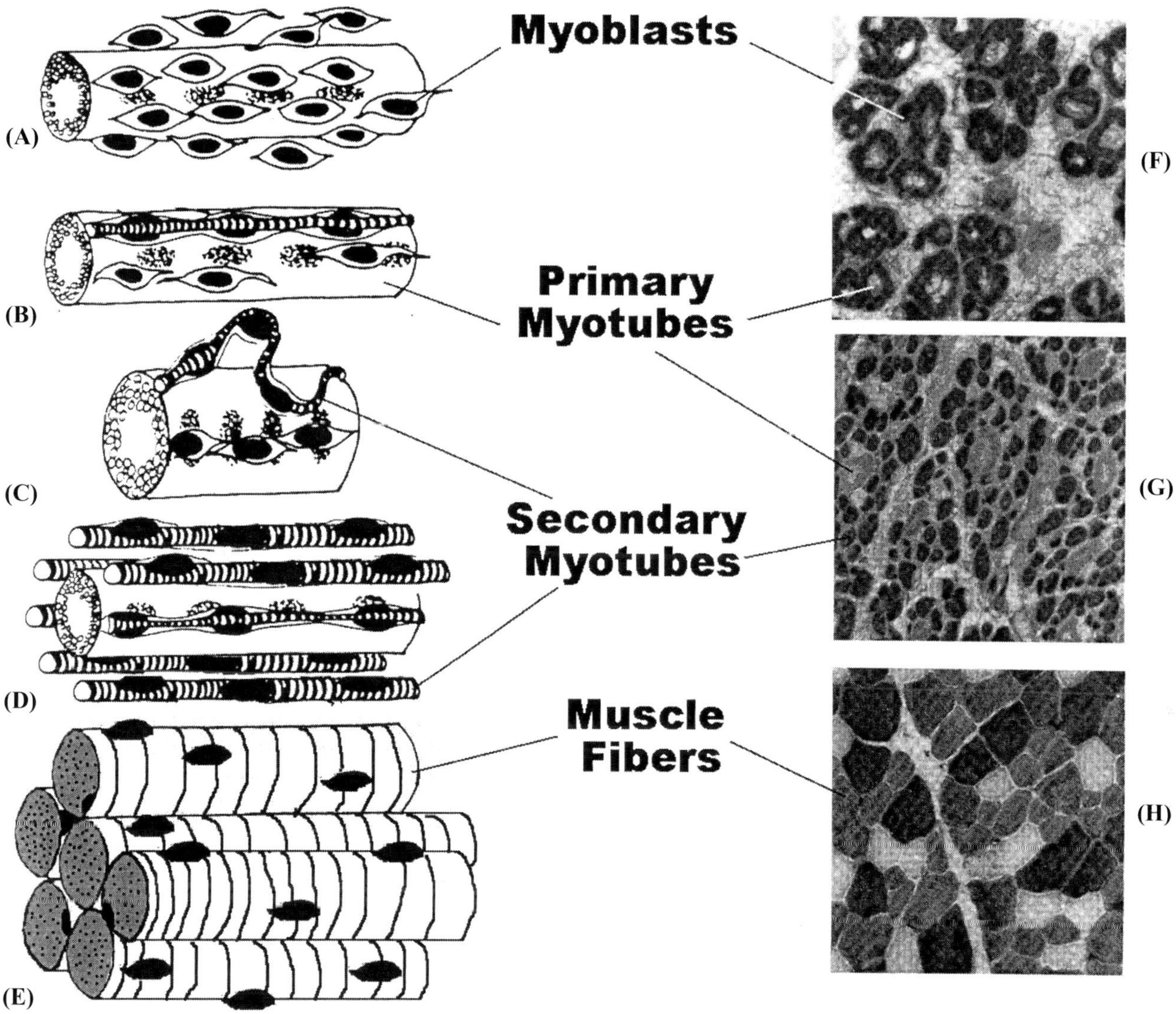

Figure 4.11 Illustration of embryonic and fetal development of muscle fibers. (A) Myoblasts surrounding a primary myotube. (B and C) Secondary myotube forming on and detaching from the surface of a primary myotube. (D) Secondary myotubes surrounding a primary myotube. (E) Several fully formed muscle fibers. (F) Cross-section of early embryonic skeletal muscle showing the predominance of primary myotubes. (G) Cross-section of early fetal skeletal muscle showing secondary myotubes surrounding primary myotubes. (H) Cross-section of fully formed skeletal muscle fibers. Reprinted with permission from *Structure and Development of Meat Animals & Poultry,* by H. J. Swatland. Copyright CRC Press, Boca Raton, Florida.

artifact of muscle sectioning and light microscopy. In fact, they appear as hollow tubes because the nuclei of newly formed myotubes are centrally located. Therefore, in any muscle cross-section, it is possible to observe primary myotubes with or without centrally located nuclei. If nuclei are not present in myotubes, the substance appearing in the hollow tube is sarcoplasm of the muscle cell.

As already stated, muscle fiber formation or myogenesis is not an event that occurs all at once (Figures 4.9 and 4.10). In most large mammals the process takes months to occur. Again, muscle fiber formation is a biphasic phenomenon, in which primary fibers form first from a population of myoblasts referred to as *embryonic myoblasts,* followed by a second wave of secondary myotube formation from a population of myoblasts referred to as *fetal myoblasts.* The time of secondary myotube development is species dependent and, in some cases, depends on the plane of nutrition of the dam. However, the morphological development of two distinct waves of muscle fibers seems to be highly conserved or similar across all species.

By integrating biochemical and molecular aspects of myogenesis with those just outlined above for morphological development of muscle, one may conceptualize that cells surrounding the primary myotubes are likely a population of determined myoblasts; i.e., replication competent and dividing as directed to increase cell number (Figure 4.11a). Then, upon receiving the appropriate signaling cue(s), for example changes in growth factor signaling, they begin to differentiate (Figure 4.4). During this secondary myotube formation, fetal myoblasts use the primary myotubes as a template for aiding in their own alignment and organization. This alignment of fetal myoblasts with primary myotubes allows them to align closely with each other, which facilitates the fusion process (Figure 4.11c). At this stage, myoblasts are loosely associated and these cells even may become attached to primary myotubes through membrane associated proteins that form tight *gap junctions* between the cells. These gap junctions serve as a means of communication between the cytoplasms of closely associated cells and may aid in the fusion process. Regardless of the exact mechanism, these fetal myoblasts fuse to form secondary muscle fibers that remain rudimentarily associated with the primary myotubes. However, after innervation and contraction of primary muscle fibers, secondary muscle fibers or myotubes are liberated or "splinter away" from the primary fiber (Figure 4.11c). After separation of muscle fibers, individual cells become encapsulated by their own basal lamina. This process of developing a secondary population of myofibers, although not completely understood, results in the formation of up to 90% of the muscle fibers found in some domestic animals (Figure 4.11g). The number of secondary fibers around each primary fiber varies from 5 to 9 in mouse and rat to over 20 in larger species, such as pig. For reasons not well understood, secondary muscle fibers possess centrally located nuclei for only a short time while nuclei of primary myotubes remain central for a much longer period of time before migrating to the periphery of the sarcoplasm where they remain for the rest of their existence in some mammals.

The complete process of developing primary myotubes is self-governing or *autonomous* in nature, at least regarding muscle tissue. In other words, development of primary myotubes proceeds with very little input from other sources. In contrast, development of secondary myotubes is predicated on the contraction of primary myotubes that occurs via innervation. This was demonstrated in studies where innervation was blocked and formation of primary muscle fibers was not affected, but secondary myotube formation was severely hampered. Therefore, the role of innervation in modulating muscle fiber development cannot be understated and plays a major role in controlling myogenesis *in vivo.*

Recall the "unidentified cells" surrounding the primary myotubes. As stated earlier, these cells could be myogenic, fibroblastic, or simply undetermined *mesenchymal cells.* Morphologically, myogenic precursor cells are undistinguishable from the aforementioned non-muscle cells. The only means currently available to know exactly whether these are myogenic is by their ability to express muscle cell-specific genes.

Micrographs of muscle fiber development are incredibly elegant and show the importance of primary muscle fibers in aiding the formation of secondary fibers. In fact, primary muscle fiber formation may be a point of critical importance in controlling muscle development. In particular, animals with the genetic propensity of developing more muscle have more primary myotubes, yet control of this phenomenon is far from being understood.

Depending on species, the process of muscle fiber development occurs during the first two-thirds of prenatal development and for the most part, is completed before birth. The total number of myofibers is generally considered established by 90 to 95 days post-conception in pigs (Figure 4.12). The total number of muscle fibers increases up to

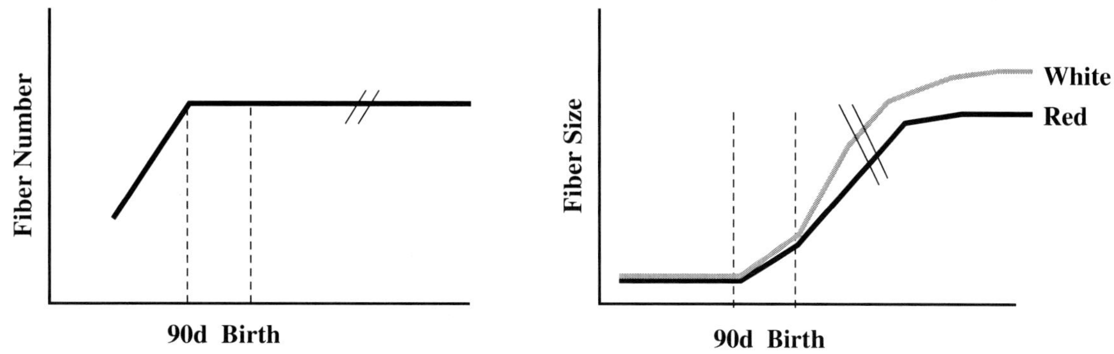

Figure 4.12 Evolution of muscle fiber number and size in pig muscle.

90 day post-conception and remains unchanged thereafter.

There is considerable difference between the size of primary and secondary myotubes, which allows for easy identification of each population during muscle development. As pointed out earlier, there may be as many as 20 secondary muscle fibers developed for each primary muscle fiber. However, as muscle continues to growth, so does muscle fiber diameter. At some point, the diameter of the secondary fibers reaches a point in late fetal development at which the difference between primary and secondary muscle fiber diameter is so little that size is not useful in distinguishing them from one another. Ultimately, in later postnatal stages, diameters of secondary muscle fibers may exceed those of primary fibers.

Myofibrillogenesis

The primary function of muscle is to contract and the principal means of accomplishing this task is the addition of myofibrils to each cell during development. Development of myofibrils is termed *myofibrillogenesis* (Figure 4.13). Even though the exact mechanism is not precisely known, much is known and merits consideration in the context of understanding how muscle develops.

One of the defining characteristics of myofibrils is that they are highly organized and consist of many different proteins (see Chapter 2). Therefore, the process of organizing such a structure within a

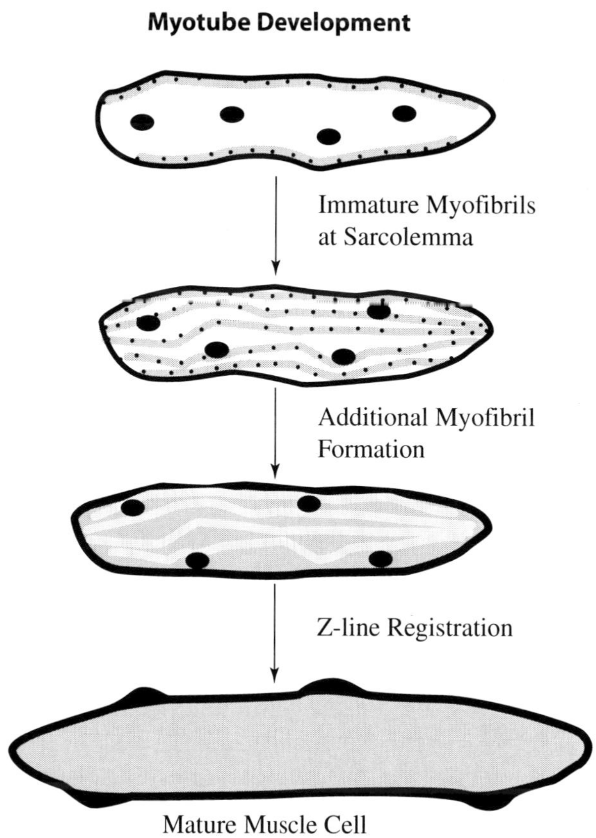

Figure 4.13 Formation of myofibrils and their alignment during myofibrillogenesis. Courtesy of D. R. Swartz, Department of Anatomy and Cell Biology, Indiana University.

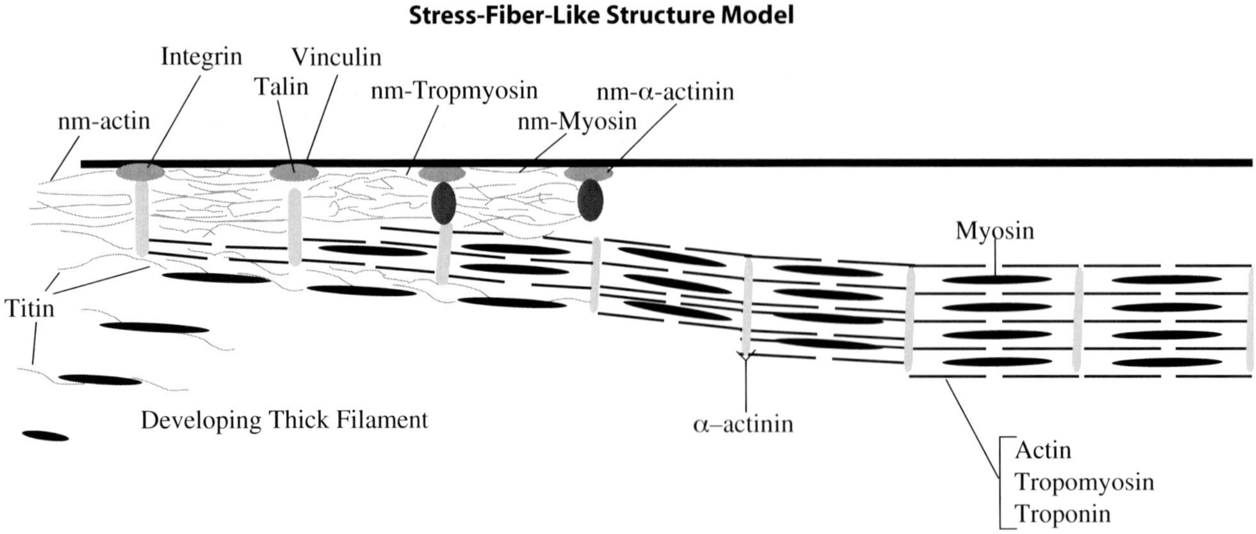

Figure 4.14 Model for the formation of myofibrils from stress fibers (nm, non-muscle). From *Proceedings of the American Society of Animal Science,* 1998. Copyright © 1998 by American Society for Animal Science. Reprinted by permission.

cell resides in the development of an organizing structure. As outlined in Chapter 2, all cells contain structural proteins that help maintain the physical structure of cells, the cytoskeleton. In the absence of such reinforcement, the structure of the cell is rendered spherical in nature. One component of the cytoskeletal system is *stress fibers,* which are bundles of microfilaments with contractile capabilities. These fibers, in most all cells, are located at the periphery of cells and interact with membrane bound proteins called *extracellular* and *cell-adhesion molecules. Integrins* (extracellular cell adhesion molecules) are transmembrane-bound proteins that interact with extracellular (outside the cell) proteins of connective tissues (fibronectin, laminin and collagen) and intracellularly with proteins of the cytoskeleton. The two families of proteins that are primarily responsible for holding cells in close contact with each other are the *adherins* (N- and M-adherins) and the *NCAMs* (neural cell adhesion molecules). Development of stress fibers in close apposition to the cell membrane likely represents a template for development of *nascent myofibrils* or those fibrils initially produced in developing muscle cells.

Two models have been proposed for explaining the mechanisms of myofibrillogenesis, the *stress fiber model* and the *premyofibril model.* In the stress fiber model, myofibrillogenesis occurs by using stress fibers as a template (Figure 4.14). Under mechanisms still unknown, non-muscle myosin begins to associate with adhesion molecules that reside in periodicity along the inner surface of the membrane. As this process continues, skeletal myosin is ushered in between the Z-line precursors by titin. As more and more skeletal myosin molecules are added to the myofibril, the Z-lines consisting of non-skeletal myosin are replaced with the mature myofibril equivalent alpha-actinin. Further maturation of the myofibril occurs by substitution or addition of mature myofibril proteins. This process is repeated many times until the cell contains its full complement of mature myofibrils.

In the competing hypothesis, the premyofibril model, stress fibers detach from the membrane (Figure 4.15). Then, there is a systematic replacement of nascent myofibril proteins or non-skeletal muscle proteins with their mature counterparts. Regardless of the precise mechanism employed in myofibrillogenesis, the principles underlying each mechanism involve stress fibers that are intimately associated with the cell membrane and act as template for the development of mature myofibrils. Even though not completely understood, myofibrillogenesis is imperative for generating and organizing the contractile proteins in muscle cells.

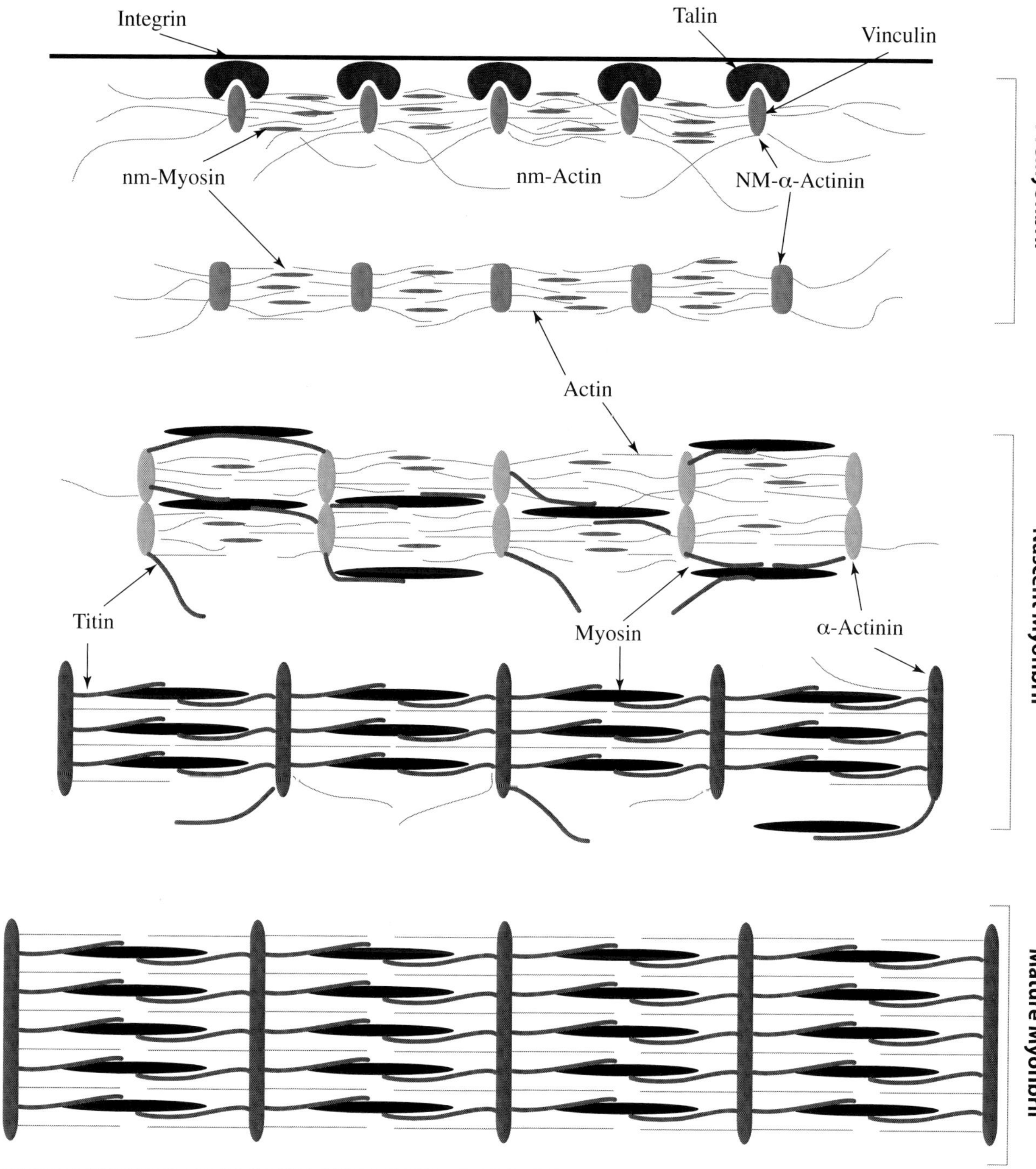

Figure 4.15 Model showing the organization of nascent myofibrils during myofibrillogenesis (nm, non-muscle). From *Proceedings of the American Society of Animal Science*, 1998. Copyright © 1998 by American Society for Animal Science. Reprinted by permission.

Chapter 5

Muscle Growth

Introduction

Muscle consists of individual muscle fibers; therefore, the collective mass or volume of a muscle can be dramatically altered by changing either the number or size of those muscle fibers (Figure 5.1). In fact, there is relationship between muscle fiber growth characteristics (number and size) and the overall growth of muscle. As illustrated in Figure 5.1, muscle hypertrophy is related to an increase in the number of cells (r = 0.3 to 0.5) or size (r = 0.6) of muscle fibers present in the muscle. In other words, muscles that experience hypertrophy generally do so via increasing fiber size or by having more muscle fibers compared to those muscles not experiencing much hypertrophy. (Note the distinction between muscle hypertrophy and muscle cell hypertrophy). Curiously, however, there is a negative relationship (r = –0.3 to –0.8) between muscle fiber size and muscle fiber number; meaning that the greater the number of muscle fibers in a muscle, the smaller they are generally. Although the reason for this relationship is not known, it shows the importance of both muscle fiber size and abundance in modulating muscle growth.

Growth and maturation of the muscle fibers formed during prenatal development must occur if an animal is to experience maximal muscle growth. To understand fully how muscle grows, one must have an appreciation for factors that impact muscle fiber number. Second, and equally important in postnatal growth, are the mechanisms underlying changes in the size of muscle fibers. Appreciation must be developed for muscle protein synthesis and degradation because the balance between these two processes is a primary determinant of skeletal muscle growth. Finally, the unique mechanism by which muscle is capable of introducing additional DNA into its individual cells to maintain growth and repair will be discussed. Understanding these critical processes and their regulation within the confines of muscle is paramount to understanding meat animal growth.

Muscle Fiber Number

The absolute number of muscle fibers in any given muscle is rather difficult to ascertain using even the most sophisticated technologies. One reason for this difficulty is that muscle fibers are not always oriented parallel to the long axis of the muscle (Figures 5.2 and 5.3). An example of such a muscle is the *longissiumus dorsi* muscle, located on each side of the vertebral column of mammals. This muscle is of particular interest to those studying meat animal growth as it is one of the economically important muscles of the animal, primarily because many expensive meat cuts (steaks and chops) include large portions of the muscle (Figure 5.4). As for the function of this muscle in living animals, it extends the vertebral column. This function is possible because the muscle fibers are oriented in

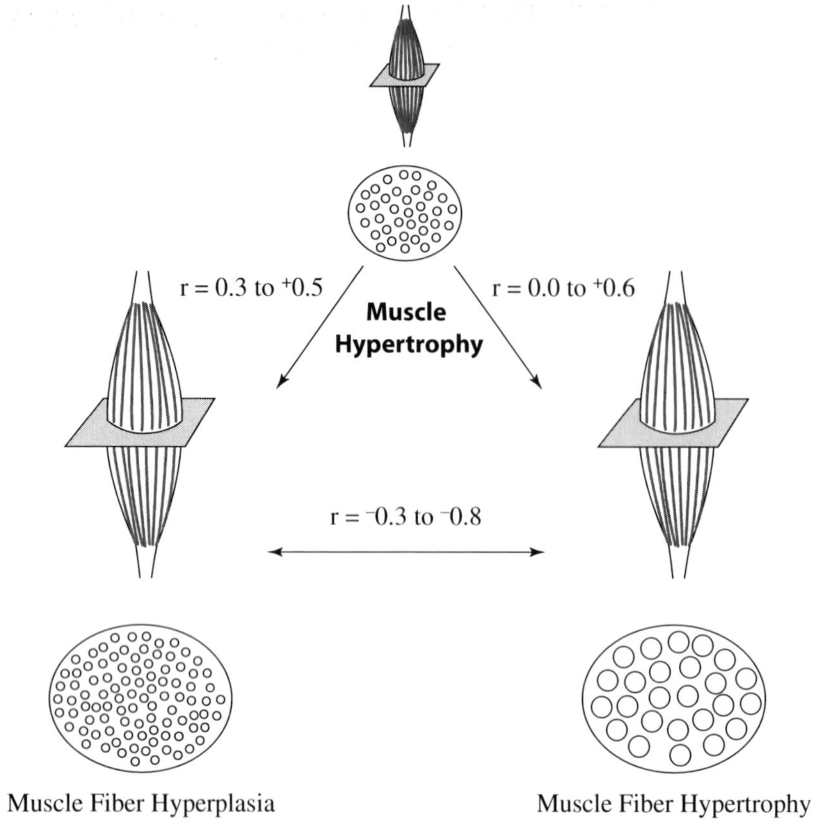

Figure 5.1 Effect of muscle cell hyperplasia and hypertrophy on whole muscle hypertrophy. Data from C. Rehfeldt, I. Fiedler, G. Dietl and K. Ender, Myogenesis and postnatal muscle cell growth as influenced by selection (2000), *Livestock Production Science,* vol. 66, page 177. Note: As selection for muscle hypertrophy occurs, it results in either muscle fiber hyperplasia (left) or muscle fiber hypertrophy (right), thus a positive relationship (r) exists between both forms of hypertrophy. Simply stated, animals selected for greater muscling possess either more muscle fibers or fewer fibers that are larger. This latter issue results in a negative relationship between hypertrophy and hyperplasia.

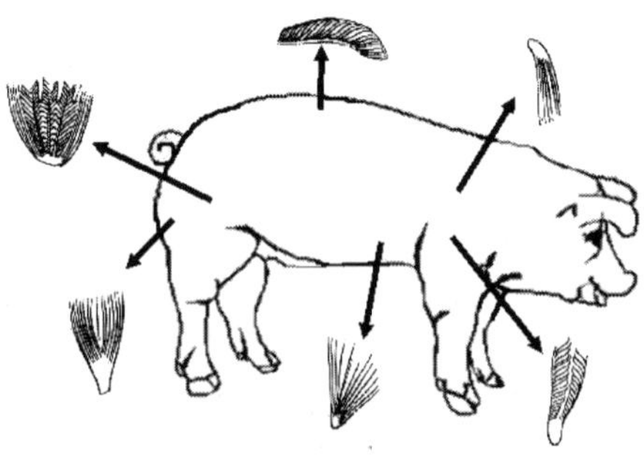

Figure 5.2 Various types of muscle in pigs.

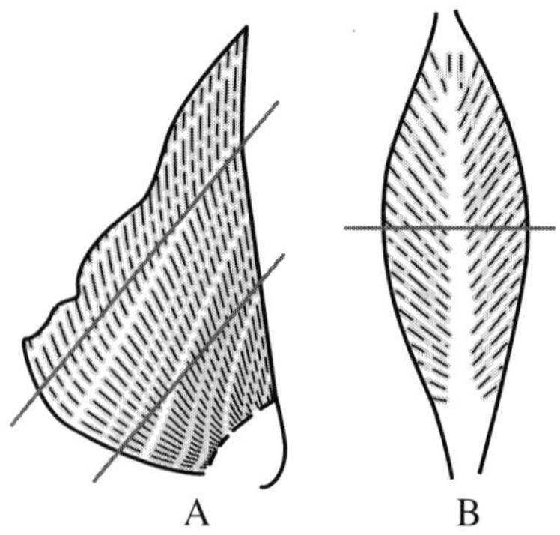

Figure 5.3 Effect of muscle fiber arrangement on assessing muscle fiber characteristics. From H. J. Swatland, *Structure and development of meat animals,* Prentice-Hall, Inc., Englewood Cliffs, New Jersey.

an oblique fashion rather than parallel to the long axis of the muscle. Unfortunately, from a research standpoint, this oblique fiber arrangement makes determination of total number of muscle fibers in this muscle very difficult. Under normal conditions, an increase in total body muscle mass is not muscle-specific and is generally realized in other muscles, so studies of muscle fiber number are best conducted using muscles in which fibers are aligned parallel with the long axis of the muscle, such as the *biceps brachii* and the *semitendinosus* muscles.

In addition to muscle fiber orientation, the fact that some muscle fibers do not extend the entire length of the muscle (Figure 5.5) contributes to some error of muscle fiber number measurements (Figure 5.6). Muscle fiber number is generally reported as the *apparent muscle fiber number,* as a means of predicting *absolute muscle fiber number,* for any given muscle and is calculated from the

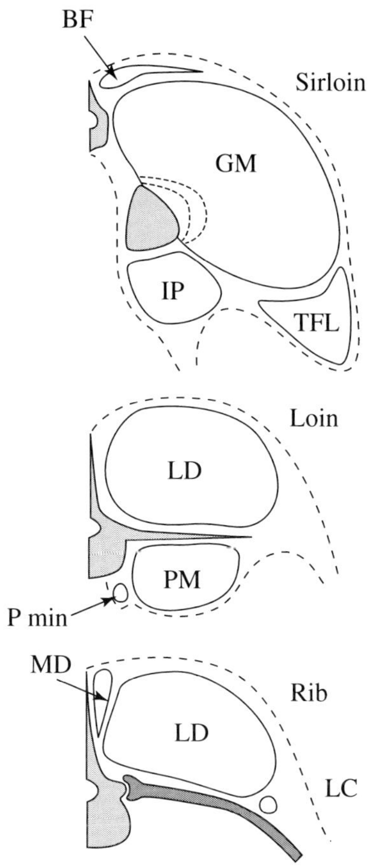

Figure 5.4 Location and type of muscle present in economically important meat cuts. LD, longissimus dorsi; GM, gluteus medius; PM, psoas major; IP, iliopsoas; BF, biceps femoris; TFL, tensor fascia lata; P min, psoas minor; MD, multifidus dorsi; longissimus costarum. From H. J. Swatland, *Structure and development of meat animals,* Prentice-Hall, Inc., Englewood Cliffs, New Jersey.

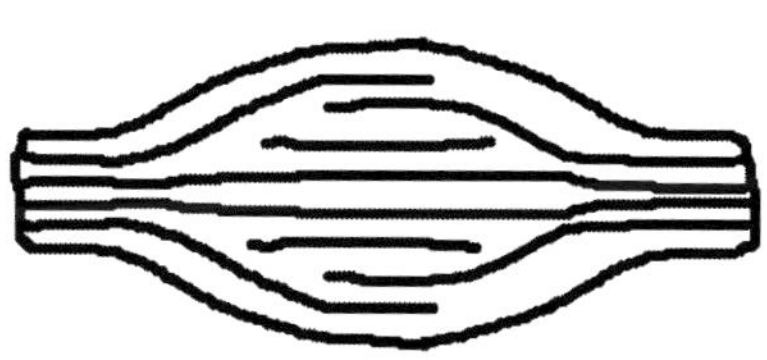

Figure 5.5 Organization of muscle fibers in a whole muscle.

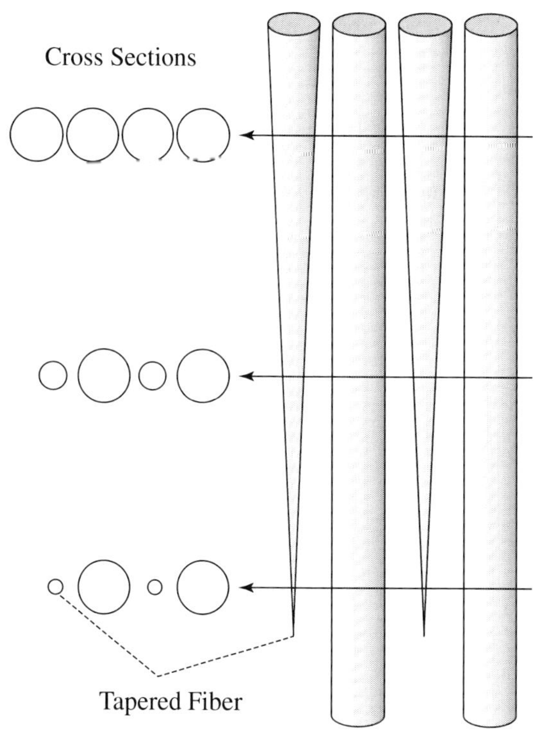

Figure 5.6 Effect of intrafascicularly terminating muscle fibers on muscle fiber parameters. Reprinted with permission from *Structure and Development of Meat Animals & Poultry,* by H. J. Swatland. Copyright CRC Press, Boca Raton, Florida.

cross-sectional area of the muscle and the *mean packing density* (number of muscle fibers per unit cross-sectional area).

Mechanisms for Increasing Muscle Fiber Number

The means by which muscle fibers are developed in the muscle tissue has been discussed in the previous chapter. Given that most of this occurs *in ovo* or *in utero,* minimal increases in fiber number, if any, occur after birth. In fact, most of the observed increases in muscle fiber number reported to occur after birth were attributed to muscle fiber lengthening of *intrafascicularly terminating fibers.* In some muscles, fibers may not extend the entire length of a muscle, but may extend various distances from the ends towards the central portions of the muscle. Other fibers may exist within the muscle with no attachment to either the origin or insertion of the muscle. These intrafascicularly terminating fibers increase in length during muscle growth and therefore, depending on where and when the muscle is cross-sectioned, muscle fibers growing in length may be mistaken for new fibers (Figure 5.6).

New muscle fibers may be generated during postnatal growth, but only during muscle regeneration following an injury that results in myofiber death. In such cases, *muscle satellite cells* located between the sarcolemma and basal lamina are activated to proliferate and ultimately fuse with other satellite cells to form new muscle fibers. The function that these special cells play in augmenting growth of existing muscle fibers, however, is profound and will be discussed in more detail later in this chapter. Even though generation of new fibers can occur during postnatal growth, it is rare and does not contribute appreciably to normal muscle growth parameters.

Factors Affecting Muscle Fiber Numbers

Animal Variation

Muscle fiber number is quite variable among animals within a species. For example, muscle fiber numbers in the *semitendinosus* of pigs can vary twofold. However, this occurrence may be related to prenatal nourishment and will be discussed to greater lengths later in this chapter. Nonetheless, there is substantial animal-to-animal variation in number of muscle fibers developed.

Muscle

Muscles exist in a myriad of sizes and shapes (Figure 5.2). The primary difference in the size of various muscles is the number of muscle fibers contained within each. Some of the smallest muscles, such as those responsible for eye movements, contain as few as a few thousand muscle fibers, whereas others, such as the *semimembranosus* muscle of the hind leg, contain several billion fibers.

Species

Body size is related to the number of muscle fibers. The difference between the muscle mass of elephants and mice is due largely to the number of muscle fibers in each animal. This mechanism for increased muscle mass has evolved because the excess growth in fiber diameter is not favorable from a metabolic standpoint. Diffusion of nutrients and metabolites in and out of a muscle fiber is restricted if radial growth of fibers is excessive. Porcine *semitendinosus* muscle contains roughly one-third the number of muscle fibers of the bovine *semitendinosus* muscle.

Nutrition

Nutrition plays an important role in the growth of all organisms. In the context of muscle, there is an interaction between nutrition and age, meaning nutrition has a greater impact on muscle at some stages of growth than at other stages of growth. Nutrition alters muscle fiber number only during prenatal development. Throughout the animal kingdom, the developing conceptus has priority over other tissues of the dam for nutrients that are partitioned among tissues and organs within the body. With few exceptions, most maternal tissues will undergo dramatic atrophy in order to mobilize enough nutrients to maintain a viable pregnancy.

In litter-bearing species, there is also a priority that exists across embryos and fetuses. This is particularly evident in pigs for which extreme restriction of nutrient absorption occurs during fetal development. Severe fetal growth retardation manifests itself by dramatically suppressing birth weight and muscle fiber formation (Figure 5.7). These pigs are termed *runt* pigs by producers. Runt pigs are those piglets weighing less than 2/3 the weight of the mean piglet weight of a given litter. Many runt pigs die shortly after birth because they are unable to compete with larger littermates for adequate nutrient intake during nursing. If, however, runts are for-

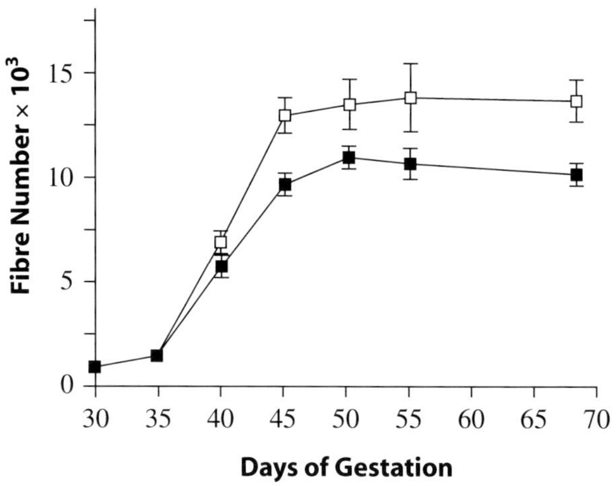

Figure 5.7 Effect of restricted nutrient intake (closed boxes) in gestating guinea pigs on fetal muscle fiber number in the biceps femoris muscle. From S. S. Ward and N. C. Stickland, Why are slow and fast muscle differentially affected during prenatal undernutrition (1991), *Muscle & Nerve,* vol. 14, pg. 259.

tunate to survive the nursing phase and successfully enter the growing phase, they become fatter than their littermates and produce carcasses with less muscle at comparable live weights (Chapter 11). This inferior carcass composition is the result of fewer muscle fibers in runt pigs than in normal littermates. The source of this variation lies in the fact that runt pigs fail to develop secondary muscle fibers during the early-middle portion of prenatal development (depending on species) as discussed in Chapter 4. Even though runt piglets seem to reflect the impact of fetal undernutrition to a maximum extent, smaller piglets in a litter have fewer muscle fibers than those large piglets. In light of the information covered earlier in this text regarding fetal muscle development (Chapter 4), it is obvious that adequate nutrition is needed during formation of muscle fibers, but once fibers are developed, postnatal nutrient deprivation has little effect on muscle fiber number.

Age

The number of muscle fibers that an animal develops is fixed at birth because muscle fiber hyperplasia (increase in muscle cell number) occurs *in ovo* or *in utero* (Figure 4.12). As pointed out in Chapter 4, muscle fiber number increases rapidly during secondary muscle fiber development. Therefore, age-related changes in muscle fiber number occur only during early to middle phases of prenatal development and may vary with species.

Most mammalian and avian species are *determinate* growers and reach a given size (mature size), provided adequate nutrients are available. In contrast, fish are *indeterminate* growers, meaning they have no predetermined mature weight. Many fish grow to a size that reflects their environment or availability of food. As a result, fish must have a means to increase tissue mass. For reasons yet unknown, fish are capable of generating new muscle fibers after hatching. Figure 5.8 shows the number of muscle fibers (y-axis) in the muscle of rainbow trout with increasing length (x-axis), which in this case can be considered age. It is clear from this graph that muscle fiber number increases significantly after hatching and contributes substantially to the overall hypertrophy of the muscle, at least during the earliest periods of growth (Figure 5.9). Notice the amount of hyperplastic growth is related to the type (red versus white) of muscle. This ability to increase muscle fiber number after birth or hatching is rather atypical in nature. Whether muscle fibers are lost during periods of reduced nutrient availability in fish is less clear. The uniqueness of fish hyperplastic growth as compared to mammals may support the time honored axiom that ontogeny recapitulates phylogeny.

Even though it is commonly held that muscle fiber number does not change with age, this is not true during *senescence* when tissue mass is no longer maintained in its mature state and muscle fibers are lost. Although senescence is not relevant to most meat-producing animals because market age is usually younger than physiological maturity, senescence is apparent in some breeding stock because superior males and females may remain in breeding herds for many years. The link between senescence and muscle fiber loss is not clear, but may be indirectly related to losses in neuronal input. This is an important area of research given the significance of muscle wasting in elderly humans.

Breed and Genetic Selection

Faster-growing breeds or strains have more muscle fibers than their slower-growing counterparts. This is particularly true regarding the relationships of muscle fiber number versus average daily gain and

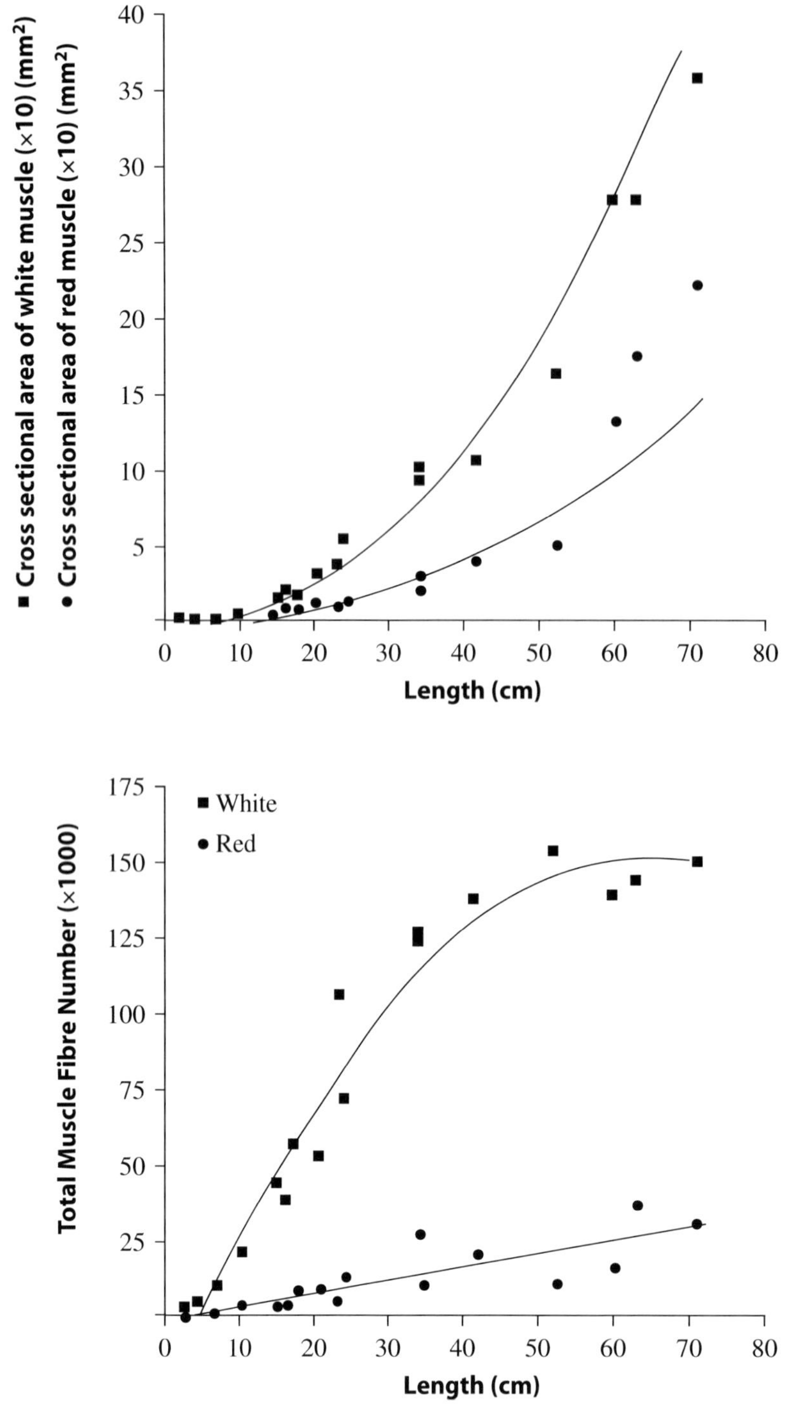

Figure 5.8 Changes in muscle fiber size (top) and number (bottom) in growing rainbow trout. From N. C. Stickland, Growth and development of muscle fibers in the rainbow trout (salmo garidneri) (1983), *Journal of Anatomy,* vol. 137, pg. 323.

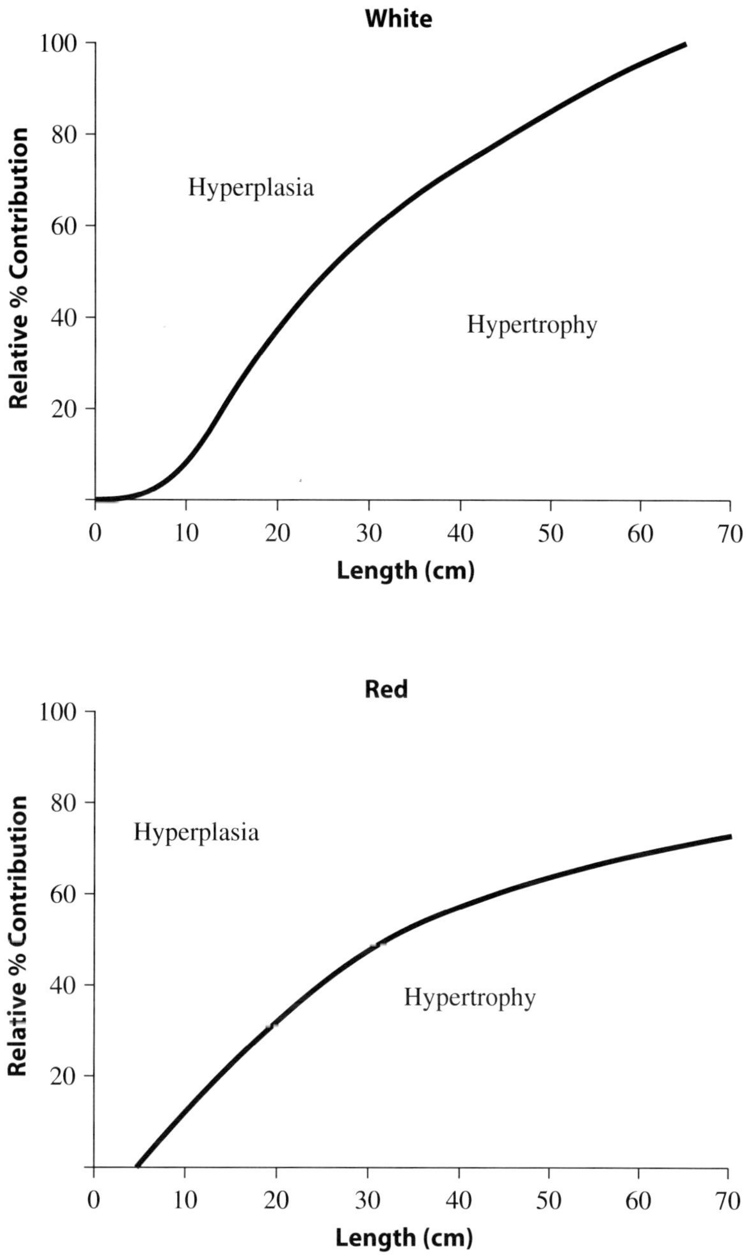

Figure 5.9 Relative contribution of muscle fiber hyperplasia versus hypertrophy to overall growth of white (top) and red (bottom) muscle in growing rainbow trout. From N. C. Stickland, Growth and development of muscle fibers in the rainbow trout (salmo garidneri) (1983), *Journal of Anatomy,* vol. 137, pg. 323.

feed efficiency (Figure 5.10). Tables 5.1 and 5.2 show differences in muscle fiber number among several breeds of cattle and pigs, respectively. Muscle fiber number is different among breeds, especially those with extreme phenotypes. Pietrain pigs, for example, are extremely muscular and possess more muscle fibers than other pig genotypes, even though their cells are generally larger in size as well (discussed later in this chapter). Improvement within breeds has led to an increased muscle fiber number and the differences in fiber number across breeds have been minimized as selection for protein accretion has occurred. This is further illustrated by studies comparing wild and domestic pigs. Wild pigs

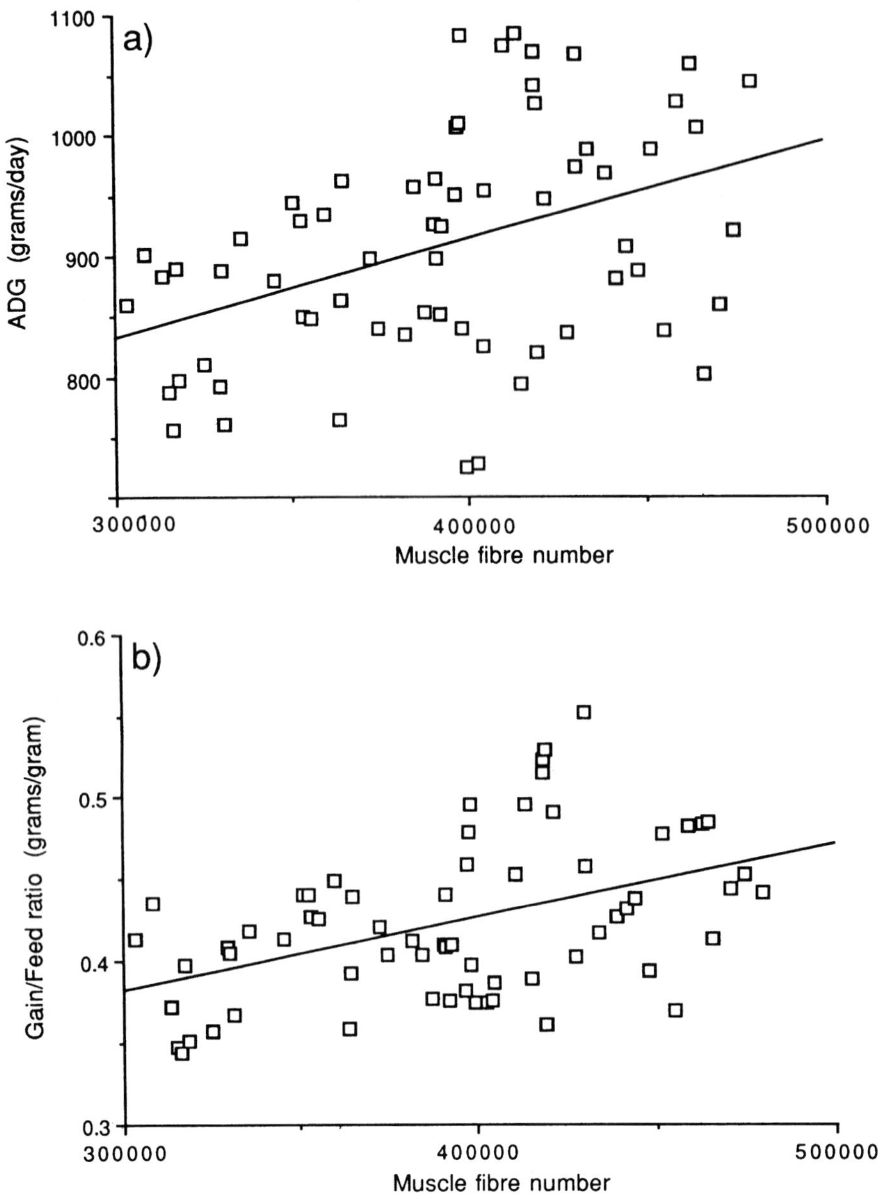

Figure 5.10 Relationship between muscle fiber number and performance traits (average daily gain, ADG; and feed efficiency, gain/feed ratio) in growing (25–80 kg) pigs. From N. C. Stickland, Muscle growth (1995), *Meat Focus International,* vol 6, pg. 241.

(i.e., the unselected ancestors to modern, domesticated breeds of pigs) contain fewer muscle fibers than domestic pigs indicating that muscle fiber number has increased with domestication (Figure 5.11). On the other hand, feral pigs (i.e., domesticated pigs that have been released or have escaped captivity) have equal numbers of muscle fibers but they are smaller due to lack of adequate nutrient intake. Similarly, selection of mice for increased body size also has resulted in increased muscle fiber numbers showing that selection for muscle size can result in an increase of muscle fiber numbers.

Sex

Males tend to have a greater number of muscle fibers at birth than females but this seems to be

TABLE 5.1 MUSCLE FIBRE NUMBER IN CATTLE OF DIFFERENT BREEDS (IN MILLIONS)

	Semitendinosus	Longissimus
Galloway	1.70	—
German Angus	1.72	—
Holstein-Friesian	1.83	2.69
White-Blue-Belgian	3.36	3.77

From J. Wagner, E. Albrecht, I. Fiedler, F. Teuscher, H.J. Papstein, and K. Ender, Growth- and breed-related changes of muscle fiber characteristics in cattle (2000) Journal of Animal Science, v. 78, pg. 1485. Savoy, Illinois.

TABLE 5.2 EFFECT OF BREED ON MUSCLE FIBER NUMBER IN PIGS

Animals	Fiber Number
Pietrain (P) gilts	1,078,000
Pietrain barrows	984,000
Danish Landrace (DL) gilts	1,124,000
Danish Landrace barrows	988,000
P × DL gilts	1,005,000
P × DL barrows	934,000
Duroc high fat line	659,000
Duroc low-fat line	802,000
Yorkshire high-fat line	738,000
Yorkshire low-fat line	797,000
Veredeltes Landschwein gilts	869,000
Veredeltes Landschwein barrows	940,000
Deutsches weisses Edelschwein gilts	758,000
Deutsches weisses Edelschwein barrows	717,000

From H.J. Swatland, Structure and development of meat animals, Prentice-Hall, Inc., Englewood Cliffs, New Jersey.

species dependent. For example, sex has little effect on muscle fiber number in pigs whereas muscles of male cattle and chickens have more muscle fibers. This is probably the result of the exposure of male fetuses to higher concentrations of androgens that exist *in utero*. Myogenesis is altered at these early stages of development in such a way that increased

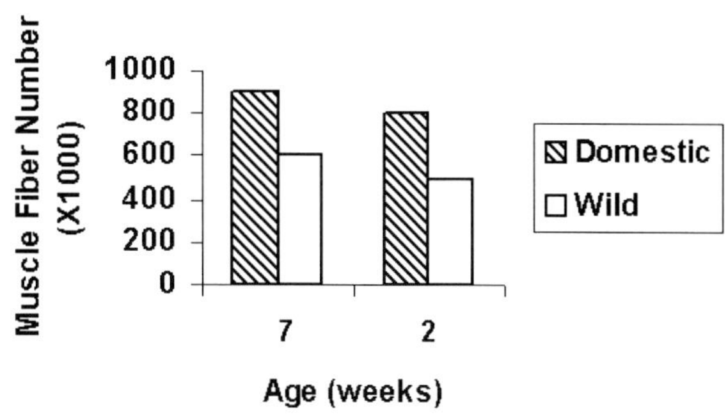

Figure 5.11 Muscle fiber number in domestic and wild pigs. Data from C. Rehfeldt, I. Fiedler, G. Dietl and K. Ender, Myogenesis and postnatal muscle cell growth as influenced by selection (2000), *Livestock Production Science*, vol. 66, page 177.

number of myofibers are produced. Ultimately, this results in a greater potential for postnatal muscle hypertrophy, resulting in males being heavier muscled than females. The hormonal regulation of growth will be discussed in a later chapter.

Genetic "Conditions"

Recall that mechanisms controlling muscle development are very complex and consist of a myriad of proteins that coordinate this process. Each protein is encoded by a specific gene in the *genome*. During DNA replication, if DNA is not replicated exactly, the organism may die because an essential gene is mutated to a point where it is becomes lethal. In other cases, a mutation may not be lethal, but may result in a protein or proteins with altered function. In some cases, the mutation may result in a more desirable trait i.e., one that actually results in improved animal performance. This advantage obviously occurs in the wild, but is occasionally amplified during selection of superior breeding stock in domestic meat-producing animals. This mechanism probably propagated *double muscled* cattle and *callipyge* sheep.

Double muscled (*culard* or *doppellender*) cattle are extremely muscular animals. These cattle do not have double the number of muscles as the name implies. Rather, they develop muscles that contain nearly twice the number of muscle fibers as those lacking this condition (Figure 5.12). As a result of this hyperplastic growth, muscles of adult animals become very large and bulging (Figure 5.13). Double muscled cattle are not larger than normal cattle; they

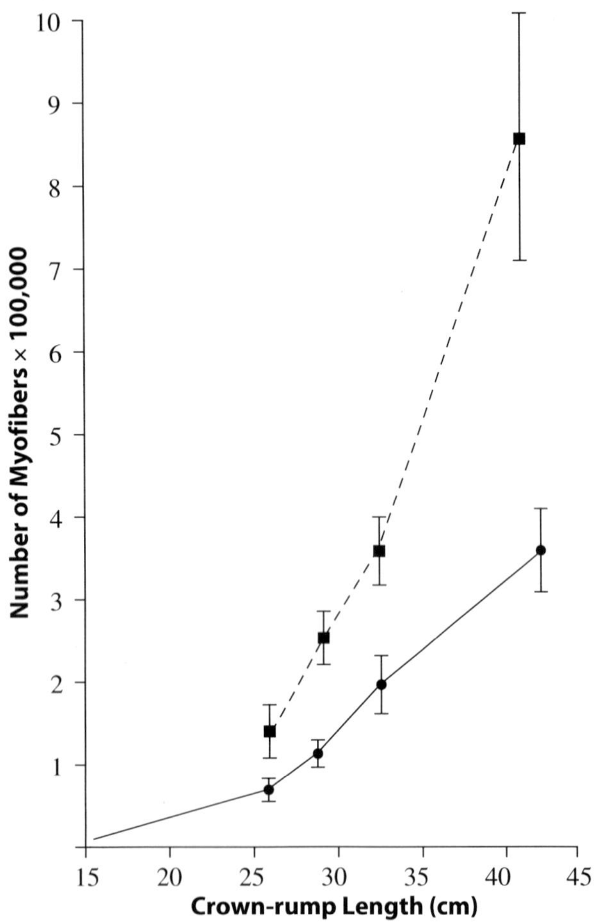

Figure 5.12 Muscle fiber number in developing double-muscled (■) and normal (•) bovine fetuses. From H. J. Swatland and N. M. Kieffer, Fetal development of the double muscled condition in cattle (1974), *Journal of Animal Science*, vol. 38, pg. 752.

simply possess more skeletal muscle. In fact, the amount of muscle relative to the amount of bone in cattle possessing the double muscle condition may reach nearly 6:1, compared with ratios for normal beef and dairy cattle of 5:1 and 4:1, respectively.

Because of the benefits of producing cattle with such economically positive attributes (increased muscle and decreased fat), these cattle were propagated over several years, especially in Europe where butchers coveted the lean carcasses. However, the double muscled condition is not compatible with many of the management practices of the US cattle industry.

As a result of elegant studies, the gene responsible for double muscled cattle was identified. The *myostatin* gene belongs to the transforming growth factor family and has a profound effect on myogenesis (Chapter 4). As postulated by many prior to its discovery, the myostatin gene is mutated in cattle that possess the classical double muscled phenotype. Mutations in the gene that result in the double muscle phenotype in cattle range from changes in one nucleotide in the gene (*point mutation*), to various *deletions* of larger regions of the myostatin gene. The *gene product* or protein encoded by the mutated gene lacks normal function. This altered function affects myogenesis (e.g., tightly controlled regulation of proliferation and differentiation of myoblasts) and results in increased muscle fiber number. The exact mechanism by which myostatin affects muscle development is still being carefully studied.

Extreme muscular development exists in other meat animal species (Figures 5.14 and 5.15). On occasion, selected lines or breeds of pigs, sheep, and poultry exhibit more muscle development relative to their normal counterparts; however these conditions have not been attributed to changes in the myostatin gene, nor have they been due to such dramatic increases in the number of muscle fibers as that known to occur in double muscled cattle. Unfortunately, as is the case with double muscled cattle, extremes in muscle size are often accompanied by physiological conditions that interfere with animal reproduction, stress resistance, and locomotion. These factors are crucial to the economic viability of animals in the meat industry.

Muscle Fiber Size

Within a species, breed or line, and muscle, there are very few factors that change muscle fiber number other than those which exist during embryonic and fetal development. With the exception of a response to an injury and subsequent repair of a muscle, there is no known means for developing muscle fibers other than during primary and secondary myotube formations that occur during fetal stages of development. Conversely, muscle fiber size increases dramatically during the last third of fetal development and this continues during postnatal phases of growth (Figure 5.16). In addition, muscle fiber length can

(A)

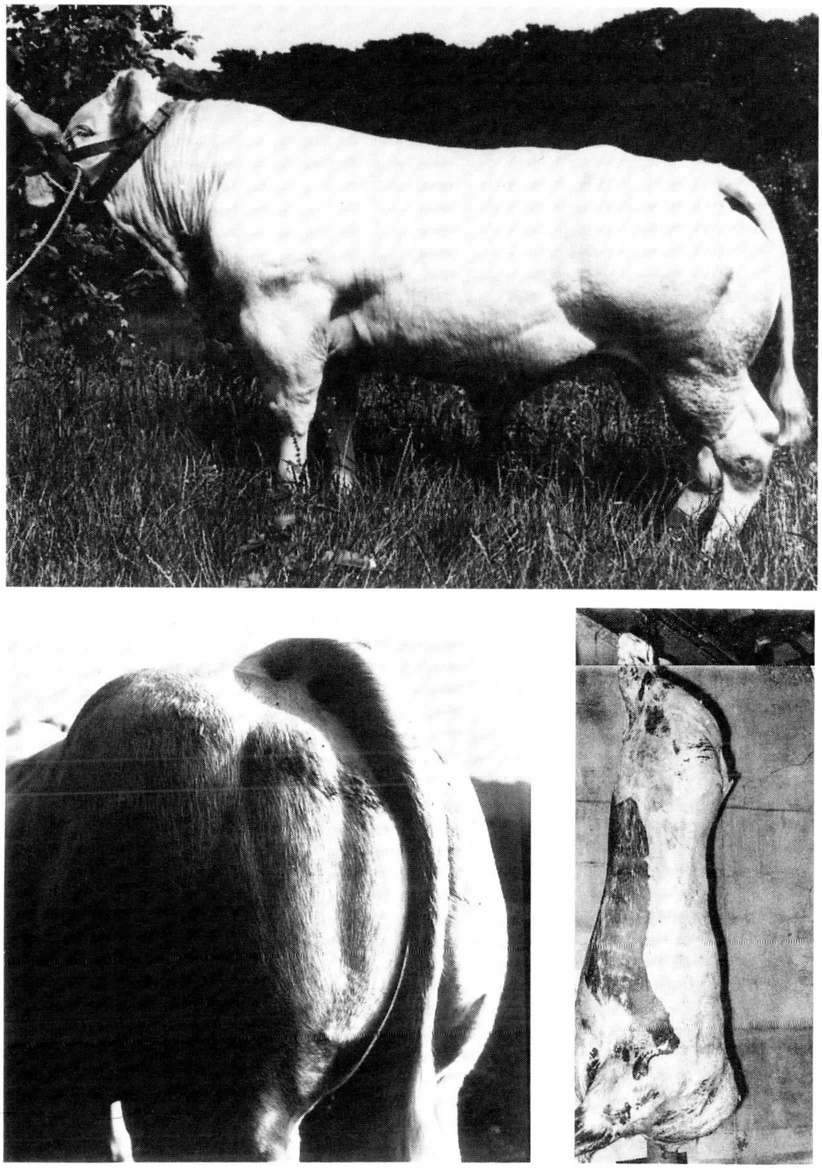

(B)

(C)

Figure 5.13 Bull and carcass that exhibit double muscling. Note the two depressions in the muscling of the round (B), and in the forequarter of the live animal (A), and the bulging round and depression of the carcass (C). From *Principles of Meat Science, Fourth Edition* by Aberle et al. Copyright © 2001 by Kendall/Hunt Publishing Company. Reprinted by permission.

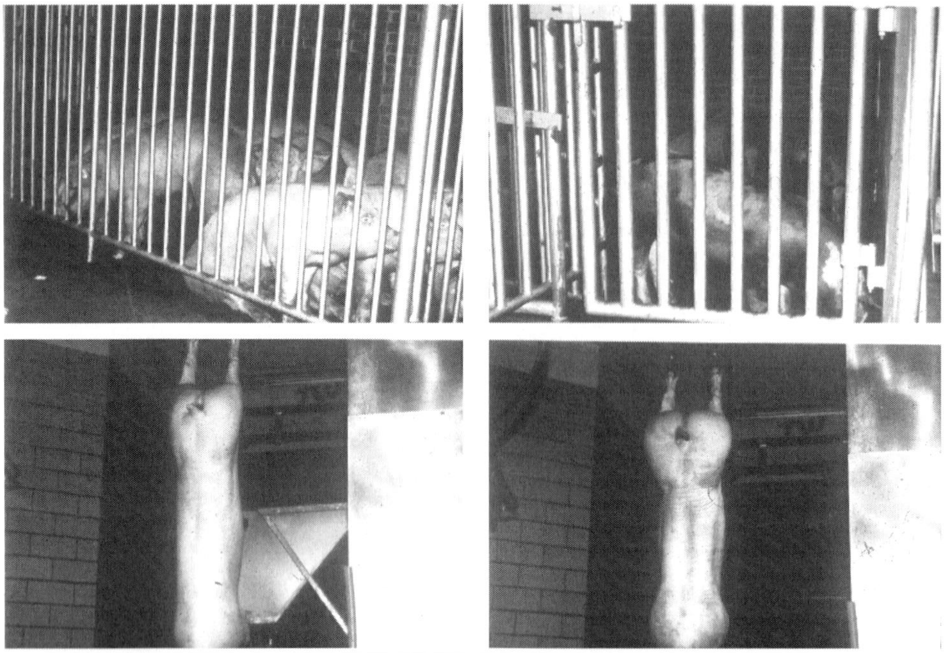

Figure 5.14 Comparison of pork carcasses from pigs possessing (right) and lacking (left) the halothane or "stress" gene. Note that many pigs lacking the halothane gene can be penned with many other pigs prior to stunning compared to those positive for the gene which require less crowded conditions. Further, note the extreme muscularity of carcasses from halothane positive pigs.

impact muscle growth. Therefore, both longitudinal and radial growth of muscle fibers contribute significantly to increased muscle growth. Moreover, as pointed out in Figure 5.1, increasing muscle fiber size is another means of stimulating muscle hypertrophy and its role in improving growth performance is illustrated in Table 5.3 by comparing the size of muscle fibers in the muscle of fast- and slow-growing lines of chickens.

Mechanisms for Increasing Muscle Fiber Size

Structural Changes

At the time muscle fibers form, they have diameters ranging between 10 to 20 μm. In some cases, mature muscle fibers can reach 100 μm in diameter as in the case of *giant muscle fibers* found in some highly selected lines of pigs and poultry. There is speculation about the physiological relevance of these giant fibers and their existence may be related to a pathological or degenerative condition in the muscle. Therefore, normal mature muscle fibers are considered to range from 50 to 80 μm in diameter. The mechanisms responsible for this increased radial growth are components of the system of generation of newly synthesized myofibrils as occurs in myofibrillogenesis (Figures 4.13, 4.14 and 4.15). Existing myofibrils can split longitudinally (Figure 5.17) to form new myofibrils, which may subsequently enlarge through the addition of myofilaments. Splitting of existing myofibrils probably accounts for the majority of increased numbers of myofibrils during the major stages of muscle fiber hypertrophy. The stimulus for increasing radial growth of muscle fibers through splitting is work or exercise. Therefore, growth resulting from such stimulus is referred to as *work-induced hypertrophy* or *exercise-induced hypertrophy*.

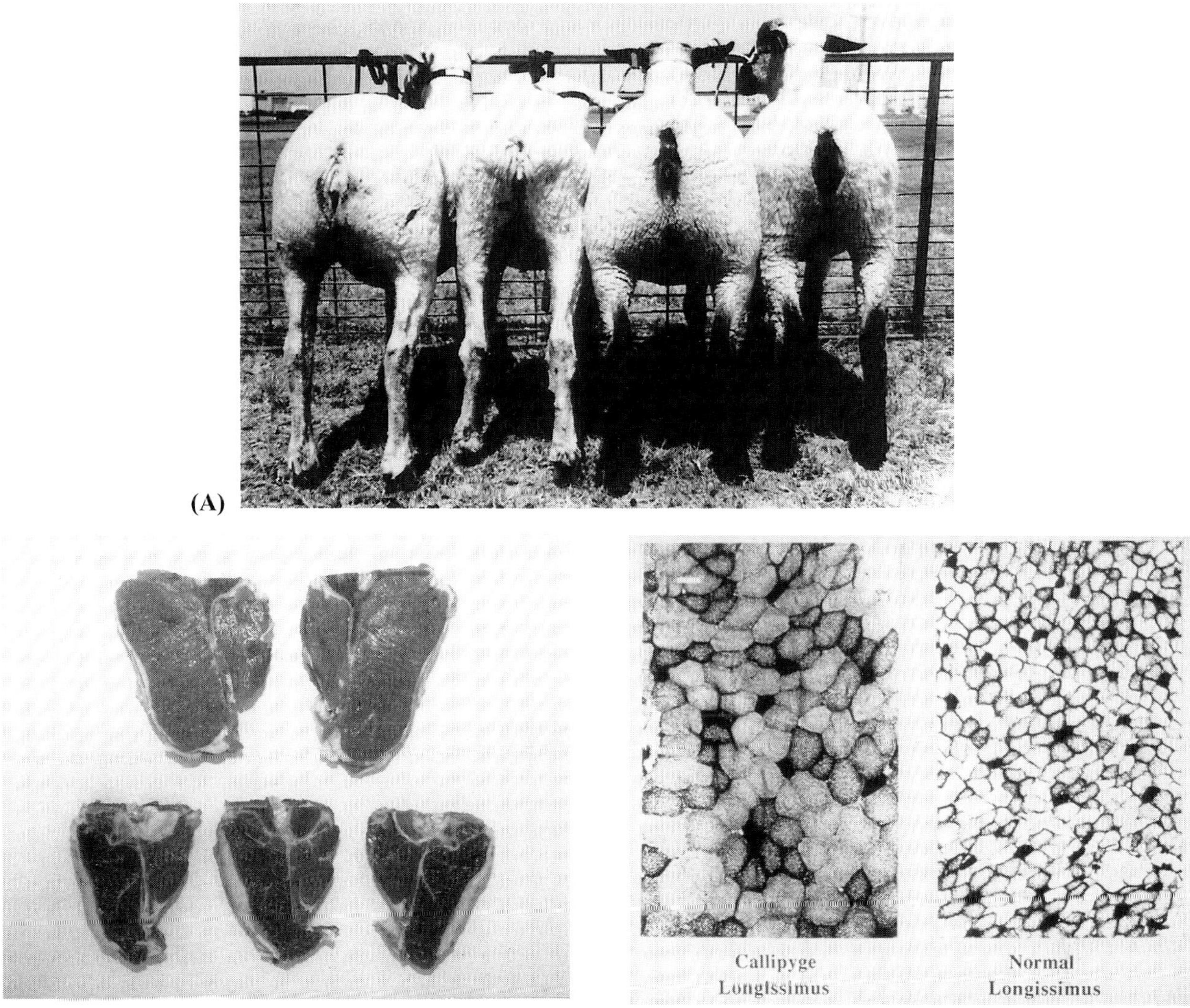

(B) (C)

Figure 5.15 Comparison of callipyge and normal genotype lambs. (A) Callipyge lambs (first and third, left to right) contrasted to normal lambs (second and fourth) of the same breeds. Photo courtesy of Sam Jackson, Texas Tech University. (B) Loin chops from callipyge (top) and normal (bottom) lambs. (C) Cross-section of *longissimus* muscle from callipyge (left) and normal (right) lambs showing muscle fiber hypertrophy of the callipyge genotype. Courtesy of C. E. Carpenter, Utah State University.

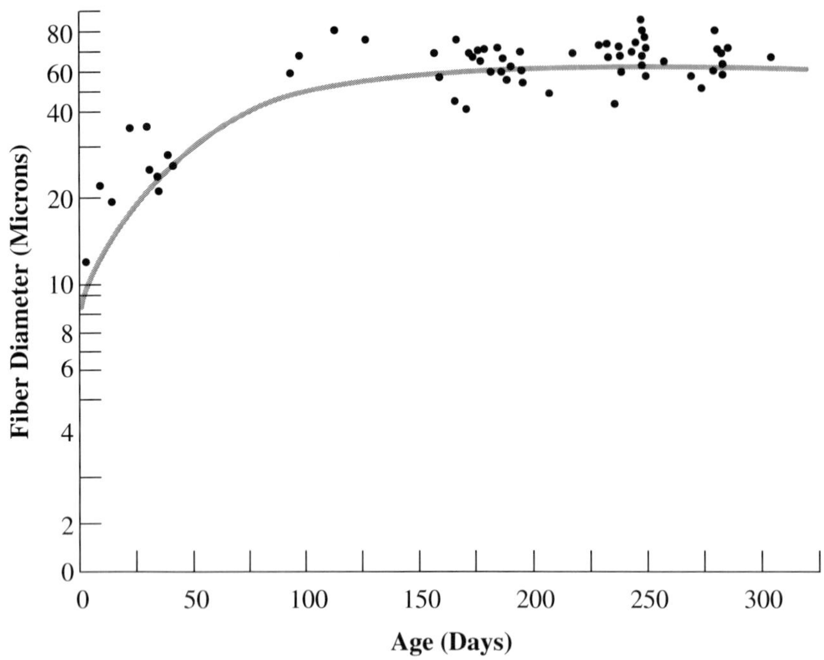

Figure 5.16 Change in longissimus muscle fiber diameter with increasing age in growing pigs. Data from B. B. Chrystall, University of Missouri. Courtesy of W. H. Freeman and Company.

TABLE 5.3 GROWTH AND MUSCLE FIBER DIAMETER IN RAPID AND SLOW GROWING CHICKENS[1]

Age	Live Weight		Fiber Area	
	Rapid	**Slow**	**Rapid**	**Slow**
(wk)	(g)		(μm^2)	
0	36	31	20	23
11	1,882	675	1,256	664
55	3,285	1,883	2,755	1,946

[1]Values are the means of the *Pectoralis* muscle for rapid- and slow-growing strains of male chickens. Data from Rémignon *et al.* (1993).

The stimulus for increasing the length of muscle fibers is related to stretch. Therefore, growth in length is a result of *stretch-induced hypertrophy* as imposed by bone lengthening during skeletal growth. Muscle fiber development in muscles occurs when the fetus is only millimeters long. Because most of the muscles of a fetus are only a fraction of the length of a fetus, growth in the length of muscle fibers is perhaps more important than radial growth during prenatal development. The mechanism that allows muscle, individual muscle cells, and myofibrils to increase in length is *sarcomere addition.* These contractile units are generally added to the end of myofibrils, although they can insert in less distal aspects of existing myofibrils. Sarcomere numbers can increase more than three times from birth to maturity in some animals. Lengthening of individual sarcomeres also can

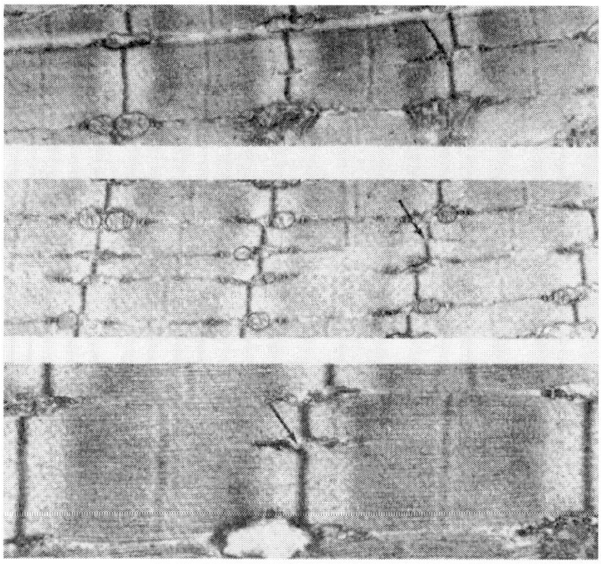

Figure 5.17 Micrographs showing longitudinal myofibrillar splitting in adult muscles. From G. Goldspink, *The structure and function of muscle,* vol 1, G. H. Bourne, editor, Academic Press, Inc., New York.

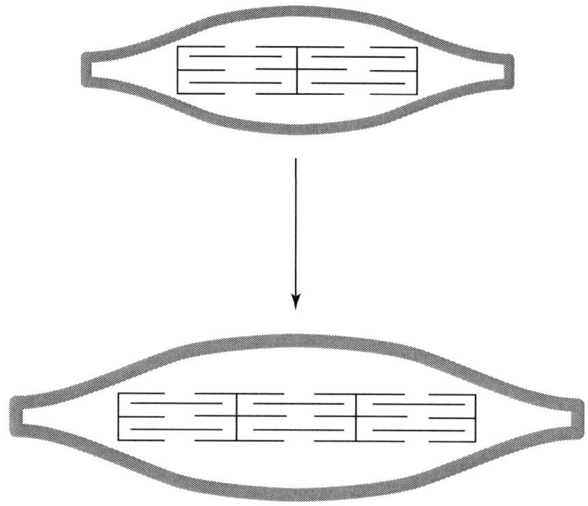

Figure 5.18 Drawing illustrating the role of sarcomere addition in muscle lengthening. Modified from B. Russell, D. Motlagh, and W. W. Ashely, Form follows function: how muscle shape is regulated by work, *Journal of Applied Physiology,* vol. 88, pg. 1127.

affect the overall length of muscle fibers (Figure 5.18), due to their arrangement in series; however, sarcomere lengthening accounts for less than 20% of the increase. During the lengthening of myofibrils, the cellular components of the myofiber e.g., sarcoplasmic reticulum, sarcolemma, etc. must accommodate the growth.

Protein Synthesis and Degradation

The increased longitudinal and radial growth of myofibers described above requires significant amounts of newly synthesized proteins. Other cellular components of the cell must accommodate this growth. Many different proteins, including *sarcomeric* (those associated with the sarcomere) and non-sarcomeric proteins are needed for the entire fiber to grow. Taken together, it is logical that the ability of muscle to manufacture or *synthesize* these proteins may affect the entire growth of muscle. Furthermore, muscle proteins exist in a dynamic state and an equilibrium between synthesis and degradation exists. The process of building protein, then systematically replacing it with a newly synthesized protein is referred to as *protein turnover.* Protein turnover is most rapid for proteins of the sarcoplasm, slowest for stromal proteins, and intermediate for myofibrillar proteins.

Protein synthesis refers to the process by which cells assemble amino acids into proteins. This involves:

- Transcribing DNA into messenger RNA (mRNA).
- Movement of mRNA out of the nucleus into the cytoplasm.
- Translation of mRNA into protein.
- Post-translational processing of the protein.
- Positioning of the protein to specific locations (Figure 5.19).

Protein degradation or *breakdown* is proteolysis of the proteins into polypeptides and individual amino acids by various proteinases within the cell. The *net protein accretion* or *deposition* is the difference between the amount of protein synthesized and degraded. Muscle fiber hypertrophy will occur when the accumulation of protein occurs, i.e., when

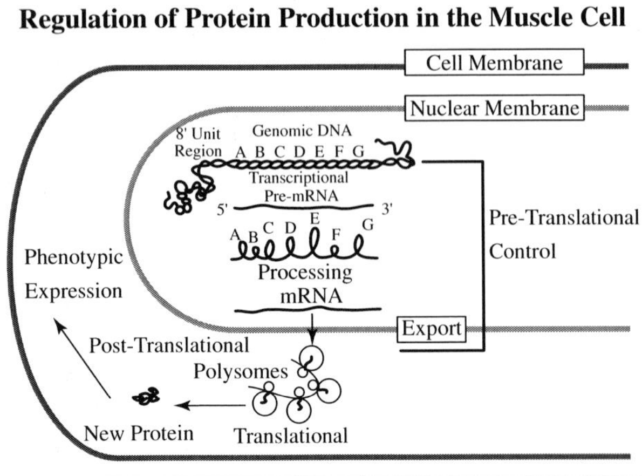

Figure 5.19 Molecular events occurring during protein production in a muscle cell.

amount of protein synthesized exceeds that degraded. *Muscle fiber atrophy,* on the other hand, occurs if protein degradation exceeds synthesis. Because there is an upper limit to the amount of protein that can be synthesized, efforts to increase protein accretion often involve exploration of methods for inhibiting protein degradation. Therefore, an understanding of protein degradation systems in skeletal muscle is important in developing strategies to maximize protein accretion.

Three *proteolytic* systems in muscle that have been studied extensively are:

- Lysosomal system
- Calpain system
- Ubiquitin-proteosome proteolytic pathway.

The *lysosomal system,* as the name implies, requires the lysosomes of cells. These cellular organelles are literally packed with various enzymes capable of degrading numerous proteins. One family of enzymes most often associated with lysosomal proteolysis is termed *cathepsins.* Cathepsins are non-specific proteases that work best in a relatively low pH environment, hence in the lysosomes, and degrade proteins to small peptides or to individual amino acids. *Cystatin,* an inhibitor of cysteine-lysosomal proteases, is a member of this system that is responsible for prevention of deleterious degradation of proteins. The lysosomal system is likely responsible for 25 to 30% of the intracellular protein turnover in muscle cells, most of which is turnover of sarcoplasmic proteins.

The *calpain system* consists of at least four major components. Three components of this system are specific intracellular proteases that are ubiquitous among cells: *μ-calpain* (or *Calpain 1*), *m-calpain* (or *Calpain 2*), and *skeletal muscle calpain* (*Calpain 3* or *p94*). The fourth component of the calpain system is *calpastatin,* which is an inhibitor of calpains. Other than the presence of a targeted protein, the calpain system requires calcium at varying concentrations to activate the system to different levels. *In vitro,* the μ-calpain protease is active at micromolar (μM) calcium concentrations, whereas the m-calpain requires nearly millimolar (mM) concentrations. Calpastatin also requires calcium for activation, and its inhibition of calpain activity is a complex process. *Autolysis* of the calpains, which reduces the concentration of calcium required for activity, is altered by phosphorylation of calpains and by their binding to proteins and phospholipids, which may explain the presence of calpain activity in cells with free calcium concentrations of less than 0.5 μM. The calpain system may be important in myofibril assembly and myogenesis and is responsible for the majority of protein turnover in living muscle cells as well as postmortem proteolysis that contributes to meat tenderization. Degrading a structure as large and as complex as a myofibril requires multiple steps, involving initiation of myofibrillar disassembly followed by further breakdown of proteins to individual amino acids. Calpains play a major role in disassembly of the contractile protein apparatus (Figure 5.20) and other proteolytic systems are responsible for further proteolysis of sarcomeric proteins.

The *ubiquitin-proteosome proteolytic pathway* contains multiple enzymes, ATP, and the polypeptide co-factor, ubiquitin (Figure 5.21). Ubiquitin is covalently attached to proteins that are to be targeted for breakdown via the action of at least three enzymes:

- E1, *ubiquitin activating enzyme*
- E2, *ubiquitin conjugating enzyme*
- E3, *ubiquitin ligase.*

Once ubiquitin molecules attach to proteins, i.e., proteins become *polyubiquitinated,* they are trafficked to *26S proteosome* complexes for degradation into small peptides. This pathway plays a primary role in degradation of short-lived regulatory

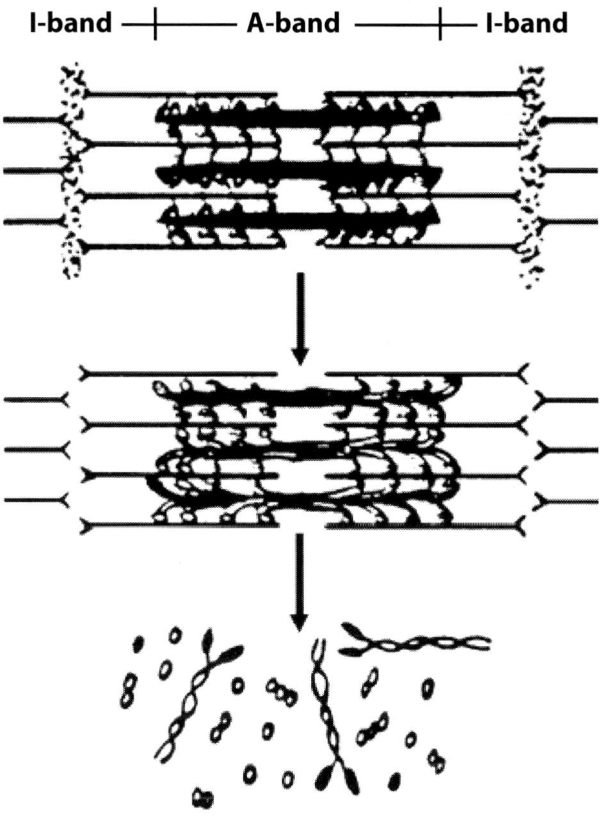

Figure 5.20 Effects of the caplain proteases on muscle degradation postmortem. From Goll et al., 1992. From Proc. Annual Rec. Meat Conf. Amer. Meat Sci. Assoc. (44th) pp. 25–36.

proteins and abnormal proteins, and has been implicated in the degradation of actin and myosin.

The exact means by which muscle cells are capable of synthesizing proteins and selectively inserting them into functioning myofibrils is not known. It is, however, likely that all three of the aforementioned pathways, and possibly others that have yet to be identified, participate in concert to perform this complex task.

Protein Accretion Rates

The rate of protein accretion or deposition is the difference between the rates synthesized and degraded. These rates are important in characterizing muscle protein metabolism and they may be expressed in different ways, two of which are as *absolute rates* and *fractional rates.* Absolute rates are used to express rates as mass per unit time (g/day), whereas fractional rates are used to express rates as relative to the amount of absolute protein present, i.e., relative to the mass of the existing protein pool, usually expressed as percent per unit time (%/day). Fractional rates often are used in research settings when comparing synthesis, degradation, and accretion rates across muscles of different sizes, i.e., different protein pool sizes. Converting a fractional rate to an absolute rate is accomplished by multiplying the fractional rate by the mass of the protein pool. Developmental changes in fractional rates are shown in Figure 5.22. As with absolute rates, the accretion rate is the difference between the rates of synthesis and degradation. As animals approach maturity, this difference becomes smaller, and eventually becomes zero, at which point no further muscle growth occurs. In a senescent state and in some disease states, the rate of degradation may exceed synthesis, resulting in muscle atrophy. Because muscle protein deposition is a major component of total body weight gain, a relatively small change in muscle protein degradation rate may have a dramatic impact on body weight gain. Consequently, much research is directed at developing strategies that reduce muscle protein turnover. Nutritional status can have major effects on muscle protein synthesis and degradation and, consequently, the energetic efficiency of protein deposition. Many of the factors affecting protein metabolism also are mediated by the endocrine system. Nutritional and hormonal regulation of protein metabolism will be discussed in later chapters.

Satellite Cell Recruitment

Increases in myofiber size are accomplished by increased accretion of protein. Although the *protein: DNA ratio,* i.e., *DNA unit* or *nuclear domain,* of the growing fiber increases, it does not increase to the extent expected, given the significant amount of protein deposited during growth. This is the result of an increase in the number of myonuclei located within the fiber during hypertrophy. Recall from Chapter 4, myonuclei, i.e., nuclei within muscle fibers, are not able to replicate, yet mature muscle fibers are still capable of recruiting new nuclei. The source of these new nuclei resides solely in the population of muscle satellite cells. These quiescent or dormant mononucleated cells, residing in the G0 phase of the cell cycle (Chapter 4), are strategically located between the sarcolemma and the basement membrane of muscle fibers (Figures 5.23). Although it is not well documented, the membranes of the satellite cell and the mature muscle fiber are in close association, and may allow for communication between the separate entities. Satellite cells are not post-

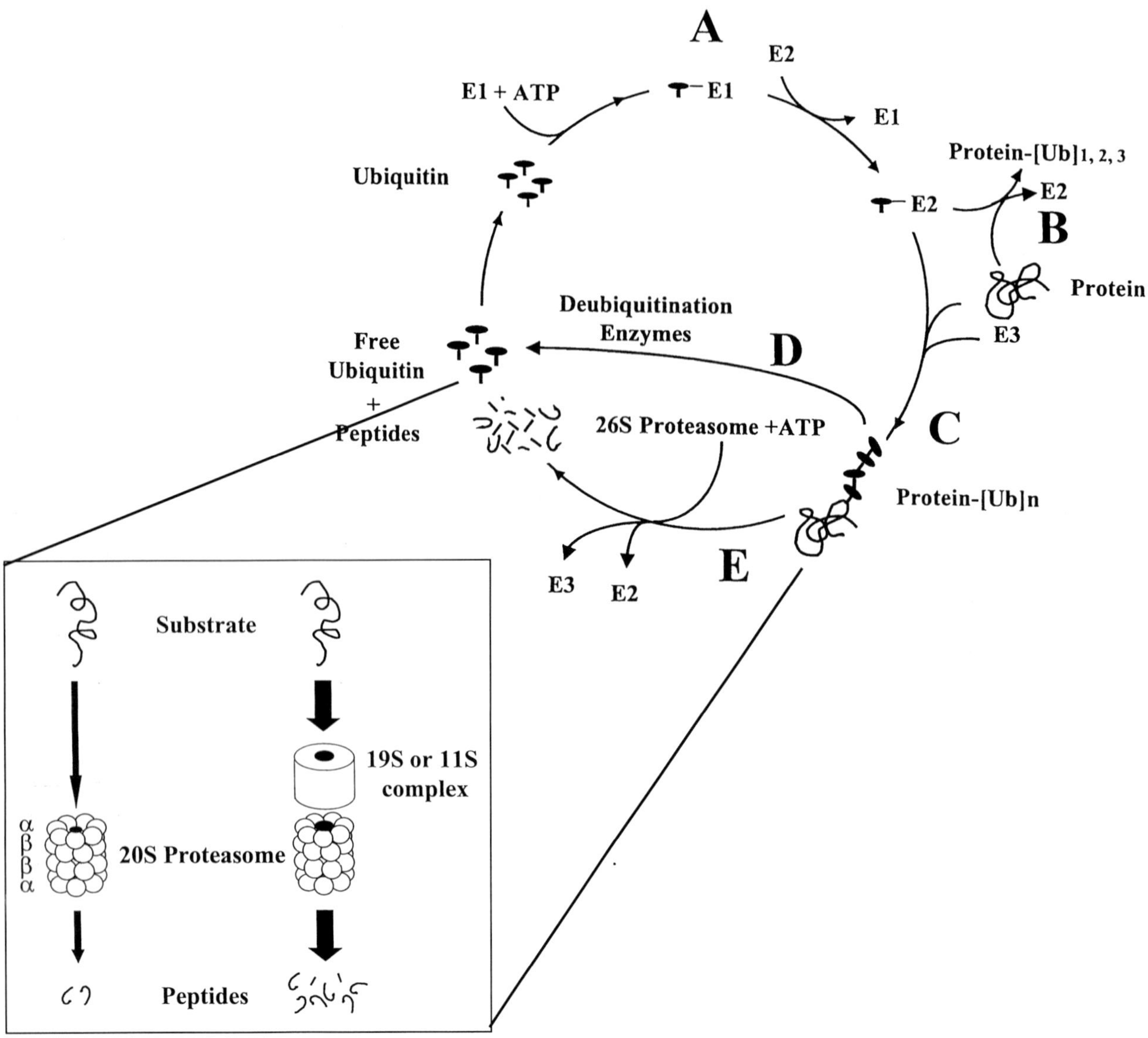

Figure 5.21 Ubiquitin-proteosome pathway for protein degradation. Modified from D. Attaix, L. Combaret, M.-N. Pouch, and D. Taillandier, Cellular control of protein degradation (2001), Proceedings of the Biennial Growth Symposium, *Journal of Animal Science,* Savoy, Illinois.

mitotic, and therefore, retain the capacity to divide, unlike those nuclei in existing muscle fibers (Chapter 4). As mentioned earlier, satellite cells are similar to myoblasts in that most are determined and restricted to the muscle lineage. Once *activated* by various environmental stimuli related to growth, however, these cells begin proliferating and their progeny are capable of differentiating and fusing with the adjacent muscle fibers, thereby contributing additional nuclei to the existing population of myonuclei. Satellite cells provide growing muscle fibers with added DNA that increases the capacity of fibers to synthesize greater and greater amounts of protein during cellular hypertrophy.

The origin of muscle satellite cells, or those occasionally termed *adult myoblasts,* is not clear. Traditionally, it has been accepted that satellite cells are derived from the somite, possibly as remnants of the embryonic or fetal myoblasts, or perhaps as a third lineage distinct from embryonic and fetal myoblasts (Figures 5.24 and 5.25). However, recent evidence suggests that some satellite cells may be of non-somitic origin. It is quite possible that other cell types serve as precursors for satellite cells and that under appropriate stimuli are able to become myogenic in skeletal muscle. These findings could serve as an explanation for the existence of subpopulations of satellite cells that appear to respond differently to growth factors and to exhibit different growth kinetics. This area of satellite cell research is very active but many questions remain unanswered.

Although existing nuclei and protein synthetic machinery of the muscle fiber are capable of producing a significant amount of protein, demand for increased protein synthesis as the muscle fiber undergoes hypertrophy exceeds capabilities of

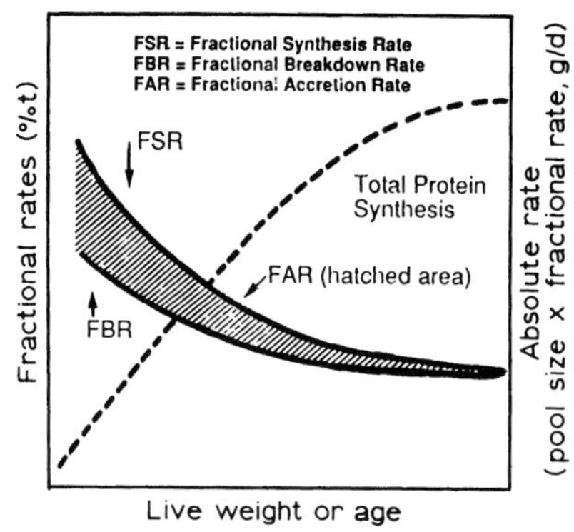

Figure 5.22 Fractional protein synthesis, degradation and accretion rates and absolute protein accretion rate in growing animals. From W. G. Bergen and R. A. Merkel, *Protein accretion, in Growth Regulation in Farm Animals, Advances in Meat Research* vol 7, A. M. Pearson and T. R. Dutson, editors, Elsevier Applied Science, London.

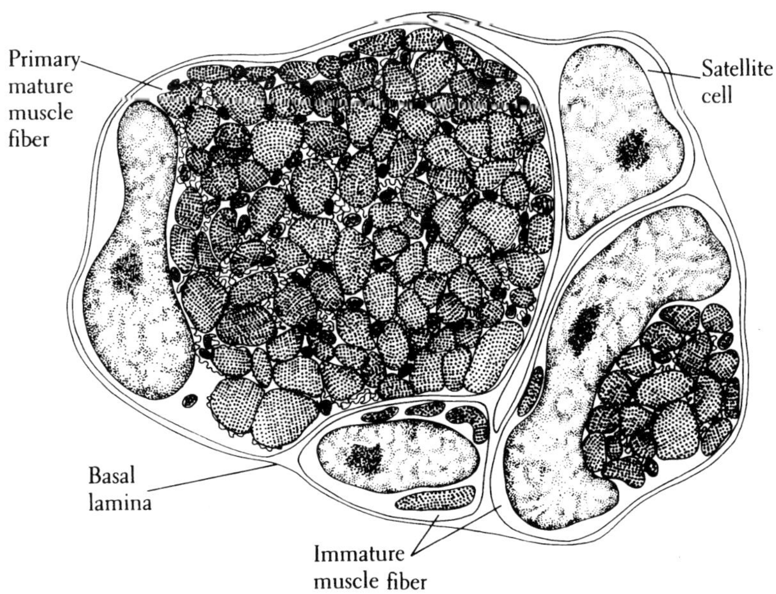

Figure 5.23 Location of satellite cells in relation to a mature muscle fiber. From *Developmental Biology* by N. Holder and V. Walbot. Copyright © 1987 by McGraw-Hill Companies, Inc. Reprinted by permission of McGraw-Hill Companies, Inc.

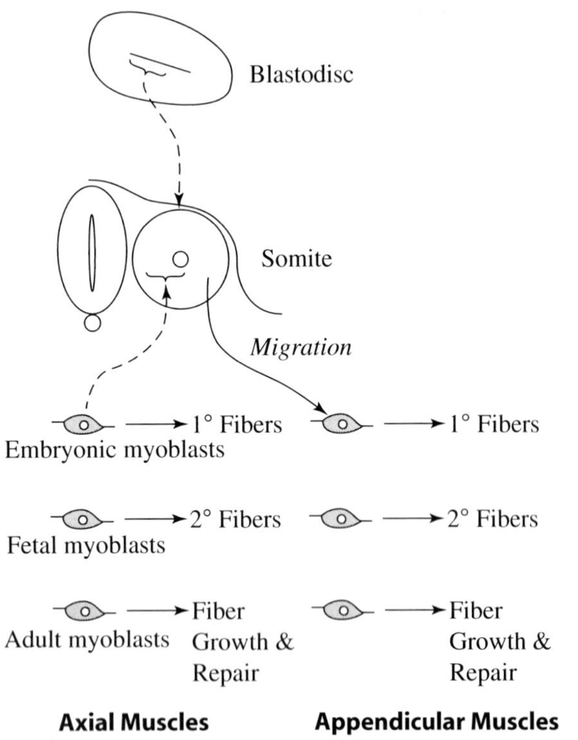

Figure 5.24 Origin of embryonic, fetal and adult myogenic lineages in developing birds. From F. E. Stockdale, Myogenic cell lineages (1992), *Developmental Biology,* vol. 152, pg. 284.

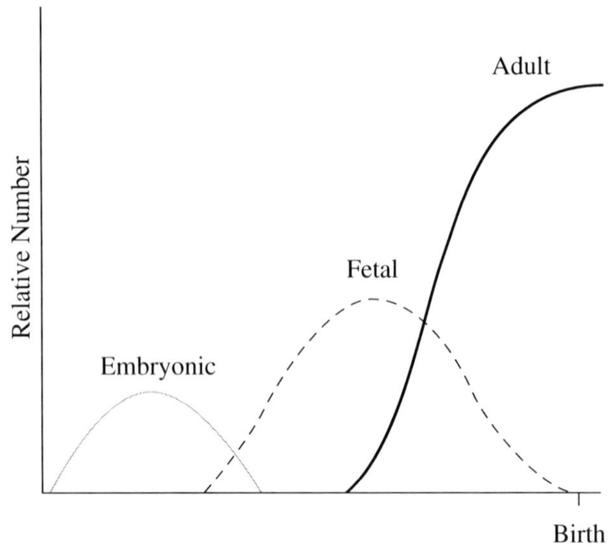

Figure 5.25 Sequential appearance of embryonic, fetal and adult myoblasts in developing muscle. From F. E. Stockdale, Myogenic cell lineages (1992), *Developmental Biology,* vol. 152, pg. 284.

existing nuclei. Thus, recruitment of satellite cells by muscle fibers is required for normal muscle growth. To illustrate the relative importance of satellite cells to postnatal muscle growth further, it is estimated that the majority of the DNA present in mature skeletal muscle is accumulated during postnatal growth (greater than 95% in some species) (Table 5.4). The majority of this DNA is presumably derived from satellite cells. Formation of new muscle fibers by satellite cells is negligible, except in cases of severe muscle injury and regeneration (Figure 5.26). The number of muscle satellite cells is highest at birth, and decreases as the muscle matures (Figure 5.27). Reasons for this decrease in number are not completely understood, but the recruitment of large numbers of satellite cells early in life coupled with loss of proliferative capacity with increased animal age contribute to this decline in number. Consequently, much research is being directed at understanding the biology of satellite cells in an attempt to enhance muscle growth and repair and to prevent muscle atrophy.

The capacity of muscle to synthesize protein can be evaluated by using the *RNA:DNA ratio* as an indicator. Efficiency of protein synthesis, on the other hand, is best assessed by the *protein:RNA ratio* or the ability of RNA to deposit protein. Both indicators tend to be greater in faster growing animals. As stated above, satellite cell recruitment by muscle fibers results in an increase in DNA accumulation in muscle as an animal matures. As a result, this can ultimately affect these ratios. Nucleic acid quantity and concentration are developmentally regulated (Figures 5.28 and 5.29). Increases in DNA and RNA content with age are important for accretion of abundant quantities of protein that contribute to the increased muscle mass. The accumulation of muscle protein is so great during postnatal growth that it results in significant dilutions (decreased concentrations) of DNA and RNA (Figure 5.29). This hypertrophy-associated dilution may be a factor contributing to the decreased fractional rates of protein synthesis.

TABLE 5.4 ACCUMULATION OF DNA IN POSTNATAL MUSCLE OF DIFFERENT SPECIES[a]

Species and Muscle	Increase in DNA Context (*X*-fold)	% of DNA Accumulated Postnatally
Male rat. 21–133 days, gastrocnemius[b]	8.4	88
Male rat. 16–86 days, gastrocnemius[c]	3.8	74
Male rat. 7–56 days, quadriceps[d]	4.8	79
New Hampshire female chicken, 0–266 days, breast muscle[e]	16.9	94
New Hampshire male chicken, 0–266 days, pectoralis[f]	96	99
New Hampshire male chicken, 0–266 days, gastrocnemius[f]	45	98
Pig, 23–118 kg. total muscle[g]	2–2.7	50–63
Sheep, 0–120 days, gastrocnemius[h]	3.1	66

[a] From Allen *et al.* (1979).
[b] Winick and Noble (1966).
[c] Enesco and Puddy (1964).
[d] Cheek *et al.* (1965).
[f] Moss *et al.* (1964).
[g] Harbison *et al.* (1976).
[h] Johns and Bergen (1976).
[i] Buttery *et al.* (1975).
From A.M. Pearson and R.B. Young, Muscle and Meat Biochemistry, Academic Press, Inc., San Diego.

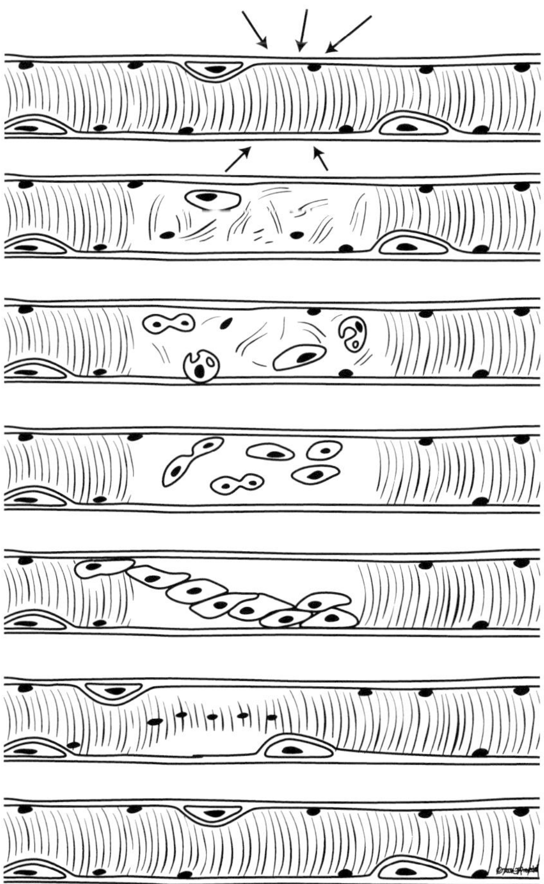

Figure 5.26 Participation of adult satellite cells in regenerating muscle fibers. Courtesy of Faye C. Prevedell.

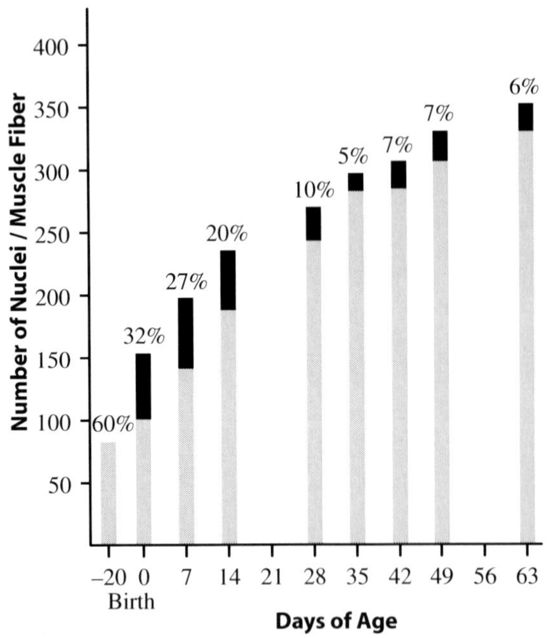

Figure 5.27 Estimation of the contribution of satellite cells to muscle growth in growing mice. The height of the bars represents total nuclei within the basement membrane of muscle fibers, while the number and shaded portion of the bars, represent the proportion of the total nuclei representing satellite cell nuclei. Note satellite cell nuclei percentages decrease with age. From A. M. Pearson and R. B. Young, *Muscle and Meat Biochemistry*, Academic Press, Inc., San Diego.

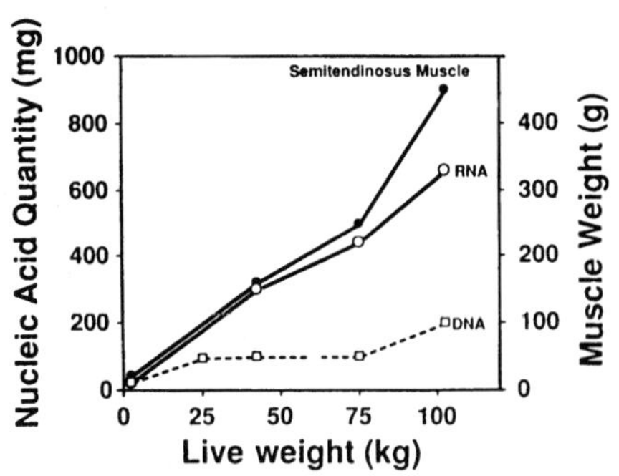

Figure 5.28 Changes in muscle weight and nuclei acid content in growing pigs. From W. G. Bergen and R. A. Merkel, *Protein accretion, in Growth Regulation in Farm Animals, Advances in Meat Research* vol 7, A. M. Pearson and T. R. Dutson, editors, Elsevier Applied Science, London.

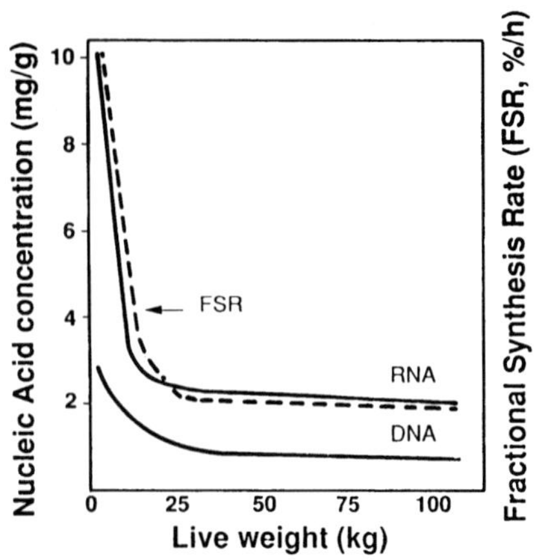

Figure 5.29 Changes in nuclei acid concentrations in muscle and fractional synthesis rate in muscle of growing pigs. From W. G. Bergen and R. A. Merkel, *Protein accretion, in Growth Regulation in Farm Animals, Advances in Meat Research* vol 7, A. M. Pearson and T. R. Dutson, editors, Elsevier Applied Science, London.

TABLE 5.5 EFFECT OF SEX ON MUSCLE FIBER SIZE (μm^2)

Breed	Bulls	Steers	% Change
Angus	4668	3894	+ 20
Simmental	4431	3810	+ 16

* Seideman & Crouse (1986), Meat Sci. 17:55.

FACTORS AFFECTING MUSCLE FIBER SIZE

Sex

Table 5.5 shows the size of muscle fibers in the bovine longissimus of intact and castrated Simmental and Angus males. Regardless of breed, there is about 20% greater cross-sectional area of muscle fibers of bulls than castrates. This is likely a consequence of the powerful androgen, testosterone, that enhances protein accretion. The action of this and other androgens will be covered later in Chapter 10. Similar differences are observed in other species.

TABLE 5.6 EFFECT OF FEED INTAKE ON MUSCLE FIBER SIZE (μm²)

Diet	Ad Libitum	R1	R2	% Change
Mean Fiber Area	3824	3024	3024	–21

R1 = feed restricted to gain 1.64 kg/day
R2 = feed restricted to gain 1.38 kg/day
Solomon *et al.* (1988) JAS 66:3279.

TABLE 5.7 EFFECT OF DIETARY PROTEIN ON MUSCLE FIBER SIZE (μm²)

Dietary Protein Content (%)	LM	SM
11	3002	2645
15	3324	3732

LM = longissimus muscle
SM = semimembranosus muscle
Solomon *et al.* (1994) JAS 72:615.

Nutrition

Nutrition clearly alters muscle fiber size. Table 5.6 shows muscle fiber size of growing (25 to 55 kg) pigs allowed unlimited access (ad libitum) to feed compared to pigs restricted to gain 1.64 or 1.38 kg per day. Pigs allowed ad libitum access to feed have muscle with larger fiber sizes than those on limited diets. In addition, a similar reduction of muscle fiber size can be induced if adequate protein is not available in the diet (Table 5.7). As a result, those animals placed on a high plane of nutrition have muscle fibers that are larger in diameter than those of animals fed a low plane of nutrition. The effect of nutrition on muscle fiber size is less obvious with increased animal age. For example, if animals are placed on restricted diets during early stages of growth, muscle fiber size will be reduced compared with counterparts allowed ad libitum access to feed. This difference, however, will not be apparent if animals are maintained for long periods of time. Muscle fiber size ultimately reaches the same level in both fed and restricted animals. As a result, feed-restricted animals require more time to attain the same muscle growth. Of course, this area of nutrition and animal growth will be covered in more detail in Chapter 11, but it must be emphasized that the effect of nutrient restriction on postnatal muscle mass is mediated through changes in muscle fiber size.

Age

Muscle fiber size increases until muscle maturity is attained. As discussed earlier, muscle fiber number tends to decrease in the senescence stages of life, perhaps from changes in innervation patterns. As a result, the muscle fibers that remain increase in size as a result of work-induced hypertrophy. This phenomenon is not observed in animals harvested at typical market weights, but is likely to occur in older animals used for breeding purposes.

TABLE 5.8 EFFECT OF BETA-ADRENERGIC AGONISTS ON MUSCLE FIBER SIZE (μm²) IN VARIOUS MEAT ANIMAL SPECIES

	Control	Treated	% Change
Cattle[1]	2565	3443	+ 34
Pigs[2]	3134	3671	+ 17
Sheep[3]	1476	1191	+ 30
Chickens[4]	2150	2925	+ 36

[1]Vestergaard *et al.* (1994) JAS 72:2298.
[2]Aalhus *et al.* (1992) Meat Sci. 31:397.
[3]Beermann *et al.* (1987) JAS 65:1514; small fiber size may be related to methodologies.
[4]Gwartney *et al.* (1992) JAS 70.2144.

Growth Promotants

Many of the growth promotants that result in increased muscle growth act by increasing muscle fiber size. Because the actions of growth promotants may be hormonal in nature or are mediated by the endocrine system, mechanisms for increasing myofiber size will be discussed in Chapter 10. Compounds in the *beta-adrenergic agonist* family *(cimaterol, isoproterenol, clenbuterol,* and *ractopamine),* when added to growing diets, markedly increase muscle fiber size (Table 5.8). Likewise, growth hormone administration to growing pigs increases muscle fiber size but in a sex (Table 5.9) and muscle-dependent (Figure 5.30) manner.

Selection and Genetic Anomalies

In addition to the double muscled condition in cattle discussed earlier in this chapter, there is a condition in sheep in which aggressive muscle development ensues after birth. Similarities with the double

Table 5.9 Effect of Exogenous Growth Hormone (pST) Administration on Muscle Fiber Size (μm²) in Boars, Gilts and Barrows

	Con	pST	% Change
Boar	4042	4232	+ 5
Barrow	3557	4610	+ 30
Gilt	3659	4488	+ 23

Con = control
pST = growth hormone
Solomon *et al.* (1990) JAS 68:1176.

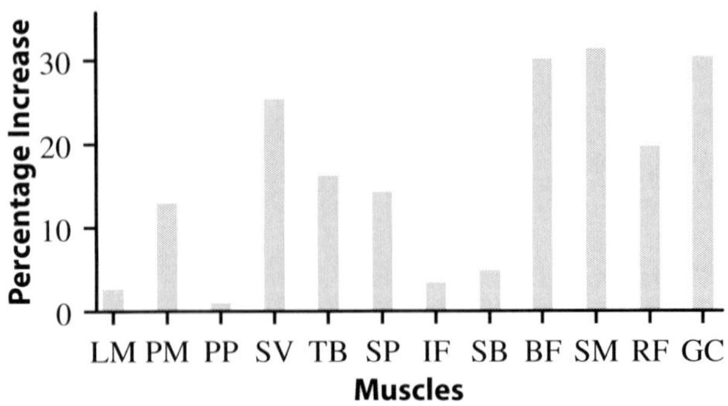

Figure 5.30 Percentage increase in the area of muscle fibers in various muscles of pigs treated with somatotropin. LM, longissiumus; PM, psoas; PP, pectoralis profundus; SV, serratus ventralis; TB, triceps brachii; SP, supraspinatus; IF, infraspinatus; SB, subscapularis; BF, biceps femoris; SM, semimembranosus; RF, rectus femoris; GC, gastrocnemius. From Y. Ono, M. B. Solomon, C. M. Evock-Clover, N. C. Steele, and K. Maruyama, Effects of porcine somatotropin administration on porcine muscles located within different regions of the body (1995), *Journal of Animal Science*, vol. 73, pg. 2282.

muscling of cattle exist, but the condition is distinctly different. Lambs possessing this condition, known as *callipyge* sheep, have larger muscle fibers than normal sheep (Figure 5.15). Callipyge is a Greek word meaning "beautiful buttocks," and likely originated because of extreme muscling in the hind legs of lambs. Lambs possessing this genotype have greatly exaggerated muscling and less fat than lambs lacking this genotype. At the cellular level, muscle of callipyge lambs has greater protein to DNA ratios and greater muscle fiber diameters. Research to determine the precise mechanism causing muscularity in callipyge sheep has lagged behind that of double muscled cattle because the genotype was identified only recently.

Species

Although some subtle differences exist in the size of muscle fibers among species, large differences are not generally observed unless the species differ in muscle fiber type, which will be covered in Chapter 6. For example, the mean cross-sectional area of muscle fibers in longissimus muscle of Montbéliard bulls (French dual-purpose cattle breed) weighing about 500 kg is about 3,000 μm². Similarly, the cross-sectional area of longissimus fibers in large pigs and rabbits is 3,400 μm² and 2,800 μm², respectively. Given the differences in the mature size of the animals, large differences in muscle fiber size are not evident.

Other Contributors to Increased Muscle Size

Orientation of Muscle Fibers

Radial growth of muscle fibers is a major contributor to the increased cross-sectional area of a muscle and may contribute significantly to the muscle-to-muscle variation in muscle fiber size (Figure 5.30). However, length of myofibers also contributes to increased cross-sectional area of some muscles, such as the *longissimus* muscle. In this muscle, muscle fibers are oriented at an angle of 25° to the longitudinal axis of the vertebral column and may contribute substantially to the overall size of the muscle.

Intrafascicularly Terminating Muscle Fibers

Cross-sectional area of muscle is increased by increased growth of intrafascicularly terminating muscle fibers (Figure 5.5). As intrafascicularly terminating muscle fibers grow in length, those originating from either end of the muscle begin to overlap in the central portions of the muscle. As a result the *girth* or belly of the muscle is increased. It has been estimated that nearly 10% of the muscle fibers in a muscle are intrafascicularly terminating muscle fibers and consequently represent a significant means by which muscle enlargement occurs.

Stromal Proteins

Muscle mass is the result of more than just increased diameter and length of the fibers that were formed during early stages of development. Muscle is not simply made of contractile proteins, but rather is a collection of different proteins from different tissues. Of particular prominence is connective tissue as outlined in an earlier chapter. As muscles grow, so does the connective tissue, albeit at a slower rate. In other words, the absolute weight of connective tissue increases in all muscles as they grow but the percentage of connective tissue in the muscle decreases because muscle fiber protein deposition is occurring at a greater rate than stromal protein deposition. Nonetheless, the area that connective tissue occupies in muscle contributes to the overall size of the muscle.

Adipose Tissue

In addition to connective tissue proper, adipose tissue contributes significantly to muscle size. This specialized form of connective tissue accumulates as marbling fat (or intramuscular fat) in the perimysium that surrounds the bundles. It displaces other tissues and subsequently increases the size of the muscles.

Chapter 6

Muscle Fiber Plasticity and Growth

Introduction

Myogenesis does not cease after muscle fiber hyperplasia ends, but it continues during all stages of animal growth, well into what is defined as the late stages of myogenesis (Chapter 4). Myogenesis includes muscle fiber *maturation*, which refers to the process immature muscle fibers undergo on their way to assuming the properties of adult muscle fibers. As outlined in earlier chapters, muscle is a collection of individual muscle cells, but these cells are not identical. Rather, muscle tissue is a very heterogeneous population of *muscle fiber types*. Muscle fiber typing, by definition, is a classification of muscle fibers according to their metabolic and contractile capabilities (inherent properties). These properties are not fixed throughout the lifetime of an animal, but they exist in a dynamic state and can respond to a number of physiological states and challenges. The factors may be as simple as repeated stimulation to contract, as experienced by runners training to improve their endurance or speed, or may be as complex as hormone signals changing body metabolism and growth. Regardless of the stimuli, muscle is capable of modifying its contractile protein and enzyme compositions during muscle fiber maturation to function under a variety of circumstances. As a result, muscle retains great *plasticity* because individual cells can change to meet the ever changing demands placed on them by the animal.

Muscle Fiber Type Organization

To understand the functional capabilities of muscle, one must understand the arrangement of muscle fiber types. Great contrast is seen in poultry wherein leg muscles of chickens and turkeys appear red, while breast muscles of the same animals appear nearly white. The explanation for this difference lies in the ratio of fiber types in muscles. Muscles that appear red contain more *red fibers,* while those white in appearance contain more *white fibers.* With few exceptions, muscles are a mixture of red and white fiber types. The shades of color between deep red and off-white depend on varying ratios of the two fiber types (Figure 6.1).

Occasionally, some muscles, such as the *semitendinosus* in pigs, exhibit regions that are predominately red while other portions of the same muscle are distinctly whiter in appearance, and contain corresponding concentrations of red and white fibers when viewed microscopically (Figure 6.2). Roughly 80% of the muscle fibers in the superficial portion of the muscle are white fibers, and the remaining 20% of those fibers are classified as red fibers. In contrast, the deep semitendinosus muscle consists of 80% red muscle fibers and 20% white fibers giving the muscle its overall red muscle designation. The *semitendinosus* muscle, often referred to as the 'eye of the round' to many in the meat industry, is located in the caudal aspect of the rear legs of most mammals. This muscle originates on the tuber ischii of the pelvis and inserts on the medial aspects of the tibia and acts to flex the knee or stifle.

Another muscle, also located in the hindlimb that has different muscle fiber types in different regions is the *biceps femoris.* As the name implies, this muscle has two heads (bi-ceps), one possessing

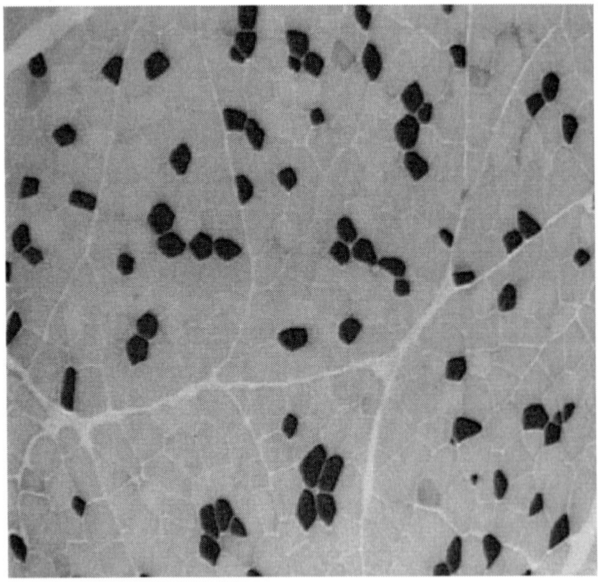

Figure 6.1 Micrograph of myofibrillar ATPase staining showing the heterogeneity of muscle fibers in the cross section of bovine longissiumus muscle.

muscle fibers that are more red and the other that is more white. The reason for co-existence of different muscle fiber types in different portions of the same muscle is related to inherent functional requirements by those portions of the same muscle.

In most species, the spatial organization of red and white muscle fibers is more or less random throughout the entire muscle. However, in porcine skeletal muscle, red muscle fibers are located near the center of fascicles (bundles) in discrete groups that are surrounded by white fibers that form the periphery of the bundles (Figure 6.3). Organization of muscle fiber types in this *rosette* or *islet* fashion probably serves little functional purpose because differences between porcine skeletal muscle and other species with more random distribution of muscle fibers show little difference in overall functional properties. More likely, this unique characteristic may be a result of differences in innervation of muscles during development, a process that dramatically impacts *muscle fiber diversification* or plasticity. Regardless of underlying mechanism, the existence of quite different muscle fiber types is readily apparent on microscopic examination of skeletal muscle (Figures 6.1, 6.2 and 6.3).

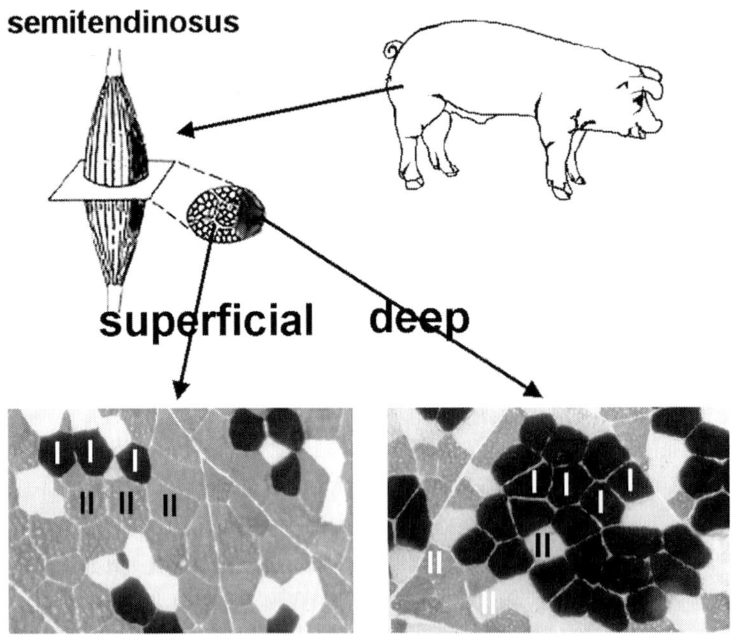

Figure 6.2 Myofibrillar ATPase stained muscle cross sections showing the vast differences in muscle fiber type composition in the superficial and deep semitendinosus muscle of pigs.

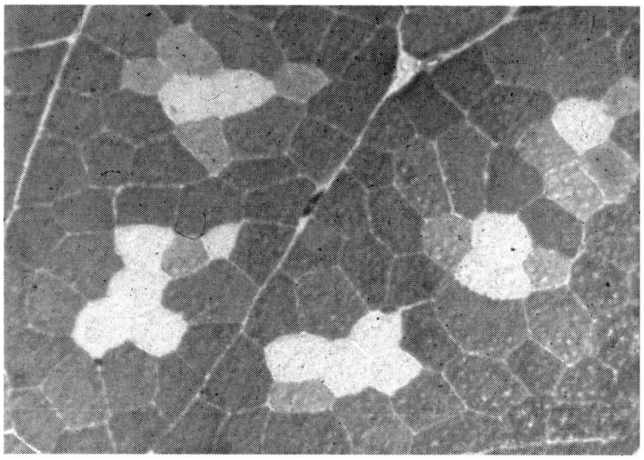

Figure 6.3 Micrograph of a myofibrillar ATPase stained muscle cross section showing the unique arrangement of muscle fiber type in porcine muscle.

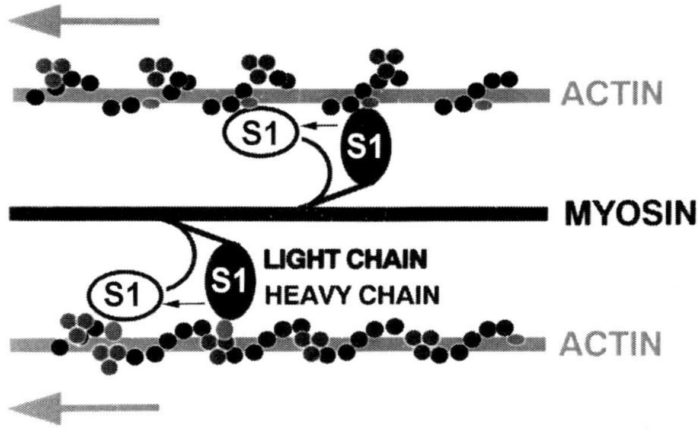

Figure 6.4 Location of the S1 myosin head that is the target of ATPase-mediated muscle fiber typing.

Muscle Fiber Types and Classification

As established above, skeletal muscle consists of a heterogeneous population of muscle fibers and individual muscles are classified as being red or white, based principally on their fresh meat/muscle color intensity. This does not imply that red muscles have a majority of red fibers; rather, muscle with only a small percentage of red fibers nevertheless appears red.

Muscle fiber typing is performed by using two characteristics that differ among muscle fiber types:

- The *speed of contraction*
- The major type of *energy metabolism* in each type of fiber.

Muscle fiber typing, per se, categorizes fiber types into discrete groups but their association with contraction speed and type of energy utilization is inexact. However, the existence of muscle fiber types is some justification for their inclusion in this text in spite of their imprecise relationships with functional properties.

Determination of Muscle Fiber Type

Muscle composition exists in a dynamic state, which means that each muscle fiber is capable of modifying or adjusting its own composition to function and operate according to the needs of the animal. Before assessing the effects of various stimuli, methods for assigning muscle fiber type must first be established.

Acto-myosin ATPase Activity

Recall from Chapter 2 that the primary function of muscle is to contract and create tension between two anatomical locations. Obviously, not all muscle fibers contained within a muscle function or contract in the same manner. The primary protein responsible for contraction is myosin, especially when bound to actin. The portion of the myosin molecule most responsible for this activity is the S1 fragment or the *globular head* (Figure 6.4, and Chapter 2). The speed at which muscle fibers contract is related to the type of myosin contained within the muscle fiber. As with most proteins, this globular protein is especially sensitive to changes in pH. Early research showed that when muscle cross-sections are incubated in buffers containing widely different pH (pH 4.3 versus pH 10.4), then placed in buffers containing ATP and other factors needed for contraction, a subset of muscle fibers in the cross-section are capable of functioning, while others are more vulnerable, or labile, to differences in pH. Therefore, the inherent ATPase activity can be inactivated by a short incubation at a less than optimal pH. This altered ability to function in different conditions is illustrated in Figure 6.5. To visualize this difference, scientists configured assays so that functioning fibers would generate a chemical precipitate that stains the muscle fibers. As a result, classification of each muscle fiber type is based on its inherent actomyosin ATPase activity. An example of such classification scheme in pig muscle is shown in Figure 6.6. For example, type 1 muscle fibers stain extensively while type 2 fibers stain very lightly when muscle cross sections are subjected to preincubation in a solution with a pH near 4.3. In contrast, the reverse occurs when the pH is changed to near 10.4. Using this type of classification system, cross-sections from most muscles of domestic animals exhibit a montage of three distinctly staining muscle fiber types: *type 1, 2A* and *2B* (Figure 6.7). Therefore, inherent ATPase activity is related to type of myosin contained within fibers as well as the speed of contraction. However, conditions needed to elicit such fiber typing are slightly different for each species.

Enzyme Abundance

In addition to differences in type of myosin contained within muscle fibers and speed of contraction, muscle fibers differ in the type of enzymes used to supply energy for contraction. In particular, muscle fibers may be classified based on enzymes that allow muscle cells to perform *glycolytic* or *oxidative* metabolism. To differentiate muscle fiber types on a metabolic basis, abundance of *lactate dehydrogenase* which represents glycolytic metabolism, whereas abundance of *succinate dehydrogenase (SDH), isocitrate dehydrogenase* and *nicotinamide adenine dinucleotide (NADH)* and *tetrazolium reductase* represent oxidative metabolism. Results of one such staining protocol are shown in Figure 6.8.

By combining the two aforementioned histological techniques on *serial cross-sections,* i.e., sections of muscle immediately adjacent to one another, each muscle fiber type in a given muscle can be classified as being either slow (S) or fast (F) contracting fibers, and as oxidative (O), glycolytic (G) or intermediate (OG) in their biochemical composition (Figure 6.6). Using this histochemical approach in which actomyosin ATPase activity is coupled with enzymatic staining, three adult muscle fiber types may be identified in adult skeletal muscle of most animals, i.e., slow-twitch, oxidative fibers (SO) or β *red* (βR) fibers; fast-twitch, glycolytic fibers (FG) or α-*white* (αW) fibers; and fast-twitch, oxidative-glycolytic, *intermediate,* or α-red (αR) fibers. The latter contract rapidly but have some enzymes for oxidative and glycolytic metabolism (FOG).

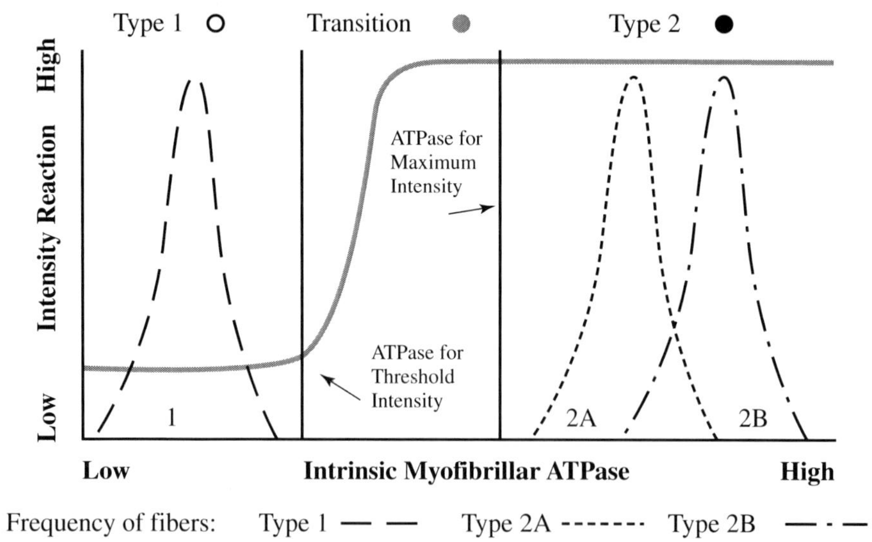

Figure 6.5 Representation of the myofibrillar ATPase staining reaction and fiber typing of various different muscle fiber types. Modified from A. Sosnicki, E. Wilson, B. Green, B. Fields, E. Sheiss, E. Olson, G. Plastow, M. Greaser, E. Pospiech, M. Szalata and R. van Laack, Biological factors underlying post-mortem pH development in porcine muscle (2000), Reciprocal Meats Conference, American Meat Science Association, Savoy, Illinois.

Unfortunately, the aforementioned systems of muscle fiber typing have created two separate nomenclatures and much confusion. To generalize, there are multiple muscle fiber types in adult skeletal muscle that differ in their ability to contract and to metabolize energy (abundance of different groups of enzymes reflecting glycolytic or oxidative capacity). In addition, oxidative βR and αR fibers are equivalent to type 1 and 2A fibers, respectively, whereas glycolytic αW fibers are similar to those classified as type 2B muscle fibers (Table 6.1).

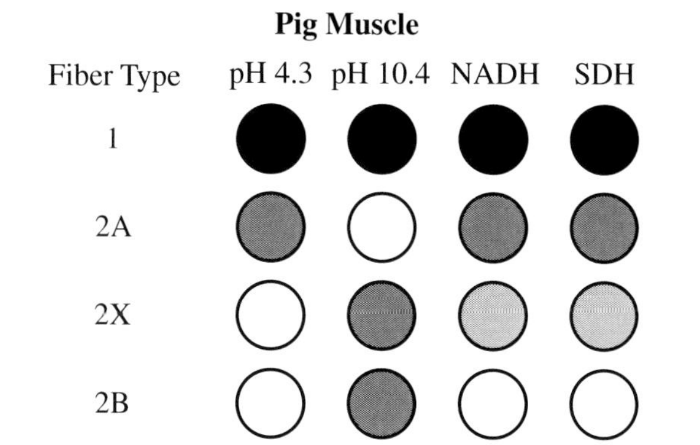

Figure 6.6 Relationship between the pH sensitivity of different muscle fiber types.

The compatibility of the two systems is not complete because there is an additional muscle fiber type in muscle of most species that was misclassified in early research as being type 2B fibers. As a result, many researchers assumed that only three fiber types existed in pigs, type 1, 2A and 2B fibers. However, after revisiting muscle fiber types in pig muscle with more sophisticated molecular techniques, scientists found that type 2B muscle fibers, which account for nearly 80% of the fibers in the longissimus muscle of pigs, actually consists of both *type 2X* (*or 2D*) and 2B muscle fibers in nearly equal proportions. Thus, much care must be taken when interpreting data regarding muscle fiber type.

Classical fiber typing of bovine muscle also may have led to erroneous results. Researchers have shown that skeletal muscle of cattle has three types of muscle fibers and have classified them as type 1, 2A and 2B. However, given the limitations of the methods, there is some skepticism that three fiber

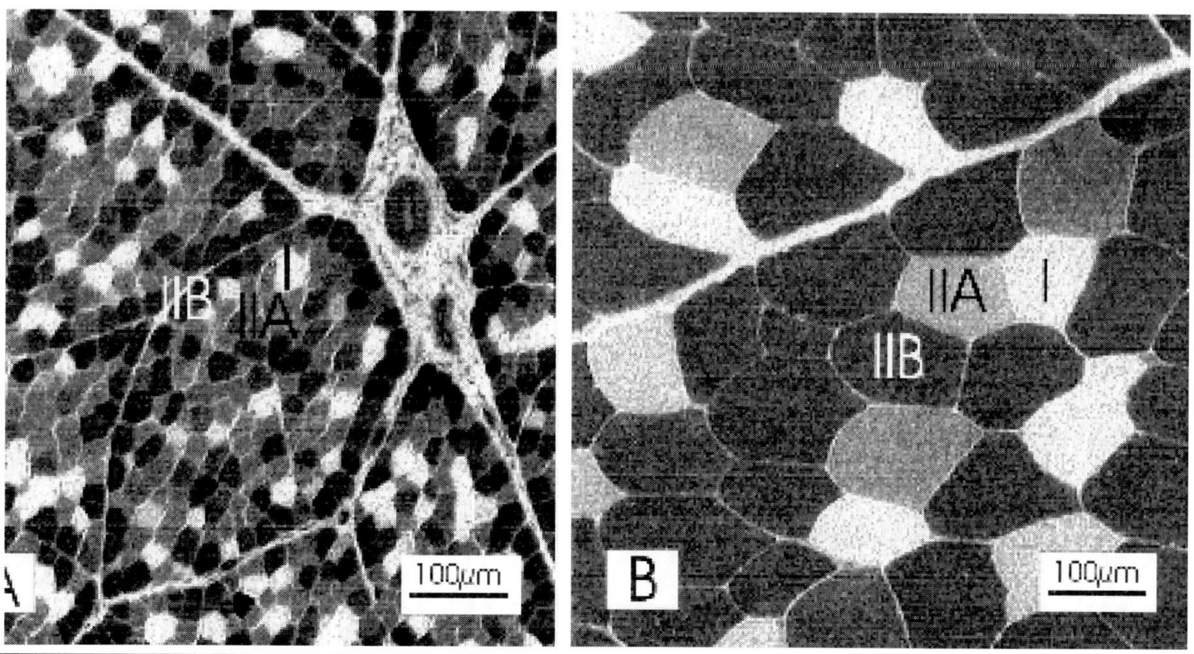

Figure 6.7 Micrographs of bovine muscle cross sections stained with a classical ATPase staining protocol. From J. Wegner, E. Albrecht, I. Fiedler, F. Teuscher, H.-J. Papstein and K. Ender (2000), *Journal of Animal Science,* vol 78, pg. 1485.

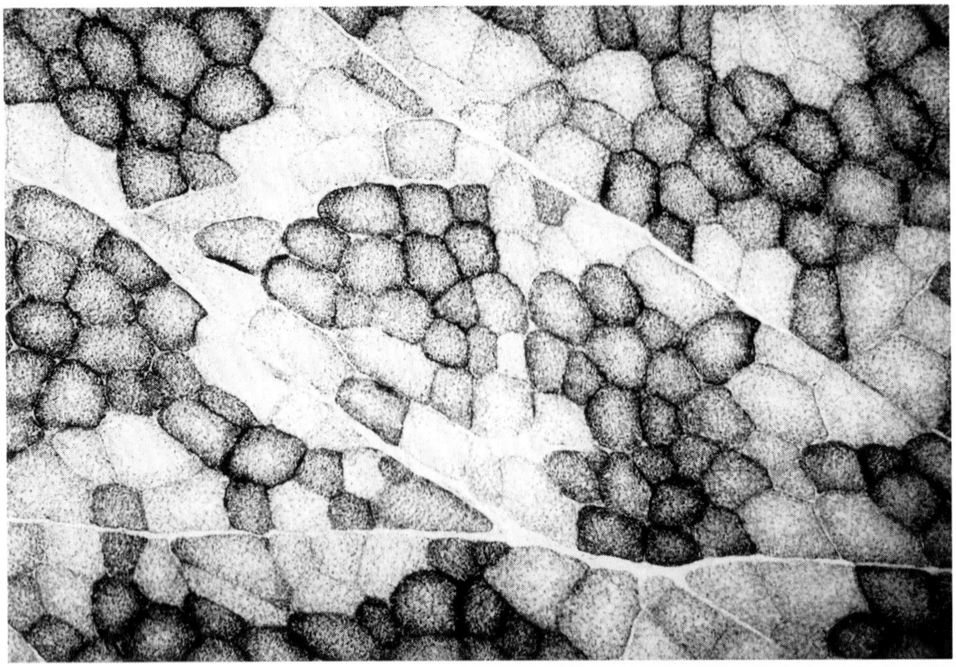

Figure 6.8 Photomicrograph of cross-section of porcine longissimus muscle, showing red and white fibers (dark and light, respectively). (× 175). From *Principles of Meat Science, Fourth Edition* by Aberle et al. Copyright © 2001 by Kendall/Hunt Publishing Company. Reprinted by permission.

TABLE 6.1 MUSCLE FIBER TYPE CLASSIFICATION

Speed Contraction	Slow	Fast		
Metabolism	**Oxidative**	**Oxidative**	**Intermediate**	**Glycolytic**
	1	2A	2X(D)	2B
	βR	αR	αW	αW

types exist instead of four as shown in pig muscle. A likely scenario is that cattle have only three muscle fiber types, but those classified as type 2B may consist of a significant number of 2X fibers. Regardless of the limitations, these protocols have firmly established the existence of different types of fibers in muscle that differ greatly in their ability to contract (fast versus slow) and metabolize energy (oxidative versus glycolytic).

OTHER METHODS OF ESTABLISHING FIBER TYPE

As mentioned above, muscle fiber typing is hampered by the fact that there is a continuum in energy metabolisms and contractile speeds across muscle fibers (Figures 6.5 and 6.9). Scientists have attempted to resolve this issue by basing muscle fiber type on the predominant *myosin heavy chain isoform* contained within the muscle fiber. A *protein isoform* refers to a protein that is closely related in structure and function to other proteins in the same family. Subtle differences in amino acid sequence create differences in optimal protein functionality or, in the case of myosin heavy chain isoforms, their ability to hydrolyze ATP when coupled to actin. To date, four adult myosin heavy chain isoforms have been identified in adult skeletal muscle of several different species; type 1, 2A, 2X (or 2D) and 2B correspond approximately to type 1, 2A, 2X and 2B muscle fiber types (Table 6.1). In addition, other myosin isoforms have been identified in developing muscle, i.e., *embryonic, fetal* and *alpha-cardiac.* Therefore, each fiber can be classified according to the predominant type of myosin contained within the fiber. As a result, there are as many different muscle fiber types as there are types of myosin heavy chain isoform. Embryonic, fetal and alpha-cardiac myosins are initially produced in developing, immature muscle fibers and will be discussed later in this chapter.

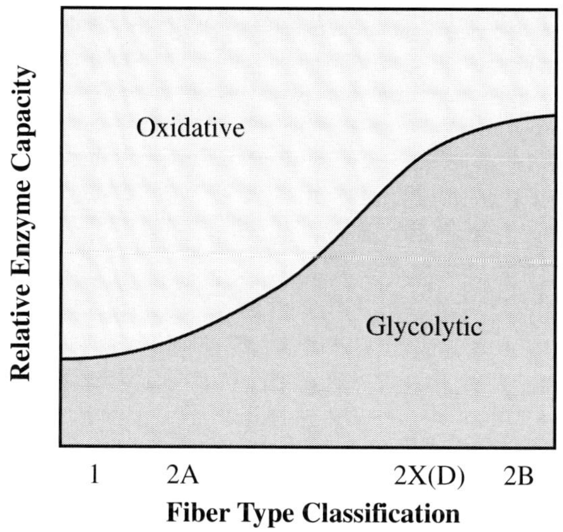

Figure 6.9 The relationship between type of energy metabolism and muscle fiber type. Note that energy metabolism is predominately either glycolytic or oxidative metabolisms. However, note that all fibers have a combination of both types of metabolisms.

Considering adult muscles, type 1 and 2A fibers are considered red muscle fibers and type 2X and 2B fibers are considered white muscle fibers (Table 6.1). All muscles contract rapidly, but differences between fast and slow fibers exist, albeit by only fractions of a second. In accordance with the aforementioned fiber types, type 1 muscle fibers contain predominately slow or type 1 myosin heavy chain isoforms and, therefore, contract slower than type 2 muscle fibers that contain predominately fast contracting type 2 myosin heavy chain isforms. Because type 1 fibers are referred to as slow contracting, red muscle fibers, errors are made by assuming all type 2 fibers are white fibers. This arises from the fact that type 2A fibers are fast contracting, but are more like type 1 fibers in terms of energy metabolism, i.e., more oxidative.

Distinguishing Features of Muscle Fiber Types

Individual muscle fiber types differ substantially in many characteristics (Table 6.2). The content of *myoglobin* as well as other *cytochromes* in muscle fibers accounts for their inherent fresh muscle color. Oxygen storage by myoglobin is consistent with the high proportion of enzymes involved in *oxidative*

TABLE 6.2 CHARACTERISTICS OF MUSCLE FIBERS IN DOMESTIC MEAT ANIMALS AND BIRDS*

Characteristics	Type 1	Type 2A	Type 2X(D)	Type 2B
Redness	++++	+++	+	+
Myoglobin content	++++	+++	+	+
Fiber diameter	+	+	+++	++++
Contraction speed	+	+++	+++	++++
Fatigue resistance	++++	+++	+	+
Contractile action	tonic	tonic	phasic	phasic
Number of mitochondria	++++	+++	+	+
Mitochondria size	++++	+++	+	+
Capillary density	++++	+++	+	+
Oxidative metabolism	++++	++++	+	+
Glycolytic metabolism	+	+	+++	++++
Lipid content	++++	+++	+	+
Glycogen content	+	+	++++	++++
Z disk width	++++	+++	+	+

* The characteristics are relative to the other fiber types.

metabolism or that energy metabolism targeted at metabolizing lipid. Along with this type of metabolic activity, the mitochondria of red fibers, which contain oxidative enzymes and cytochromes, are more numerous and larger in size than those in white muscle fibers. Red fibers also have greater lipid content, some of which presumably serves as a source of metabolic fuel and greater *capillary density,* which facilitates transfer of metabolic wastes and nutrients, particularly oxygen, to and from the vascular bed. The smaller diameter of red fibers compared with white fibers reduces the diffusion distances these materials must travel.

Conversely, white muscle contains enzymes that can readily and efficiently use glycogen as an energy substrate, i.e., *glycolytic metabolism.* Glycolytic metabolism, which predominates in white fibers, may occur in either the presence (*aerobic*) or absence (*anaerobic*) of oxygen. Thus, white fibers have a lower capillary density than red fibers. White fibers have a more extensively developed sarcoplasmic reticulum (SR) and T-tubule system, both of which are consistent with their more rapid contraction speed. White fibers also have more narrow Z disks than red fibers.

White fibers have a *phasic* mode of action; they contract rapidly in short bursts and they are easily fatigued. Red fibers contract more slowly, but for a longer time. This slower but sustained mode of action is generally referred to as *tonic* (continual tension) contraction. Red fibers are functionally important in posture and, because of their metabolism, they are less easily fatigued, so long as oxygen is available. The intermediate fibers (2X) contract faster than red fibers and are less easily fatigued than white fibers (2B).

Development of Muscle Fiber Types

Muscle fibers undergo a process of maturation, which leads to the pattern of distribution found in adult muscle. Recall from Chapter 4, muscle fiber development occurs in a very orderly fashion where primary myotubes form in the center of a fascicle or muscle bundle and secondary muscle fibers originate from myoblasts that fuse to the surface of these earliest developing muscle fibers. This process is mediated by innervation and, although the mechanism by which innervation affects muscle fiber type is not completely known, it likely involves intracellular signaling via changes in calcium concentration (Figure 6.10). Changes in cytoplasmic calcium concentrations often are used in *signal transduction* pathways, which are means of communicating a signal from cell membranes to other intracellular components, including nuclei. Nerves that innervate slow muscle fibers exert a chronic type of stimulation, whereas those innervating fast fibers are more phasic in nature. These two types of stimulation patterns provide a distinctly different calcium level in cytoplasm, which ultimately controls muscle fiber phenotype.

Fiber type diversification, or *specialization,* is accompanied by changes in myosin heavy chain content as well as changes in metabolic enzyme abundance and other cellular characteristics. Many different myosin heavy chain isoforms have been identified in mammalian skeletal muscles, including those observed in the *extraocular* muscles (muscles controlling movement of the eyes) and other super-fast contracting myosin isoforms. Each myosin isoform is the product of a different gene belonging to the same gene family. Studies of developing muscles have shown that embryonic, fetal/perinatal and α-cardiac myosins are transiently expressed during development, while the four adult isoforms (1, 2A, 2X(D) and 2B) start to appear during development and continue to be expressed in adult muscle. A simplified scheme of muscle fiber diversification in pig skeletal muscle, based on myosin isoform transitions, is shown in Figure 6.11. Primary myotubes first appear and initially express embryonic, fetal and slow type 1 myosin heavy chains. During the fetal period, these fibers subsequently mature to type 1 fibers in most muscles, but can also give rise to type 2A fibers in highly glycolytic muscles, such as the superficial white portion of pig semitendinosus muscle. Secondary fibers also express embryonic and fetal myosin during the fetal period. However, unlike primary myotubes, they do not express type 1 myosin before late gestation. Perinatally, a subpopulation of secondary fibers in the direct vicinity of primary myotubes starts to express type 1 myosin and mature into type 1 fibers. Some of these fibers transiently express the α-cardiac myosin before maturing to type 1 fibers. Adult fast type 2A myosin is present in some secondary fibers during the fetal period, whereas 2X and 2B myosin appear in muscle shortly after birth. During the first weeks postnatal, secondary fibers that do not express type 1 myosin mature into either type 2A, 2X or 2B containing muscle fibers. Of course, this process is different for fast and slow muscles (Figure 6.12).

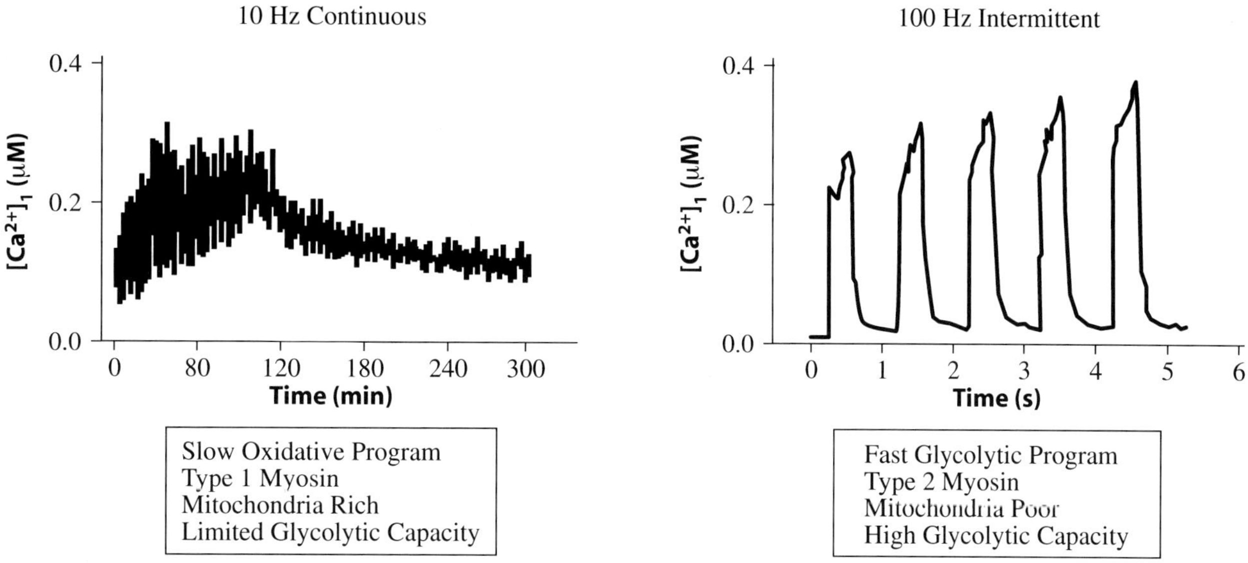

Figure 6.10 Different types of motor nerves lead to specific changes in cytosolic calcium concentrations. Note the extreme difference in basal and maximal calcium concentrations between the two types of nerve-dependent stimulation. From E. N. Olson and R. Sanders-Williams, Remodeling muscles with calcineurin (2000), *BioEssays,* vol 22, pg. 510.

Embryonic and fetal myosins are expressed during myogenesis, but disappear at various stages during the maturation process. Comparison among species (Figure 6.13) shows that fetal myosins disappear in muscle before birth in cattle, between 10 and 15 days of age in pigs, and around 30 days in rats and rabbits.

Muscle Fiber Types and Growth Performance

Muscle weight is a function of muscle fiber cross-sectional area and length (Chapter 5). By definition, glycolytic fibers are generally larger in diameter than oxidative or slow muscle fibers (Table 6.2). Therefore, muscles with more glycolytic fibers are generally larger. One aspect of muscle fiber types that deserves attention at this point is that white muscles have less protein turnover than red muscles. Recall from Chapter 5 that protein turnover can dramatically affect animal growth efficiency. Therefore, animals with a whiter muscle are more efficient because less energy is lost during the turnover process. In addition, the fact that red muscles have greater amounts of RNA is consistent with the more rapid protein turnover rates of red muscle; however, the exact relationship between fiber type composition and growth performance is still somewhat unclear. Even though meat quality, per se, is not actually a measure of growth performance, muscle fiber type composition is related to the development and ultimate quality of meat. This aspect will be further explored in Chapter 12.

Factors Affecting Muscle Fiber Type Composition

As discussed, muscle fiber types exist in a dynamic equilibrium and can change depending on a myriad of environmental, physiological and cellular stimuli. Changes in muscle fiber type occur in a well-documented pathway: 1 ↔ 2A ↔ 2X ↔ 2B. The importance of the aforementioned pathway simply means that if a type 2A muscle fiber is stimulated to become a type 2B muscle fiber, it must first become a type 2X muscle fiber, after which, it can then proceed to be a type 2B muscle fiber.

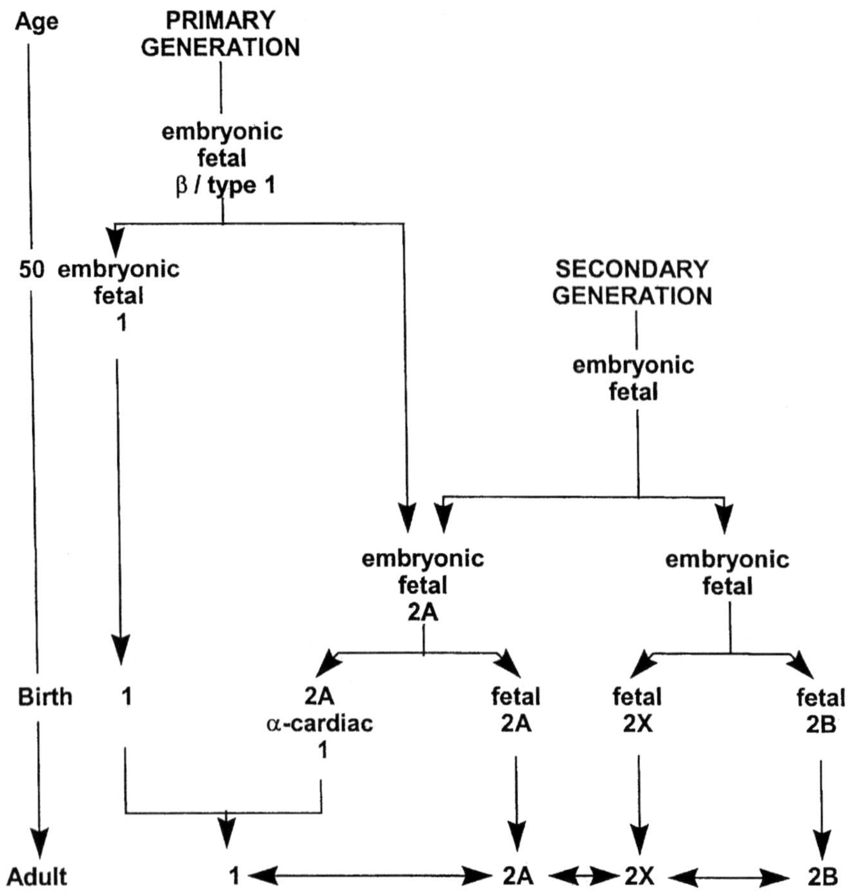

Figure 6.11 Evolution of muscle fiber types in pigs. From L. Lefaucheur and D. E. Gerrard, Muscle fiber plasticity in farm mammals, *Journal of Animal Science,* Savoy, Illinois.

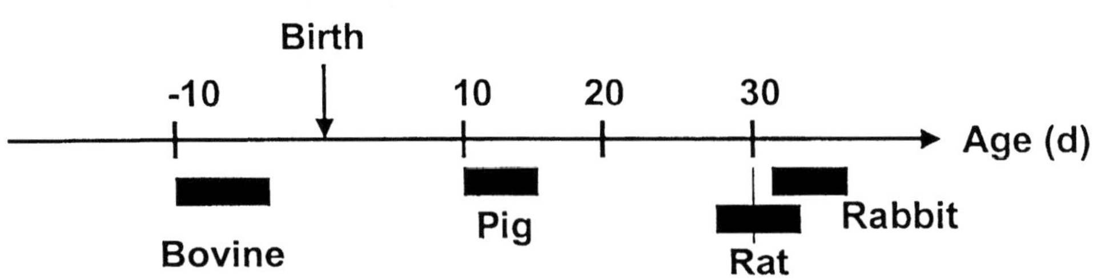

Figure 6.12 Disappearance of fetal myosin heavy chain isoforms in the muscle of various animals. From L. Lefaucheur and D. E. Gerrard, Muscle fiber plasticity in farm mammals, *Journal of Animal Science,* Savoy, Illinois.

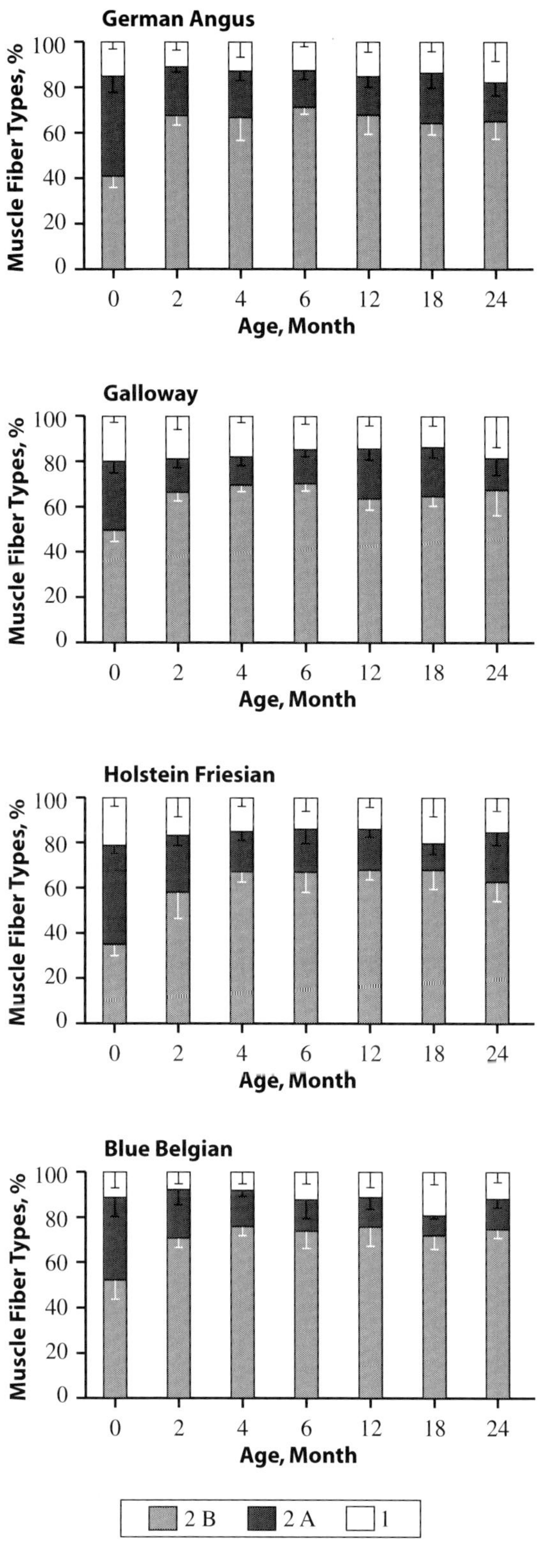

Figure 6.13 Relative muscle fiber type composition in four cattle breeds. From J. Wegner, E. Albrecht, I. Fiedler, F. Teuscher, H.-J. Papstein and K. Ender (2000), *Journal of Animal Science,* vol 78, pg. 1485.

Because transitions occur between and among muscle fiber types, they may possess two different types of myosin heavy chains. Recall from Chapter 2 that myosin exists within the myofibrillar structure of a muscle fiber and that roughly 1000 myofibrils exist in a muscle fiber. Therefore, if a type 1 muscle fiber is being stimulated to become a type 2A muscle fiber, the fiber must begin to replace the myosin in the cell. Obviously, given the amount of total myosin in a muscle cell and the size of the myosin molecule, this is a daunting task that will require several weeks to accomplish. Because of this phenomenon, these "transitory fibers," or those fibers lacking the appropriate criteria to be classified as either type 1, 2A, 2X(D) or 2B, are termed *type 2C fibers.* However, failure to detect a unique type 2C myosin isoform, or its associated gene, simply means that muscle possessing an increased frequency of type 2C fibers is often considered in a state of heightened transition.

Physiological Function

Different muscles differ in their muscle fiber type composition so they can function differently under different conditions. For example, some muscles are required to maintain tension over extended periods of time. These muscles, therefore, must be fatigue resistant or they will fail to perform the task at hand. Muscles responsible for maintaining posture are examples of such muscles and contain an increased number of red, fatigue-resistant muscle fibers (specifically type 1 and 2A). Muscle closely associated with the spinal column and located near bones, and those needed to maintain limb extension (e.g., *serratus ventralis, teres major, infra-* and *supraspinatus* muscles) continually provide the tonic tension to maintain posture. Conversely, superficial muscles that represent the bulk of muscle possessed by an animal, consist primarily of fast contracting, glycolytic fibers (specifically 2X and 2B). These muscles provide quick, powerful movements that last for a short period of time. Muscles such as the *longissimus* muscle that extends the spinal column and the *gluteal* and *biceps femoris* muscles of the hindlimb are especially active during periods when animals are running. Therefore, knowing the function of a muscle provides great insight into *muscle fiber composition,* i.e., the relative amount of each muscle fiber type present.

Species

The proportion of muscle fiber types in a given muscle varies with species. The proportion of type 1, 2A and 2B muscle fibers in longissimus muscle is respectively:

- 35, 24 and 41% in bulls
- 15, 31 and 54% in sheep
- 10, 7 and 83% in Large White pigs
- 1, 6 and 93% in New Zealand rabbits.

These proportions contribute to the overall nature and color of meat produced by each specie.

Genetics

Within a species, the most important factors that influence muscle fiber type composition of a given muscle are probably genetic factors, e.g., breed. However, because mature body size can be highly different between breeds, differences can vary depending on weight, age or degree of maturity (expressed as percentage of adult body weight).

Regardless of species, wild or unimproved breeds contain more oxidative and less glycolytic fibers than domesticated animals. This is particularly evident from studies of the wild and domestic pig (Table 6.3). Once selection, via domestication, has occurred, additional gains in lean tissue growth rate are not observed. Generally increases in lean growth are closely associated with increases in glycolytic muscle fiber types or a reduction in oxidative fibers. In accordance, Meishan pigs, a highly prolific breed from China, exhibit a low muscle growth potential, and have greater proportions of oxidative fibers.

TABLE 6.3 PROPORTION OF FIBER TYPES IN M. GRACILIS OF WILD AND DOMESTIC PIGS

Group	Fiber Type % 1	2A	2B
Domestic pigs:			
Barrows	15.4	11.0	73.6
Gilts	15.6	18.6	65.8
Wild pigs:			
Prepubertal	21.5	31.9	47.6
Mature	36.6	50.1	13.3

* Weiler et al., Anat. Histol. Embryol, 24, 1995. Consequences of Selection on Muscle Composition: A Comparative Study on Gracilis Muscle in Wild and Domestic Pigs.

In cattle, muscle fiber type differs with breed; however, similar to pigs, differences exist only among those exhibiting markedly different muscle phenotypes. Double muscled cattle, as already discussed in a previous chapter, differ greatly from ordinary beef cattle in that the observable muscularity is greatly exaggerated. Likewise, dairy type animals differ greatly from non-dairy type breeds by lacking muscularity as well. As a rule of thumb, differences in muscle fiber type composition are rather small, if evident at all, among beef breeds. However, comparing across extremes (i.e., dairy, beef and double muscled), glycolytic fiber type abundance increases with muscle mass (Figure 6.13). This is consistent across all species. The heritability estimate for muscle fiber type in pigs is ~0.46 and suggests that conventional selection strategies could modify this growth related attribute.

Major Genes

Several genes influence muscle fiber type composition of muscles. These are the halothane and Rendement Napole (RN) genes in pigs, the myostatin gene in double-muscled cattle, and the callipyge gene in sheep (Table 6.4). Halothane, myostatin and Callipyge (see Chapter 5) genes clearly result in animals with greater than normal muscle hypertrophy. As a result, presence of these genes increases the percentage of glycolytic muscle fibers. In contrast, the RN gene mainly increases the relative abundance of intermediate (2X) fibers. Curiously, muscle of RN carrier pigs exhibits a higher glycogen content in white muscles, a characteristic that leads to development of reduced meat quality characteristics. This will be discussed in some detail in Chapter 12.

Time

The bulk of changes occurring in muscle take place shortly after birth. Figure 6.14 shows the dramatic increase in the weight of the longissimus muscle, total lactate dehydrogenase and isocitrate dehydrogenase during early postnatal growth in pigs. Remember that lactate dehydrogenase and isocitrate dehydrogenase abundance are reflective of fast and slow muscle fibers, respectively. Therefore, glycolytic metabolism increases dramatically up to 2 to 3 weeks after birth, then continues to increase, even up to body weights of 100 kg. Conversely, only a small increase in muscle oxidative metabolism is observed in the first 2 to 3 weeks, followed by a gradual decrease. As a result, muscle metabolism becomes more glycolytic with increasing age. The early postnatal period is likely a critical period for muscle fiber type diversification. In bovine muscle tissue, however, changes occur during the latter

TABLE 6.4 EFFECTS OF MAJOR GENES ON HISTOCHEMICAL TRAITS OF VALUABLE MUSCLES IN DIFFERENT SPECIES

Major genes	Halothane[1]	Double-muscling[2]	Callipyge[3]	Rendement Napole[4]
Species	pig	cattle	sheep	pig
Total Number of Fibers	=	↗	=	=
Glycolytic fibers				
Number, %	↗	↗	↗	↘
CSA, μm²	↗	↗	↗	=
Rel. area, %	↗	↗	↗	↘

[1] Lindholm *et al.,* 1977; Essén-Gustavsson et al., Depreux et al., 2002.
[2] Holmes and Ashmore, 1972; Ashmore *et al.,* 1974; Swatland and Kieffer, 1974; West, 1974.
[3] Koohmaraie *et al.,* 1995; Carpenter *et al.,* 1996.
[4] Lebret *et al.,* 1998.
From L. Lefaucheur and D.E. Gerrard, Muscle fiber plasticity in farm mammals, Journal of Animal Science, Savoy, Illinois.

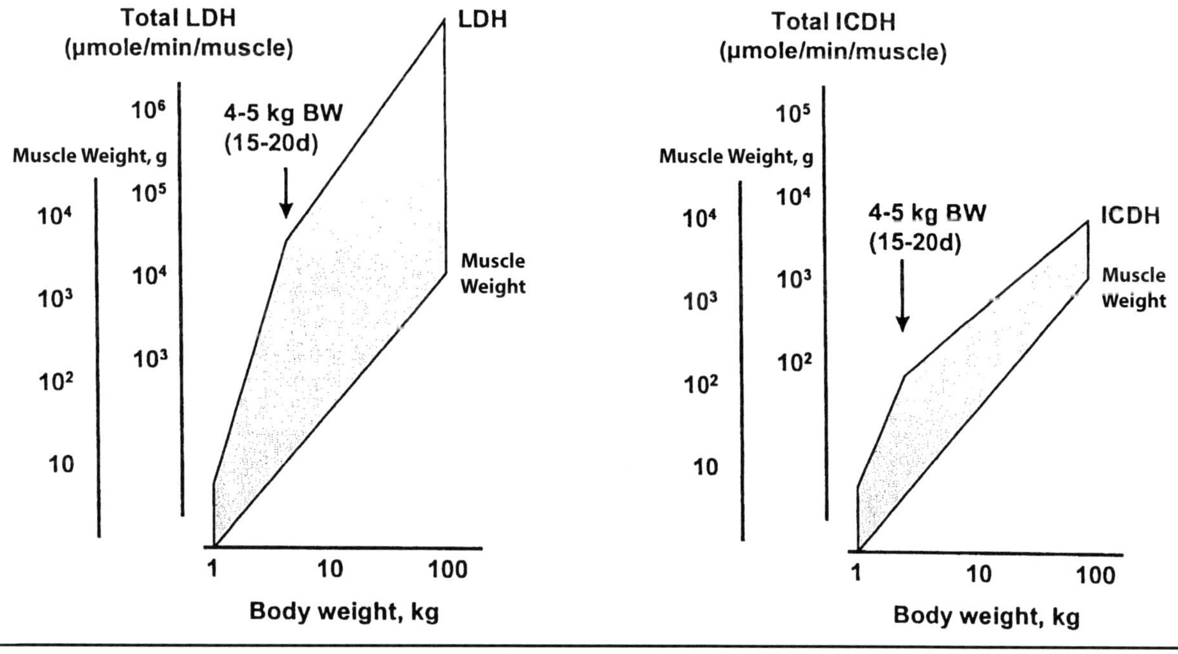

Figure 6.14 Changes in the abundance of lactate dehydrogenase (LDH) and isocitrate dehydrogenase (ICDH) in growing pig muscle. From L. Lefaucheur and D. E. Gerrard (2000), Muscle fiber plasticity in farm mammals, *Journal of Animal Science,* Savoy, Illinois.

stages of gestation, supporting the idea that cattle are more mature at birth than pigs. Nontheless, during postnatal growth in most species, muscle fiber type becomes more glycolytic.

Innervation

The extent of participation by nervous tissue in modulating muscle fiber plasticity should not be underestimated. In very early studies utilizing microsurgery, researchers conducted cross-innervation to test the hypothesis that the type of innervation affects muscle fiber type. In these studies, nerves that innervated fast and slow muscle fibers were surgically switched. Results of these studies showed that muscle fibers changed their contractile and biochemical properties depending on the type of stimulation they received from their corresponding nerve. Nerves signal muscle to contract via acetylcholine that is released at the motor end plate. Acetycholine causes the membrane of the muscle cell to depolarize and subsequently causes calcium to enter the cell and modulate contraction. All nerves, however, do not signal the cell in the same manner (Figure 6.10). The manner in which nerves deliver signals to the cell, therefore, is a function of the number of times that it depolarizes the muscle cell membrane. A *chronic* stimulation, or *high frequency, low amplitude* stimulation will result in a sustained, low level of calcium in the cytoplasm of muscle cells. This continued presence of calcium results in a sustained or tonic muscle contraction that is characteristic of red muscle fibers. In contrast, other nerves are capable of a more *static* stimulation, or *low frequency, high amplitude* stimulation. In this case the cytoplasm of muscle cells is flooded with a high level of calcium for a very short time which results in a rapid twitch of the muscle fiber, hallmark of fast fibers.

Innervation can be mimicked through the use of electrodes that can be placed in muscles to deliver stimulation patterns characteristic of each type of innervation. Experimental results of such studies firmly support the role of innervation in modulating the control of muscle fiber type.

Sex

Sex strongly influences muscle fiber type composition in cattle. In addition to a general decrease in size of muscle fibers, castration increases the proportion of glycolytic fibers at the expense of fast oxido-glycolytic (intermediate) fibers leading to a more glycolytic and less oxidative metabolism. In pigs, most studies have shown differences between females and barrows (castrate males). Generally, however, intact males possess a greater number of type 1 muscle fibers than muscle from females, again supporting the fact that removal of male sex hormones increases the glycolytic nature of the muscle.

Nutrition

Moderate feed restriction (< 30% reduction from ad libitum) does not change muscle fiber type composition. However, severe feed restriction (≥ 50% reduction from ad libitum) at early ages decreases the amount of glycolytic muscle fibers at the expense of the type 1 fibers. Protein restriction has been shown to decrease glycolytic metabolism, which is a function of reduced growth of faster contracting fibers. As a result, any animal that has reduced muscle growth, will have a higher proportion of oxidative fibers.

Embryonic undernutrition or runting will shift fiber type as well as fiber number and size (Chapter 5). Because runting preferentially affects secondary muscle fiber development, there are fewer fast fibers developed. As a result, those runt pigs reaching market weight will have muscle that is more oxidative.

Environment

Ambient temperature influences muscle fiber type composition and muscle energy metabolism. Long-term exposure to cold can increase the proportion of type 1 fibers and increase oxidative metabolism. The physiological significance of these changes is probably related to an increase in cold-induced *thermogenesis*. However, to what extent these adaptations are the result of sustained shivering or to changes in substrate availability or hormonal levels, in particular to thyroid hormones, remains to be determined. From a practical meat animal standpoint, a severe response to a cold environment on red muscles could lead to increased variability of the color among muscles, leading to reduced consumer acceptability.

Exercise

Laboratory studies using artificial electrical stimulation to stimulate muscle use have shown that contractile activity *per se* influences muscle fiber type composition of muscles. In general, this type of activity induces changes in the relative composition of muscle fiber types from one containing more fast contracting muscle fibers to one having more slow contracting muscle fibers.

TABLE 6.5 RELATIVE CHANGE (PERCENTAGE OF CONTROL) IN THE AMOUNT OF TYPE 1, 2A, 2X AND 2B MYOSIN HEAVY CHAIN (MYHC) ISOFORM IN THE LONGISSIMUS MUSCLE OF PIGS FED VARYING LEVELS OF PAYLEAN® OVER A 6 WEEK PERIOD

	MyHC			
Dose (ppm)	**Slow**	**2A**	**2X**	**2B**
0	100	100	100	100
20	100	68	52	129
60	100	51	41	160

From an animal production standpoint, the effect of exercise on muscle fiber type may be important. Production systems, especially in the US, are characterized as being very *intensive*. The theory underlying this type of production system literally is to provide enough space for the animal to exist. As a result, less energy is required or expended by the animal on unnecessary, energy-wasting movements. Animals reared in these types of intensive environments, indeed, possess less slow contracting, red muscle fibers. In contrast, those animals reared in a free-range or *extensive* production environment, have more red muscle fibers. Whether or not either of these production systems should be used for animal production is beyond the scope of this text but there is little question that physical exercise increases the overall oxidative metabolism and thereby increases the relative amount of slow muscle fibers in skeletal muscle. Conversely, a reduction in activity increases the glycolytic nature of muscle. It should be pointed out, however, that the type of exercise is more important than total amount. For example, exercise regimes that rely heavily on speed over a short period of time increase the glycolytic nature of muscle, whereas those occurring over longer periods of time, such as long distance endurance training, increase the relative oxidative nature of muscle fibers.

Hormones

Among hormones, *thyroid hormones* seem to be particularly important in regulating fiber type maturation and diversification. In particular, T3 stimulates the disappearance of fetal/neonatal myosins in favor of adult fast moysins during the peri- and early postnatal periods, and induce a slow to fast transition in adult muscles. Conversely, those animals experiencing reduced levels of circulating thyroid hormone have muscle that is more oxidative in nature.

TABLE 6.6 EFFECT OF CIMATEROL ON MUSCLE FIBER TYPE COMPOSITION IN YOUNG FRIESIAN BULLS

Fiber Type (%)	Control	Treated	% Change
Type 1	24.0	20.4	–15
Type 2A	24.2	8.6	–65
Type 2B	51.7	71.1	+38

Vestergaard *et al.* (1994) JAS 72:2298.

Growth Promotants

Growth hormone (GH) The anabolic action of GH on muscle development has been well studied in a number of domestic meat-animals and will be covered in some depth in Chapter 10. Despite increasing muscle fiber size, no major effect of GH has been observed on muscle fiber type composition.

Beta-adrenergic agonists In contrast to that shown for growth hormone, pigs treated with beta-agonists possess muscle that contains more 2B muscle fibers. Pigs fed the beta-adrenergic agonist ractopamine *(Paylean®)* develop more type 2B muscle fibers in a time and dose-dependent manner. The bulk of the increase in type 2B fibers is mainly at the expense of type 2A and 2X(D) fibers, which decline over the same timeframe (Table 6.5). The same sort of effect has been observed when cimaterol, another type of beta adrenergic agonist, is fed to cattle (Table 6.6). Thus, feeding beta adrenergic agonists to animals pushes muscle fiber type composition to a faster contracting more glycelytic phenotype.

TABLE 6.7 EFFECT OF IMPLANTS (RALGRO® AND SYNOVEX-S®) ON MUSCLE FIBER TYPE COMPOSITION IN CATTLE

Fiber Type %	Steers	♂ Bulls	♂ (Synovex-S®)	♂ (Ralgro®)
1	24	28	25	27
2A	22	27	29	29
2B	53	44	45	43

Seideman *et al.* (1986) Meat Sci. 17:75.

Steroid implants Much work has been conducted to study the effect of synthetic steroids on growth rate in ruminants. The mechanism by which these compounds elicit a response will be covered in Chapter 10. However, regarding muscle fiber type, it is generally accepted that treatment of steers with *trenbolone acetate* and *estradiol* increases the percentage of fast-twitch oxido-glycolytic fibers at the expense of fast-twitch glycolytic fibers in *longissimus* muscle. The resulting shift in muscle fiber composition, therefore, denotes a shift toward a more oxidative metabolism, or that reminiscent of sex-mediated muscle fiber type changes.

This same scenario can be observed in the data presented in Table 6.7. In this case, intact males have a muscle fiber composition that is more oxidative than steers. However, when bulls are implanted with hormones that counteract the male hormones, there is a shift towards the castrated male, albeit not completely. In other words, the intact male possesses muscle of an extreme type, relatively more oxidative when compared to castrates. Conversely, steers represent the opposite extreme, more glycolytic, compared to intact males. Therefore, any modification in the endocrine system will likely result in muscle with a fiber type composition between these two extremes.

Chapter 7

Adipose Tissue Growth and Development

INTRODUCTION

Adipose tissue is a modified connective tissue sometimes referred to simply as fat. Adipose tissue is important during growth because it serves as an energy reserve. This is particularly important during times when nutrient intake is less than nutrient expenditures. Lipid storage is a particularly useful means of storing energy because the energy density of lipid is greater than that of carbohydrate or protein. In other words, there is more energy contained within a unit of fat than there is in a unit of carbohydrate or protein and it is stored without associated water so it is 6 to 8 times more energy dense than is a protein-rich tissue such as muscle. Furthermore, adipose tissue provides insulation, support, protection of vital organs, and generation of heat (brown adipose tissue). Fat located within meat imparts a desirable flavor to some meat products, thereby increasing its value. In contrast, fat deposition in growing animals becomes a liability to producers when it results in added production costs and consumer rejection. Adipose tissue is much more than an energy storage site. Adipose tissue is a source of hormones and other signaling molecules that regulate body metabolism. Its many roles make it an extremely important tissue to animal growth biologists who are charged with improving animal growth efficiency.

Anatomical Location of Adipose Tissue

Development of adipose tissue occurs at specific sites in the body. These sites are called *adipose tissue depots* or *fat depots,* and form from an accumulation of adipocytes, the predominant cells of adipose tissue that are filled with triglyceride. There are several fat depots in meat-animals and these are known to develop at different times and at different rates. Furthermore, the relative size of the depots differs across species (Figure 7.1).

VISCERAL FAT

The earliest fat depot to form in appreciable quantities is located within the body cavity of animals. It is often called *visceral fat* because of its close association with the viscera. Even though it is often considered a single location, it is actually distributed throughout the body cavity because of the diverse milieu of organs strategically positioned in various locations that require protection and insulation. The *mesenteric fat* that is located around the intestines is one of the earliest developing fat depots in the body cavity. This fat may be called *lace fat* because of its delicate, ornate appearance that closely mimics the underlying circulatory system of the organ. Other components of the visceral fat depot include abdominal, pelvic, heart, and kidney fat. *Caul fat* is a thin sheet of adipose tissue contained in a large fold of connective tissue over the stomach and adjacent organs. *Perirenal fat* is the fat located around the kidney and is often called *kidney fat* (Figure 7.2). *Leaf fat* is located between the lining of the thoracic cavity and the ribs in mammalian meat animals, but in poultry, it usually refers to the abdominal fat pad within the body cavity and overlying the intestines. The amount of kid-

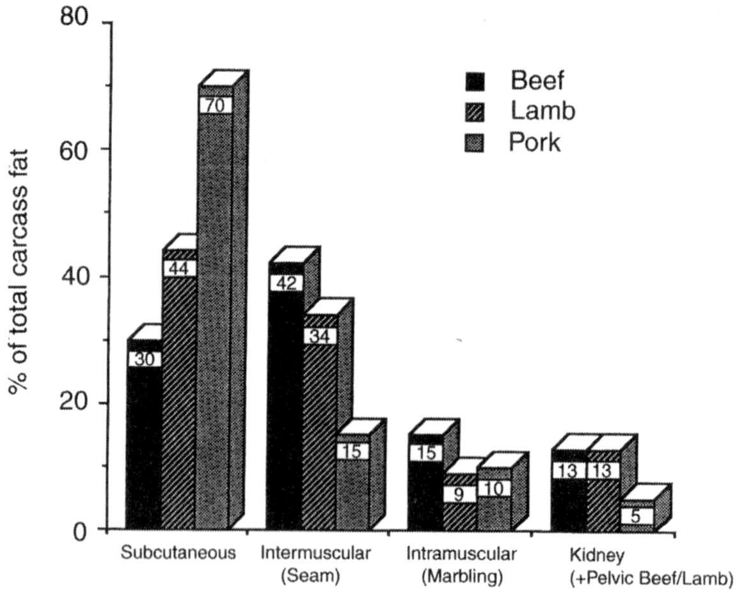

Figure 7.1 Distribution of fat between the subcutaneous, intermuscular, intramuscular, and kidney (and pelvic) fat in beef, lamb, and pork carcasses. From *Livestock & Carcasses* by Boggs et al. Copyright © 1998 by Kendall/Hunt Publishing Company. Reprinted by permission.

ney fat as a percentage of carcass weight varies across species, and in cattle, is somewhat dependent on breed. For example, as a percentage of carcass weight, dairy breeds tend to have greater amounts of kidney fat than many beef breeds.

Subcutaneous Fat

The second fat depot to form during development is *subcutaneous fat,* which is the fat located just under the hide or skin. This depot is sometimes called *external fat* and is often referred to as *backfat* or *fatback* in pigs, depending on location. As a percentage of total carcass fat, subcutaneous fat is the largest in most pork carcasses. Subcutaneous fat develops as three layers in many species (Figure 7.2).

- The *outer layer* is the first to develop and is therefore the oldest layer at any given animal age. This layer probably acts as insulation for the animal.
- The *middle layer* of subcutaneous fat is the second layer to form and is normally the thickest postnatally, thus represents the bulk of subcutaneous fat in all species. Furthermore, it is the most metabolically active layer.
- The *inner layer* is the latest developing layer and therefore is very small and thin. The inner layer is sometimes very difficult to detect in lean carcasses.

Intermuscular Fat

Intermuscular fat represents the third fat depot to develop, although the order of development of the subcutaneous and intermuscular depot may be interchanged in some species. Intermuscular fat is the fat located around and between individual muscles or groups of muscles (Figure 7.2). It is closely associated with the epimysium of muscles, and is often referred to as *seam fat.* In some species such as cattle and sheep, seam fat may contribute substantially to the total fat content of the body.

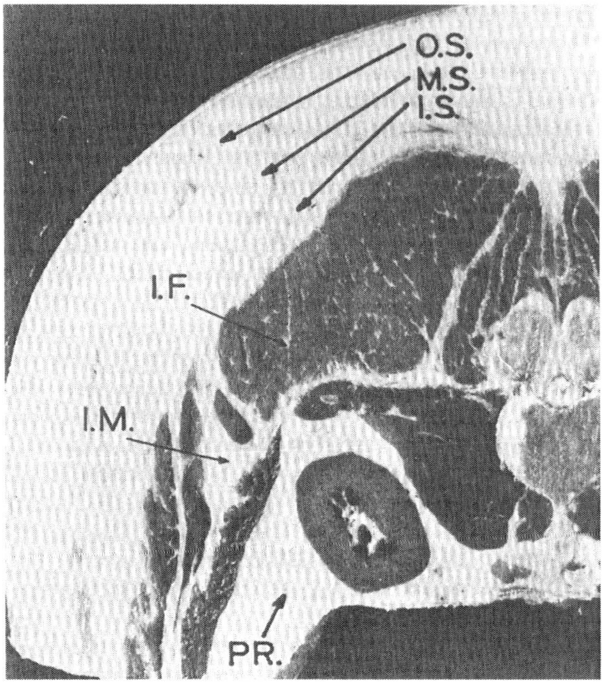

Figure 7.2 Anatomical locations of various fat depots. Transverse section through the lumbar region of a pork carcass depicting six locations of adipose tissue. The areas include: outer (OS), middle (MS), and inner (IS) subcutaneous backfat separated by two distinct layers of fascia intermuscular fat (IM) which is an extension of (IS), interfasicular (intramuscular) fat (IF) and perirenal fat (PR) surrounding the kidney. Source: [Anderson, II].

Intramuscular Fat

Intramuscular fat is the latest depot to develop and consequently contributes the least to total carcass fat (Figure 7.2). This depot represents fat deposited within a muscle, between individual muscle bundles and is specifically associated with the perimysium which surrounds the bundles of muscle fibers (muscle fasciculi). In the pathological state known as *fatty degeneration,* there is invasion of the bundles and fat is deposited in the endomysium surrounding individual muscle fibers. Intramuscular fat, also referred to as *marbling,* is related to the eating quality of meat because it is highly correlated to the flavor of intact meat products, especially steaks and chops.

Adipogenesis

Similar to myogenesis, precursor cells involved in the origination of adipose tissue, i.e., *adipogenesis,* are derived from the mesoderm of the developing embryo. The mesenchymal cells associated with connective tissue sites give rise to *adipoblasts* which are the early form of fat cells (Figure 7.3). During the earliest stages of adipose tissue development, there is an increased vascularization of connective tissue (stroma). This vascularization continues as adipose tissue matures, and may be associated with an extensive capillary network. Given that adipose tissue is responsible for storing energy for the body, it is obvious that increased blood flow to this tissue is necessary. Following the early vascularization of the connective tissue, *lobules,* or a collection of adipoblasts, form and eventually give rise to a larger *lobe* that is enclosed in a delicate sheath of collagenous fibers. As adipogenesis continues, cells of the lobules continue to proliferate and give rise to additional cells. In response to a variety of cues, adipoblasts stop replicating and may differentiate into preadipocytes, at which time small intracellular lipid droplets (*multilocular* lipid) become evident. As the lipid droplets coalesce to form one large globule (*unilocular* lipid), the adipocyte matures into the predominant cell type of adipose tissue. Further deposition of lipid in the adipocyte results in cellular hypertrophy or enlargement of each adipocyte.

Adipocyte Hyperplasia and Hypertrophy

Morphologically, adipoblasts are < 20 μm in diameter when small droplets begin to form near the center of the cell. Mature adipocytes, on the other hand, have diameters up to approximately 120 μm and can be as large as 300 μm in diameter in some very obese animals. In immature adipocytes, a high percentage of the cellular mass is cytoplasm. With maturation, the lipid droplet represents greater than 95% of the cytoplasmic volume and the nucleus is forced to the periphery of the cell.

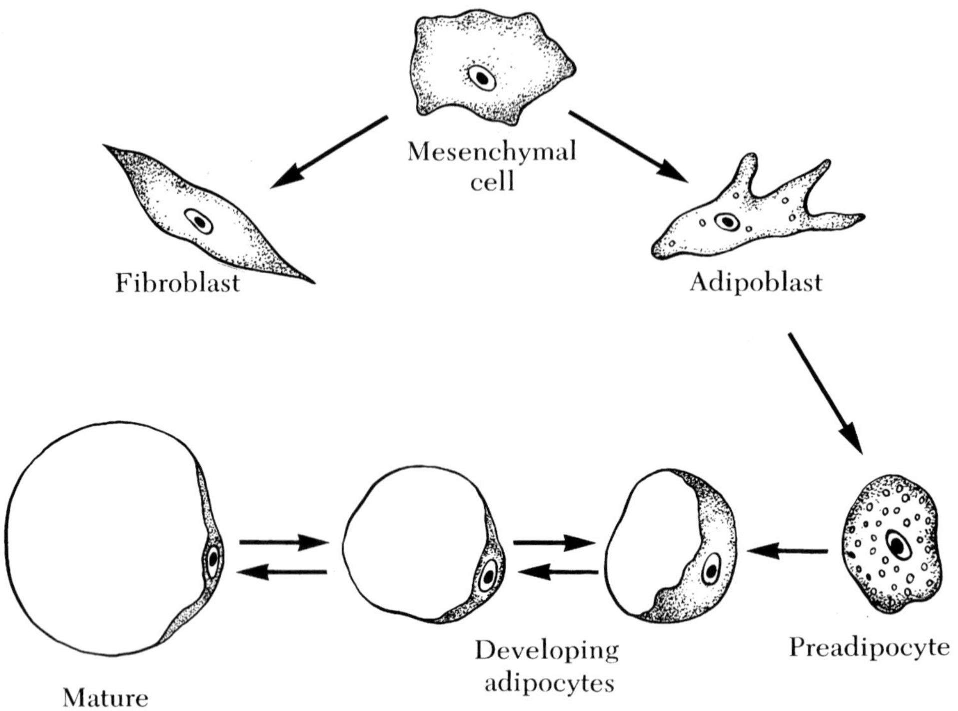

Figure 7.3 Adipocyte development.

Some species differences exist in the perinatal development of adipose tissue:

- In cattle, most of the adipocytes have unilocular lipid at the time of birth.
- In sheep, most adipocytes likewise have unilocular lipid at birth, and those with multilocular lipid have completed the transition to unilocular lipid by 2 to 3 weeks of age.
- In pigs, multilocular lipid is prevalent at birth and coalescence to unilocular lipid occurs postnatally.

It is unclear why these species differences exist, but it may be related to the stage of maturity of the animal at the time of birth.

Much fat cell hyperplasia occurs prenatally, but there is some adipoblast hyperplasia during the perinatal period. However, proliferation of the cell ceases once it differentiates into an adipocyte. Early growth biologists thought that all fat cells were developed prenatally and that any increases in *adiposity* or fatness were due solely to fat cells increasing in diameter. This has been shown to be inaccurate, as it is now known that additional cells can be recruited in adipose tissue depots. Recruitment of new fat cells results in a bimodal distribution of adipocytes (Figure 7.4). The source of these new fat cells is likely the cells of the associated connective tissue. Recall that adipose tissue begins simply as areas of loose connective tissue and that loose connective tissue contains many fibroblasts and fibroblast-like cells. These precursor cells presumably differentiate into adipocytes, and with time and adequate nutrition, accumulate lipid. As animals age, existing adipocytes exhibit hypertrophic growth in the presence of sufficient energy for fat deposition. This hypertrophy is accomplished by the increase in diameter and volume of the adipocytes. A small increase in diameter may result in a significant increase in volume because of the mathematical relationship between diameter and volume. As the mean diameter of the existing adipocyte population increases with animal age, a new population of adipocytes is eventually recruited. The small size of

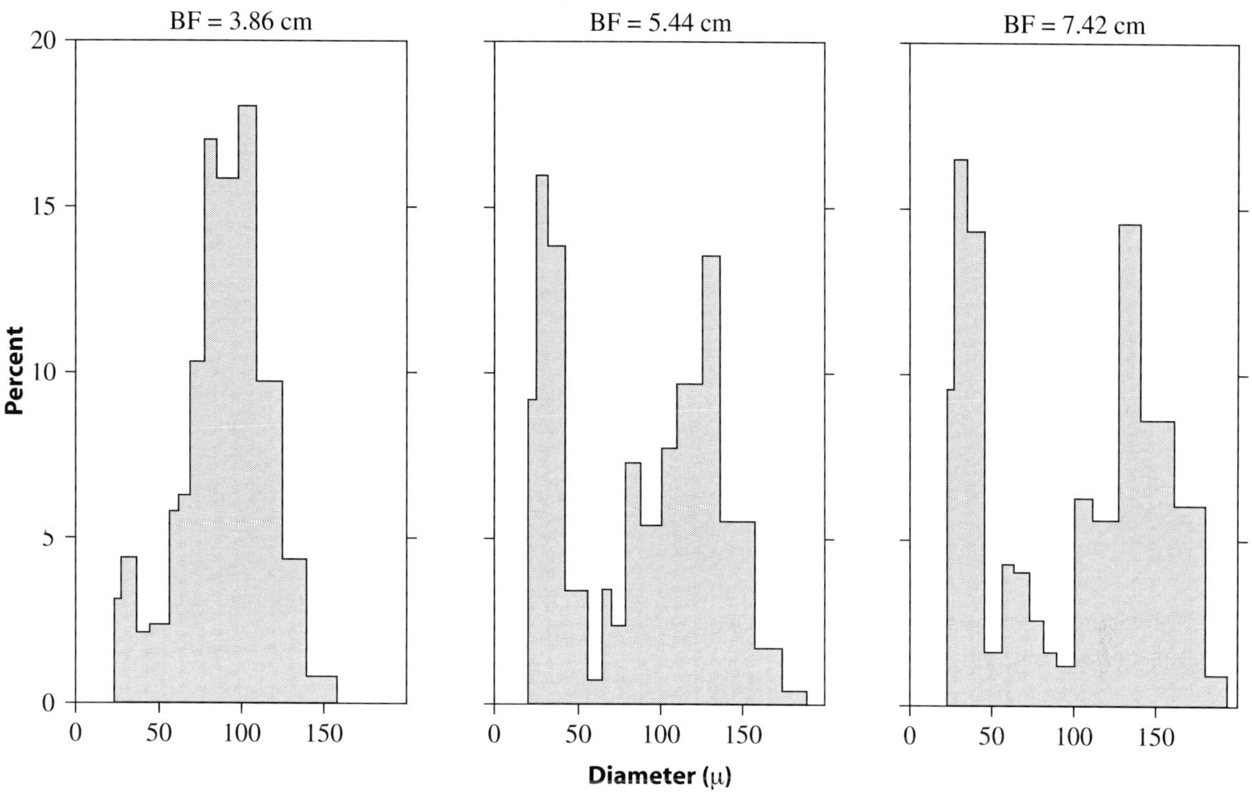

Figure 7.4 Distribution of adipocyte diameter in pigs with different backfat thickness. BF = average backfat thickness (cm). From C. E. Allen, D. C. Beitz, D. A. Cramer and R. G. Kaufman, *Biology of Fat in Meat Animals, North Central Regional Research Publication,* number 234, University of Wisconsin, Madison, Wisconsin.

this second population of adipocytes is evidence that these are newly recruited cells. Repeated phases of hyperplasia also may occur and are especially evident in obese individuals. In summary, it is clear that the increase in adipose tissue mass during animal growth is the result of the combination of increased adipocyte number and increased adipocyte size (Figure 7.5).

Molecular Control of Adipogenesis

As is the case for myogenesis, identifying those genes responsible for controlling various aspects of adipogenesis is of intense research interest. Adipocyte differentiation is initiated by a series of transcriptional events involving several transcription factors (Figure 7.6). Ultimately, genes necessary for adipocyte function and lipid accumulation, such as *lipoprotein lipase* (LPL), *stearyl-CoA-desaturase, fatty acid synthetase, acetyl CoA carboxylase,* and *glucose transporter-4* are up-regulated.

Three factors involved in differentiation of fat cells are:

1. *Peroxisome proliferator-activated receptor gamma* (PPAR γ)
2. The *CCAAT/enhancer binding protein* (C/EBP) family
3. The *adipocyte determination and differentiation factor 1* (ADD1)/*steroid regulatory element binding protein 1* (SREBP1).

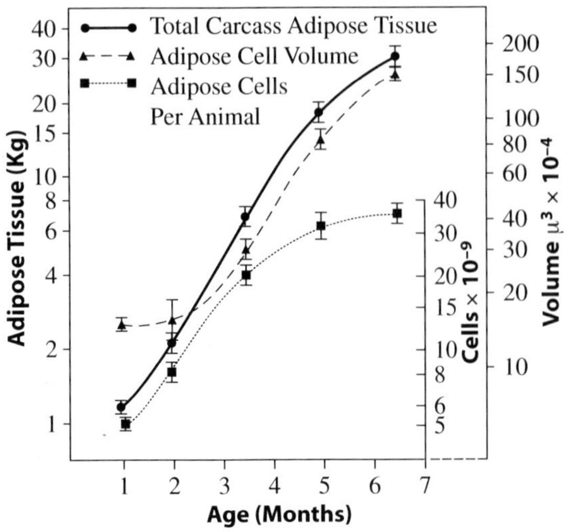

Figure 7.5 Age-related changes in carcass adipose tissue weight, volume of adipose cells in the subcutaneous depot and estimated adipose cell number in growing pigs. From C. E. Allen, D. C. Beitz, D. A. Cramer and R. G. Kaufman, *Biology of Fat in Meat Animals, North Central Regional Research Publication,* number 234, University of Wisconsin, Madison, Wisconsin.

Members of the PPAR family function as heterodimers with the *retinoid X receptor* (RXR) through binding to *peroxisome proliferator response elements* (PPRES) in the promoter regions of genes whose expression is induced during adipogenesis. Experimentally, fibroblasts can be induced to accumulate lipid by genetically-engineering cells to over-express PPAR γ. Consequently, PPAR γ has received considerable attention as a key regulator of adipose tissue development. C/EBP binding sites also are found in a number of genes that are induced during differentiation. C/EBP members belong to a *basic region/leucine zipper transcription factor family* of *nuclear transcription factors,* which are capable of developing homo- and heterodimers that can bind to the 5′ controlling elements (those DNA sequences responsible for turning on and off genes) of fat-specific genes. C/EBP β and C/EBP δ are two members of the C/EBP family that are important in initiation of adipocyte differentiation. These two members induce PPAR γ expression in cells, whereas cells lacking these two members do not differentiate into adipocytes. A third member, C/EBP α is thought to maintain the expression of PPAR δ in differentiated adipocytes. ADD1/SREBP1 is a member of the basic helix-loop-helix family of transcription factors that promotes in increased differentiation. It is evident

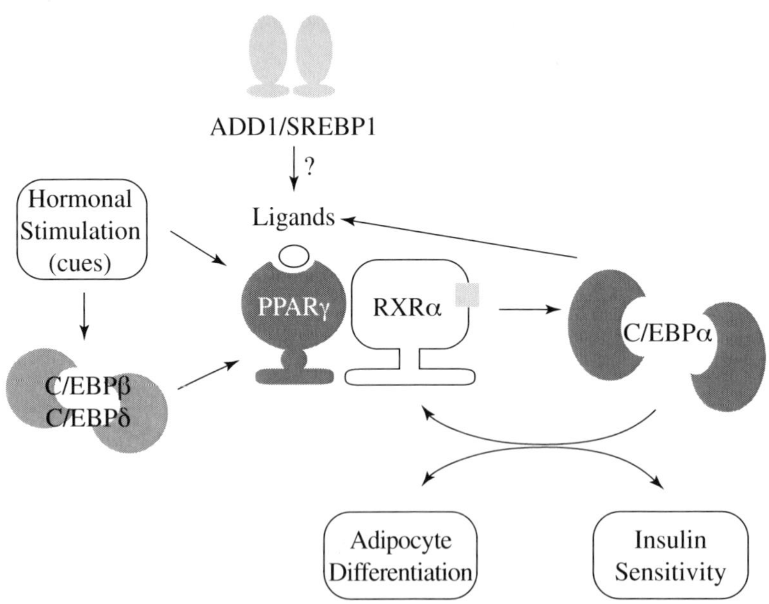

Figure 7.6 Molecular mechanisms controlling adipocyte differentiation. From Z. Wu, P. Puigserver and B. M. Spiegelman, Transcriptional activation of adipogenesis, *Current Opinion in Cell Biology* (1999), vol 11, pg. 689.

that the transcriptional cascade in adipocyte differentiation includes many activators and repressors yet to be identified. The identification of transcription factors involved in this process is an area of active investigation in both human medicine and agriculture because control of cell number may be one means to regulate lipid accumulation.

Brown Fat Differentiation

Brown fat exists in many species at birth and is composed of adipocytes called *brown adipocytes*. Adipocytes of brown fat contain more and larger mitochondria than those of white fat. Brown fat is important to neonatal survival because it is important in generation of heat. The enzymes present in brown fat are virtually indistinguishable from those in white adipose tissue, but brown adipose tissue expresses *uncoupling protein-1* (UCP1), which allows for the uncoupling of oxidation from ATP formation, thus resulting in heat production. It is unclear whether brown adipose tissue is depleted or if some is converted to white adipose tissue shortly after birth. It has been suggested that brown adipocytes may serve as precursors for white adipose tissue. Little to no brown fat exists in neonatal pigs and no brown fat is detected in poultry.

Adipose Tissue Metabolism

Triglycerides or *neutral fat* constitute the major form of lipid stored in adipose tissue even though *phospholipids* and *sterols* account for a percentage of lipid in the body. As stated previously, adipose tissue is metabolically very active and proficient at storing energy for the body, as well as mobilizing energy from its cells. The extent to which these processes occur, relative to one another, determines the overall accumulation of adipose tissue. Therefore, an understanding of how energy is moved in and out of adipose tissue is instructive to understanding animal fattening.

The rate of fat deposition in adipose tissue is a function of:

- The absorption of preformed fatty acids from the blood
- Fatty acid synthesis and triglyceride formation
- Lipolysis

All of the processes involved in absorption of nutrients from the blood, and subsequent synthesis and *esterification* to form triglycerides in an adipocyte are collectively referred to as *lipogenesis*. Conversely, lipolysis is the process of mobilization or release of fatty acids from triglycerides. Thus, the *net lipid accretion rate* is equal to the rate of lipogenesis minus the rate of lipolysis.

Lipogenesis

The major site of lipogenesis is dependent on species. Other than the mammary gland, adipose tissue is the major site of lipogenesis in cattle, sheep, and pigs. Liver is the major site in avian species. Hepatic lipogenesis likely accelerates triglyceride accumulation in liver, which can be detrimental to the bird (fatty liver), but also produces *pate du fois gois* (fatty goose liver), a delicacy common in France. The site at which fatty acids are synthesized impacts adipocytes and energy storage.

Adipocytes may accumulate lipid by:

- Extracting fatty acids from circulating triglycerides
- *De novo* synthesis of fatty acids

Fatty acids are subsequently esterified to triglycerides. The overall process of esterifying fatty acids to glycerol to make triacylglycerol and phospholipids is named glycerolipid biosynthesis (Figure 7.7).

Blood triglycerides may originate from either the food ingested or hepatic synthesis. In poultry, fatty acids produced in the liver can be transported via the circulation to adipose tissue as triglyceride packaged into *very low density lipoproteins* (VLDL). VLDL are too large to cross the endothelium of the capillary and cell membrane of the adipocyte. Therefore, fatty acids and glycerol are split from circulating triglycerides by *lipoprotein lipase* that is located on the lumen side (inner surface) of capillaries. Lipoprotein lipase is a *rate-limiting enzyme* for use of blood triglycerides. The liberated fatty acids diffuse across endothelium. Glycerol is not metabolized by adipocytes and is targeted to the liver for further metabolism. Fatty acids can diffuse into adipocytes and cytoplasmic lipid-binding proteins are involved in moving the fatty acids to the appropriate locations within the adipocyte. Once the fatty acids are in the adipocyte, they may be *esterified* into triglycerides using glucose as the precursor of the glycerol backbone. Alternatively, fatty acids may be used for other cellular processes, such as membrane synthesis, control of cellular metabolism and transcriptional control (Figure 7.8).

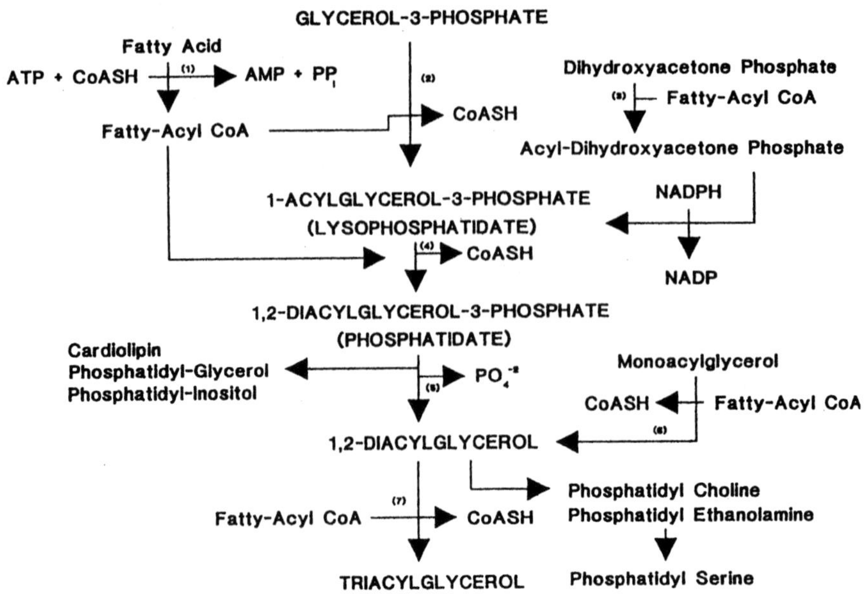

Figure 7.7 Glycerolipid biosynthesis. Numbers refer to enzymes: (1) acyl-CoA dihdroxyacetone phosphate acyltransferase, EC 2.3.1.42; (4) lysophosphatidate acyltransferase, EC 3.1.3.20; (5) phosphatidate phosphohydrolase, EC 3.1.3.4; (6) monoacylglycerol acyltransferase, EC 2.3.1.22; (7) diacylglycerol acyltransferase, EC 2.3.1.20. From *The Biology of Fat in Meat Animals, American Society of Animal Science,* Champaign, Illinois.

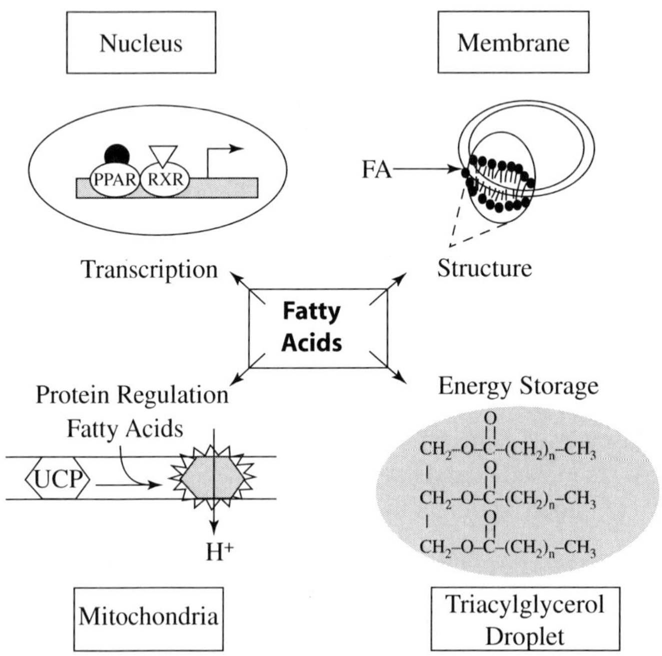

Figure 7.8 The role of fatty acids in adipose tissue. FA, fatty acids; PPAR, peroxisome proliferator activated receptor; RXR, retinoid X receptor; UCP, uncoupling protein. •, fatty acid and/or derivative; Δ, 9-*cis* retinoic acid. From D. A. Bernlohr, N. R. Coe and V. J. LiCata, Fatty acid trafficking in the adipocyte, *Cell & Developmental Biology,* vol 10, pg. 43.

Dietary fat serves as a second source of circulating fatty acids. Fatty acids derived from diet ingredients are transported from the small intestine to the adipose tissue as triglyceride packaged into lipoproteins called *chylomicrons*. In monogastrics, many of the fatty acids used for adipocyte triglyceride synthesis are the dietary fatty acids. Consequently, the fatty acid composition of monogastric adipose tissue depots is similar to that of the diet. This is why some fats, especially those that are highly unsaturated, i.e., those with high unsaturated:saturated fatty acid ratios, when fed to pigs will affect the texture of fat in resulting pork carcasses. Fats with a high proportion of unsaturated fatty acids have a lower melting point and are softer at a given temperature than more saturated fats. For this reason, high oil diets, such as those with appreciable amounts of peanut oil are not used in finishing swine diets. This phenomenon does not occur to such an extent in ruminants because the rumen microbial action modifies many of the dietary fatty acids before they are utilized. Dietary fat likely contributes to adipose fat about the same in many species; however, the unique feature of ruminants is the biohydrogenation of dietary fatty acids prior to their use.

De novo synthesis of fatty acids and subsequent triglyceride formation in adipocytes requires that precursors of triglycerides are transported into adipocytes. Enzymes required for fatty acid synthesis are shown in (Figure 7.9). The principal substrate

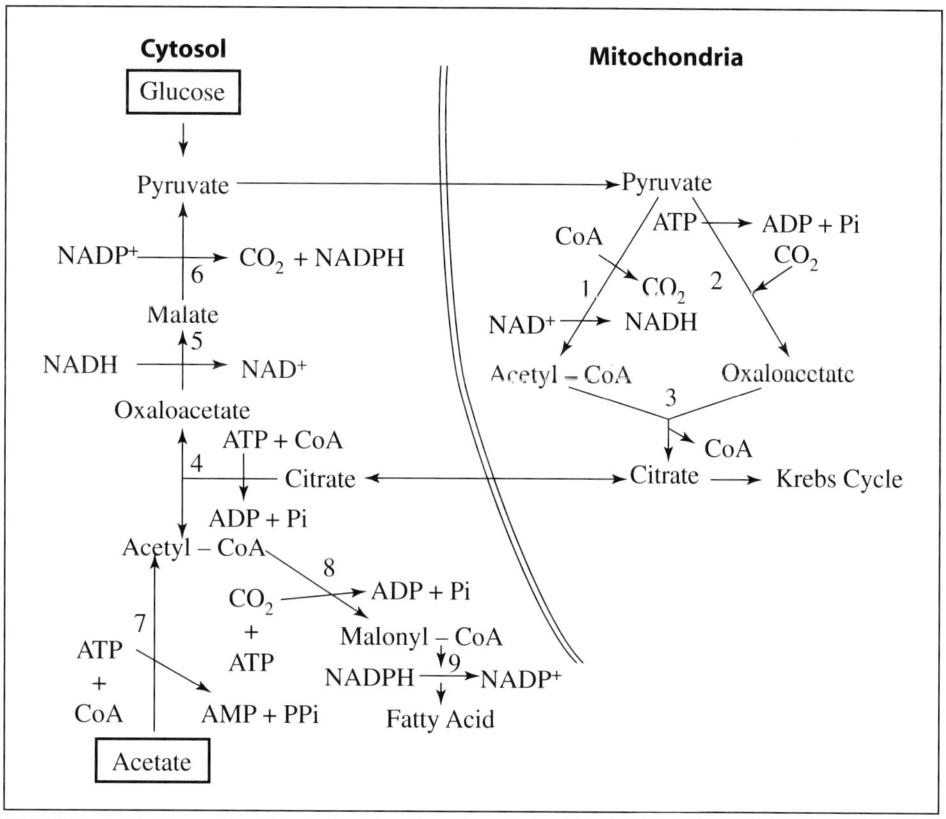

Figure 7.9 *De novo* synthesis of fatty acids from glucose and acetate. Common names for enzymes numbered in figure are: 1. Pyruvate dehydrogenase (EC1.2.4.1), 2. Pyruvate carboxylase (EC6.4.1.1), 3. Citrate synthetase (EC4.1.3.7), 4. Citrate cleavage enzyme (ATP-citrate lyase, EC4.1.3.6), 5. NAD-malate dehydrogenase (EC1.1.1.37), 6. NADP-malate dehydrogenase or malic enzyme (EC1.1.1.40), 7. Acetyl-CoA synthetase (EC6.2.1.1), 8. Acetyl-CoA carboxylase (EC6.4.1.2) and 9. Fatty acid synthetase. From C. E. Allen, D. C. Beitz, D. A. Cramer and R. G. Kaufman, *Biology of Fat in Meat Animals, North Central Regional Research Publication,* number 234, University of Wisconsin, Madison, Wisconsin.

for lipogenesis is dependent on species. Nonruminants use glucose predominantly as lipogenesis substrate, but ruminants use *acetate,* a volatile fatty acid produced in large quantities by the rumen, as the predominant substrate. Other short chain fatty acids produced in the rumen (*propionate* and *butyrate*) also can be used to some extent as substrates for fatty acid synthesis. The production of high amounts of acetate coupled with low quantities of *glucokinase* (GK), *citrate cleavage enzyme* (CCE, ATP-citrate lyase), and *malic enzyme* (ME, NADP-malate dehydrogenase) in liver and CCE, ME, and *pyruvate carboxylase* in adipose tissue are believed to contribute to the absence of fatty acid synthesis from glucose in ruminant liver and adipose tissue. The use of acetate for fatty acid synthesis in ruminants allows the limited supplies of glucose in ruminants to be spared for utilization by the nervous system, erythrocytes, and mammary tissue. Glucose is used in lipogenic tissues for generation of NADPH and production of alpha-glycerol phosphate, the latter of which forms the glycerol backbone for triglyceride synthesis. Young ruminants whose rumens are not developed to produce sufficient quantities of short-chain fatty acids appear to possess sufficient lipogenic enzymes to support lipogenesis from glucose.

Lipolysis

Removal or release of fatty acids from adipocytes, involves breakdown of triglyceride within cells and diffusion of fatty acids across cell membranes. Enzymes responsible for this process are known as *lipases.* The major lipase responsible for hydrolysis of triglyceride into glycerol and fatty acids in adipocytes is *hormone-sensitive lipase.* A number of other adipocyte proteins also are involved in lipolysis, including *protein kinase A, adenylyl cyclase, phosphodiesterase,* and *guanine nucleotide binding proteins* (Figure 7.10). These proteins, as well as hormone sensitive-lipase, are hormonally

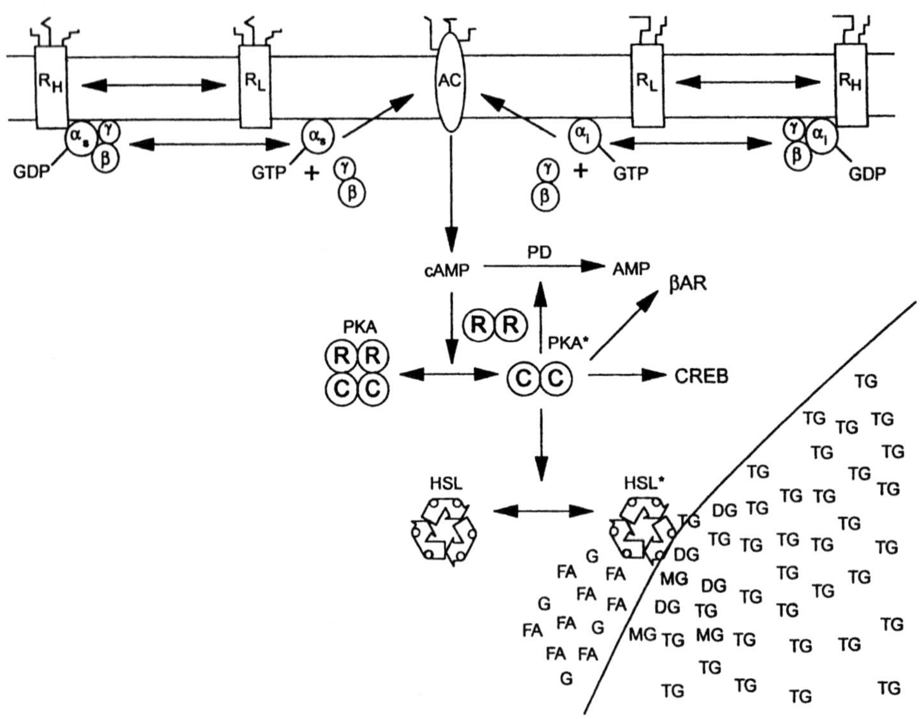

Figure 7.10 Mechanism for lipid mobilization from cells. Abbreviations: high-affinity receptor (RH); low affinity receptor (R_L); adenylyl cyclase (AC); phosphodiesterase (PD); protein kinase A (PKA); hormone-sensitive lipase (HSL); β-adrenergic receptor (βAR); cAMP response element binding protein (CREB). From *The Biology of Fat in Meat Animals, American Society of Animal Science,* Champaign, Illinois.

regulated. After the fatty acids are liberated within the adipocyte, they may be targeted to various processes or may leave the cell.

Composition of Adipose Tissue

The composition of adipose tissue is variable and is generally composed of:

- 76 to 94% lipid
- 1 to 4% protein
- 5 to 20% water.

The predominant class of lipids in the animal body are the neutral lipids and are in the form of triglycerides in the adipocytes of the various depots. The fatty acids that compose the triglycerides vary among species and depots and contribute to the differences in firmness of depot fat in carcasses (Table 7.1). One of the most abundant fatty acids in the animal body is oleic acid. The most abundant triglycerides are those that contain one palmitic fatty acid and two oleic fatty acid molecules (palmitodiolein) and those that contain one oleic, one palmitic, and one stearic fatty acid molecule (oleopalmitostearin). Palmitic and stearic fatty acids are the two saturated fatty acids that predominate in animal fats, whereas palmitoleic, oleic, linoleic, and linolenic acids are the predominant unsaturated fatty acids.

Factors Affecting Adipose Tissue Composition and Lipid Metabolism

Cellularity and Age

The composition of adipose tissue from a young animal differs from that of an older animal. The amount of lipid, as a percentage of tissue weight, is greater, and that of water and protein is less in old compared with young animals. This developmental change is highly correlated to fat cell size, which increases with age in most fattening animals. Fatty acid composition also tends to change as animals age because the proportion of triglyceride to phospholipid increases as adipocytes undergo hypertrophy.

The capacity of adipose tissue to perform lipid metabolism, either lipogenesis or lipolysis, is related to the number and size of adipocytes contained within adipose tissue, so attempts to study lipid metabolism usually are normalized to adipocyte number or size. For example, many factors known to alter lipid metabolism ultimately do so by regulating the activities of key enzymes across many adipocytes. Of particular importance are those enzymes required for NADPH generation (*glucose-6-phosphate dehydrogenase, 6-phosphogluconate dehydrogenase,* NADP-*isocitrate dehydrogenase* and *malic enzyme*). But because the number and size of adipocytes are related to these enzyme activities, measurement of enzyme activity in a given sample of adipose tissue must be normalized to adipocyte number or size. This allows researchers to determine how metabolism is regulated on a cellular basis.

Anatomical Location

Recall that numerous fat depots located throughout the body accumulate at different rates and at different times during development and therefore their composition will vary (Tables 7.2 and 7.3). As an example, the number of intramuscular adipocytes may account for nearly one half the total number of adipocytes in the body of pigs. However, that same tissue only represents just over 10% of total lipid in the body. These results show that the intramuscular depot contains less fat and more immature fat cells than those of other, earlier maturing fat depots. Another example is found within different layers of subcutaneous fat which vary in fatty acid composition. The differences in fat firmness among fat depots and among sites within fat depots are due to different ratios of unsaturated:saturated fatty acids.

Species

Species differences that exist in adipose tissue composition are due largely to differences in lipid metabolism. As discussed earlier in this chapter, hydrogenation of fatty acids by rumen microbes has a significant effect on the type of fatty acids available for triglyceride synthesis. Furthermore, the amount of tissue desaturase enzyme activity differs across species resulting in different ratios of saturated:unsaturated fatty acids in depots.

In addition to species differences in lipid biosynthesis, not all species grow at the same rate and, therefore, at a given chronological age the number of mature adipocytes will likely differ. This difference will contribute to differences in lipid metabolism and composition.

TABLE 7.1 FATTY ACID AND TRIGLYCERIDE COMPOSITOIN OF SOME ANIMAL FAT DEPOTS (PERCENT BY WEIGHT)

		Pig			Cattle		Sheep		Fish
					Subcutaneous				
Component	**Chicken**	**Outer**	**Inner**	**Perirenal**	**Subcutaneous**	**Perirenal**	**Subcutaneous**	**Perirenal**	**Herring oil**
Fatty acids									
Lauric	—	trace	trace	trace	0.1	0.2	0.1	0.1	—
Myristic	0.1	1.3	0.1	4.0	4.5	2.7	3.2	2.6	6.7
Palmitic	25.6	28.3	30.1	28.0	27.4	27.8	28.0	28.0	11.5
Stearic	7.0	11.9	16.2	17.0	21.1	23.8	24.8	26.8	1.4
Arachidonic	—	trace	trace	trace	0.6	0.6	1.6	2.6	—
Total saturated	32.7	41.5	47.3	49.0	53.7	55.1	57.7	59.5	19.6
Palmitoleic	7.0	2.7	2.7	2.0	2.0	2.2	1.3	1.9	8.6
Oleic	20.4	47.5	40.9	36.0	41.6	40.1	36.4	34.2	14.4
Linoleic	21.3	6.0	7.1	11.8	1.8	1.8	3.5	4.0	1.2
Linolenic	—	0.2	0.3	0.2	0.5	0.6	0.5	0.6	0.9
Parinaric	—	—	—	—	—	—	—	—	1.45
Gadoleic	—	—	—	—	—	—	—	—	14.3
Arachidonic plus chupandonic	0.6	2.1	1.7	1.0	0.4	0.2	0.6	0.8	1.94
Cetoleic	—	—	—	—	—	—	—	—	22.0
Total unsaturated	67.3	58.5	52.7	51.0	46.3	44.9	42.3	41.5	73.76

From C.E. Allen, D.C. Beitz, D.A. Cramer and R.G. Kaufman, Biology of Fat in Meat Animals, North Central Regional Research Publication, number 234, University of Wisconsin, Madison, Wisconsin.

TABLE 7.2 CHARACTERISTICS OF ADIPOSE TISSUE FROM SEVEN ANATOMICAL LOCATIONS IN THE PIG

Depot Location	% Lipid	Cell Diameter (μm)	g TG/Cell
Outer subcutaneous	88.8	129	1.04
Middle subcutaneous	93.4	134	1.31
Inner subcutaneous	87.7	131	1.08
Intermuscular	76.3	108	0.62
Perirenal	96.1	153	1.60
Mesenteric	84.0	123	0.92
Leg subcutaneous	70.9	106	0.58

TG = triglycerides
From C.E. Allen, D.C. Beitz, D.A. Cramer and R.G. Kaufman, Biology of Fat in Meat Animals, North Central Regional Research Publication, number 234, University of Wisconsin, Madison, Wisconsin.

TABLE 7.3 CHARACTERISTICS OF PORCINE ADIPOSE TISSUE FROM THREE LOCATIONS[a]

Trait	Subcutaneous	Intermuscular	Intramuscular[b]
Moisture, %	7.23	10.31	14.56
Ether extractable lipid, %	91.71	88.29	83.44
Lipid/moisture ratio	12.7	8.6	5.7
Cell diameter, μm	169.0	151	125
Cell volume, $\mu m^3 \times 10^5$	25.3	18.0	10.2
Cell no./g lipid $\times 10^5$	4.21	6.04	10.66

[a] All values are the mean of four animals.
[b] Determined on combined fat strands from *M. trapezius* and *M. Latissimus dorsi.*

From C.E. Allen, D.C. Beitz, D.A. Cramer and R.G. Kaufman, Biology of Fat in Meat Animals, North Central Regional Research Publication, number 234, University of Wisconsin, Madison, Wisconsin.

GENETICS

Even within species, breed and strain differences exist in adipose cellularity and resulting lipid metabolism. Obviously, different breeds or genetic lines of animals differ in their ability to deposit adipose tissues. For example, some genetic lines of pigs used in the 1950's and 1960's exhibited subcutaneous fat thickness of over 5 centimeters at market weight, yet today, some genetic lines have deposited only a few millimeters by that time. Clearly, fatter lines possess a higher percentage of lipid in backfat than leaner genotypes. Leaner genotypes are associated with changes in adipose cellularity. Because of less adipocyte hypertrophy, the recruitment of new adipocytes is less. This change in cellularity also impacts the overall composition of adipose tissue and consequently may impact the firmness of some fat depots.

SEX

Sex differences in most physiological contexts generally imply changes in hormone status, particularly estrogen and testosterone. In the case of adipose tissue composition, estrogen promotes filling of fat depots, whereas testosterone tends to inhibit lipid deposition. The increased fatness of carcasses from females is associated with increased size of adipocytes and consequently fewer adipocytes per unit of tissue (Table 7.4), whereas (for ruminants) castrated males are somewhat intermediate between females and intact males. One also expects fat of females to contain more lipid than that from a male

TABLE 7.4 EFFECTS OF BREED, SEX AND AGE ON SUBCUTANEOUS ADIPOSE TISSUE TRAITS IN SHEEP

	Breed		Sex			Age (weeks)		
Trait	**Suffolk**	**Southdown**	**Ewe**	**Wether**	**Ram**	**8**	**16**	**32**
Live weight, kg.	43.1	36.9	35.8	38.9	45.4	22.1	35.5	62.5
Lipid, %	78.8	79.2	80.0	82.2	74.6	74.9	78.3	83.6
Adipocytes/g tissue ($\times 10^5$)	20.8	20.2	18.2	20.2	23.1	29.3	19.0	13.2
Adipocyte diameter, µm	96.7	96.7	99.9	97.4	92.9	82.5	97.3	110.4
Adipocyte volume ($\mu m^3 \times 10^4$)	5.56	5.70	6.29	5.72	5.03	4.03	6.06	8.58

From C.E. Allen, D.C. Beitz, D.A. Cramer and R.G. Kaufman, *Biology of Fat in Meat Animals,* North Central Regional Research Publication, number 234, University of Wisconsin, Madison, Wisconsin.

as long as other factors such as breed, weight and anatomical location are held constant. Increased fat content of females at constant weight reflects the powerful effects of estrogen throughout the body to achieve maturity through epiphyseal closure and other effects. Effects of sex hormones on lipid metabolism, as well as gender effects on distribution of fat among fat depots will be discussed in later chapters.

HORMONES

In addition to the sex hormones mentioned above, other hormones have significant effects on adipose tissue composition and metabolism. The impact of hormones will be covered in more depth in a later chapter, but it is important to note that the endocrine system is very involved in integrating the metabolism that regulates nutrient partitioning. *Leptin* is one hormone of particular interest that appears to play a major role in this process. The leptin gene was discovered while studying *obese* and *diabetic* mice, which are genetically obese. Leptin signaling mechanisms are blunted in these mice resulting in marked *hyperphagia* (overeating), reduced energy expenditure and morbid obesity. Leptin is produced by adipocytes and circulating concentrations are related to adipose tissue mass. Circulating levels of leptin are significantly elevated in obese animals and also increase with overfeeding. Conversely, leptin levels are relatively low in leaner animals. The role of leptin in adipose tissue metabolism is not clear, but it is involved in regulating feed intake and appetite and has effects on the immune system (Figure 7.11).

NUTRITION

The composition of the diet has significant impacts on adipose tissue composition and lipid metabolism:

- High fat diets depress fatty acid synthesis, especially in nonruminants, and the fatty acid composition of the diet impacts the composition of triglycerides that are stored in adipocytes.
- Low fat, high carbohydrate diets tend to increase rates of de novo fatty acid synthesis to produce fatty acids for triglyceride synthesis.

Changes in nutrient intake significantly alter mobilization of energy in and out of fat cells. During periods of reduced energy intake, lipolysis is increased so that fatty acids are liberated for oxidation. Conversely, during high dietary energy intake, excess energy is stored as triglyceride. Depots affected by these changes in planes of nutrition depend on stage of animal development (Table 7.5). Quantifying effects of nutritional regimen on adipocyte diameter depends on a constant age or constant weight as a basis of comparison. Lambs fed a maintenance diet versus an ad libitum diet illustrate that depots respond differently to nutritional regimen (Figure 7.12). Altering nutritional status late in development at a time when the intramuscular fat depot is forming tends to first alter that depot. Consequently, presence of high degrees of marbling in carcasses generally is indicative of feeding high planes of nutrition prior to slaughter. Dietary protein:energy ratio also is important in the composition of body weight gain. For example, diets with amino acid deficiencies or

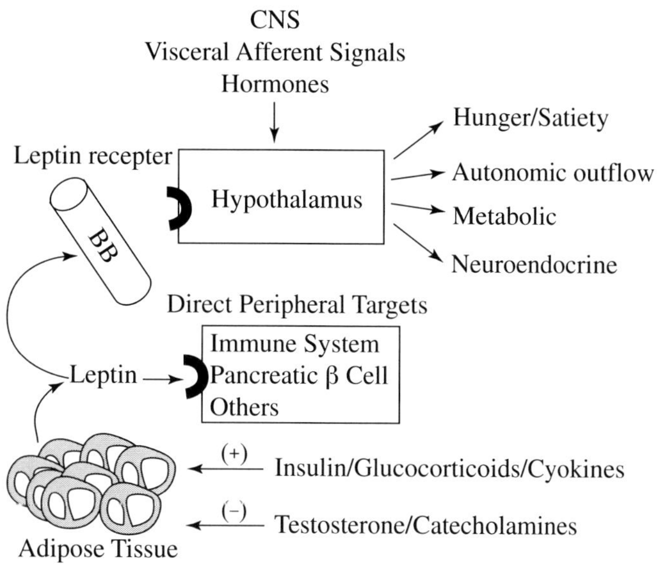

Figure 7.11 Location and mechanism of leptin action. From R. S. Ahima and J. S. Flier, Adipose tissue as an endocrine organ, *TEM,* vol 11(8), pg. 327. BB, blood-brain barrier.

TABLE 7.5 INFLUENCE OF DIETARY REGIME ON DIAMETER OF ADIPOCYTES FROM THREE LOCATIONS IN PIGS STUDIED AT CONSTANT AGE AND WEIGHT.

	Poland China		**Duroc**	
Cell Diameter, μm	**Control**	**Underfed**	**Control**	**Underfed**
Subcutaneous fat				
Age constant	119	78	108	85
Wt. constant	112	111	118	108
Intramuscular fat				
Age constant	64	51	62	49
Wt. constant	64	57	65	58
Omental fat				
Age constant	102	75	89	81
Wt. constant	100	97	98	96

From C. E. Allen, D. C. Beitz, D. A. Cramer and R. G. Kaufman, *Biology of Fat in Meat Animals,* North Central Regional Research Publication, number 234, University of Wisconsin, Madison, Wisconsin.

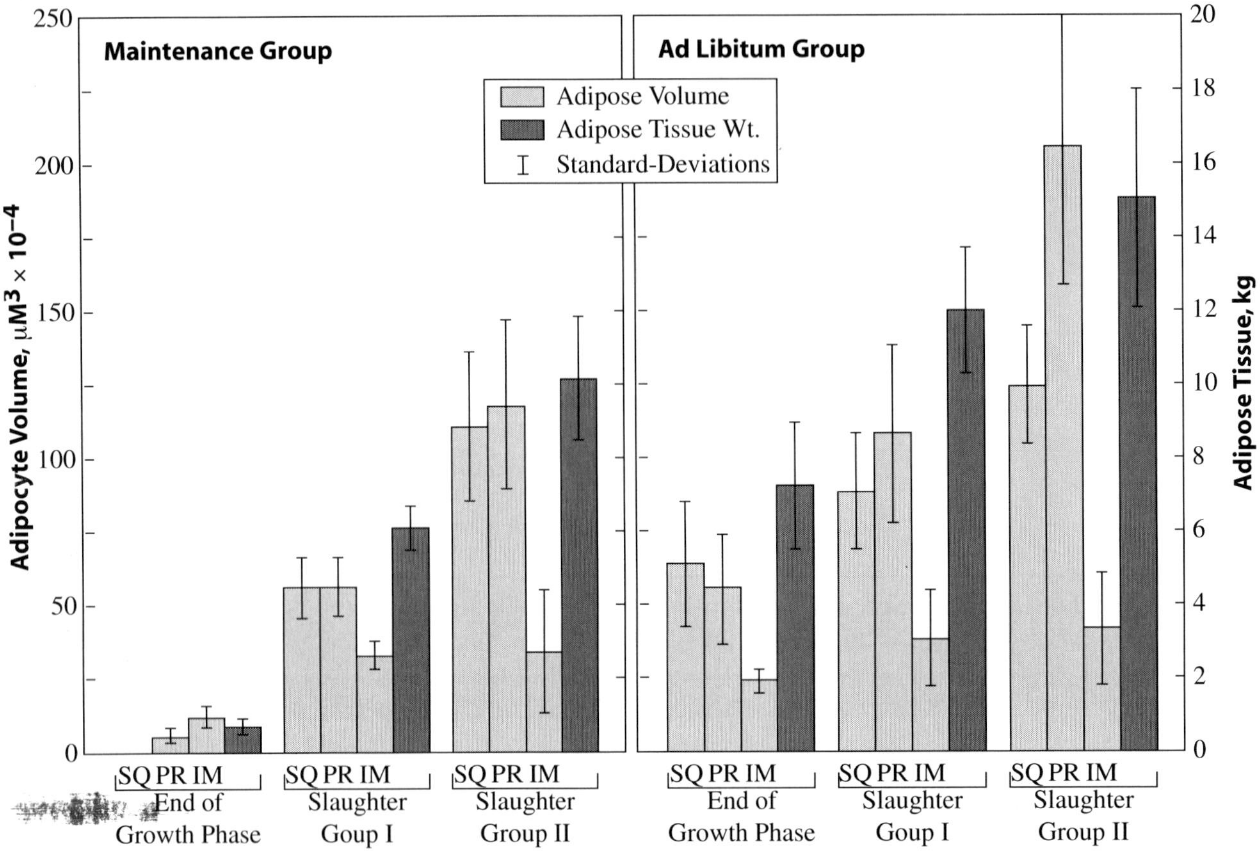

Figure 7.12 Changes in adipocyte volume in the subcutaneous (SQ), perirenal (PR) and intermuscular (IM) fat depots of lambs fed either ad libitum or maintenance diets and slaughtered at 3 different times. From C. E. Allen, D. C. Beitz, D. A. Cramer and R. G. Kaufman, *Biology of Fat in Meat Animals, North Central Regional Research Publication,* number 234, University of Wisconsin, Madison, Wisconsin.

imbalances often will result in increased lipogenesis because lean tissue accretion cannot be maximized with such unbalanced diets. Nutritional regulation of growth is discussed in greater detail in a later chapter.

Although general effects of dietary fatty acid profile on adipose tissue lipid composition have been discussed earlier, it should be mentioned that some fatty acids exhibit unique biological properties when included in the diet. Conjugated linoleic acids (CLAs) are geometric and positional isomers of linoleic acid, and, when fed to various mammalian species and poultry, reduce fat deposition and alter the fatty acid profile of the depot lipid. Although the mechanisms by which CLA exert their actions are unclear, they may involve alterations of adipogenesis, lipogenesis, and possibly desaturase activities. Reduction of fat deposition and changes in composition that favor meat quality have important implications for animal production. Consequently, CLA mechanism of action is being studied extensively.

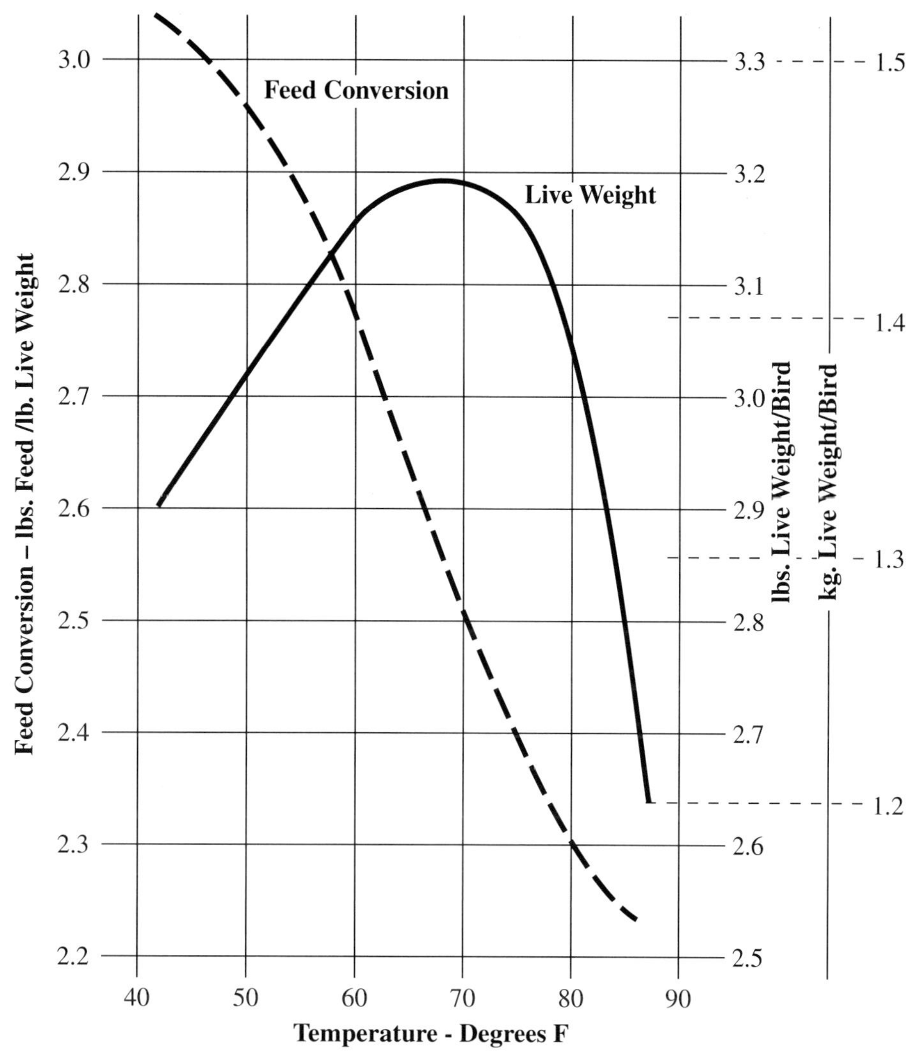

Figure 7.13 Effect of ambient temperature on animal performance traits.

Environmental Temperature

Environmental temperature may have significant effects on lipid metabolism in warm-blooded animals. Animals exposed to temperatures outside of the thermoneutral zone will have altered metabolism as they attempt to adapt to the adverse environment. If ambient temperature is lower than the thermoneutral zone, many animals will mobilize adipose tissue to support the processes involved in production of heat. Feed intake often will increase as a consequence of increased energy requirements under such conditions. Feed:gain ratios usually are altered because dietary nutrients tend to be repartitioned away from growth toward processes necessary to maintain body temperature (Figure 7.13). As ambient temperature increases above the thermoneutral zone, feed intake decreases and the processes that generate heat, including growth will be inhibited.

Chapter 8

Growth and Development of Bone

INTRODUCTION

The growth and development of bone is important to meat-animal agriculture because skeletal growth determines the frame size of an animal and the size of carcass that is ultimately produced. Normal development and maintenance of bone also is necessary to ensure the structural soundness and longevity that is important to food animals. In addition to providing structural and locomotion functions, bone also provides protection to organs and tissues, and serves as a storage depot for various compounds, such as inorganic ions, that can be mobilized in times of need. The growth and storage functions of bone are under hormonal and growth factor regulation, similar to other tissues.

Origin of Cells Involved in Bone Formation and Resorption

Osteogenesis (the formation of bone) occurs in both prenatal and postnatal life by transformation of connective tissue. Bone is formed by either *endochondral ossification* which requires a cartilaginous template or by *intramembranous ossification* in the absence of a cartilaginous template. A number of different cell types are involved in bone formation and originate from different cell lineages during embryo development. Activities of various cells result in formation of *cortical* (*compact* or *lamellar*) bone that is found in shafts of the *appendicular* skeleton and formation of *cancellous* (trabecular) bone that is found in the *axial* skeleton and in epiphyses of long bones. A third type of bone, referred to as *medullary bone,* is found in birds, and is metabolically active in that it provides a rapidly available source of calcium for egg shell development.

The axial skeleton is formed from cells derived from sclerotome cells whereas the appendicular skeleton is formed from cells derived from lateral plate mesoderm. These progenitor cells migrate to the sites of future bone formation and undergo differentiation to appropriate cell types (Figure 8.1). This differentiation at sites of future bone formation requires upregulation of various growth factors and signaling molecules and is dependent upon interaction of mesenchymal cells with epithelial cells (Figure 8.2). Mesenchymal condensations that result from these interactions determine the site of bone formation as well as the shape of bone that is formed at the site.

- ✓ *Osteoblasts* are derived from *mesenchymal* cells (also referred to as *osteoprogenitor cells*) and are directly involved in bone matrix production. Differentiation of osteoblasts requires expression of a transcription factor named *Cbfa1* (*core binding factor 1,* also known as *Runx2*). This factor is an early marker of osteoblast differentiation and is necessary, but cannot support complete differentiation without expression of other transcription factors.
- ✓ *Osteocytes* are mature osteoblasts and are responsible for maintenance of the bone matrix. In adult bone, the number of osteocytes can be ten times the number of osteoblasts. Much of the osteocyte population is involved in matrix production, but in some cases may

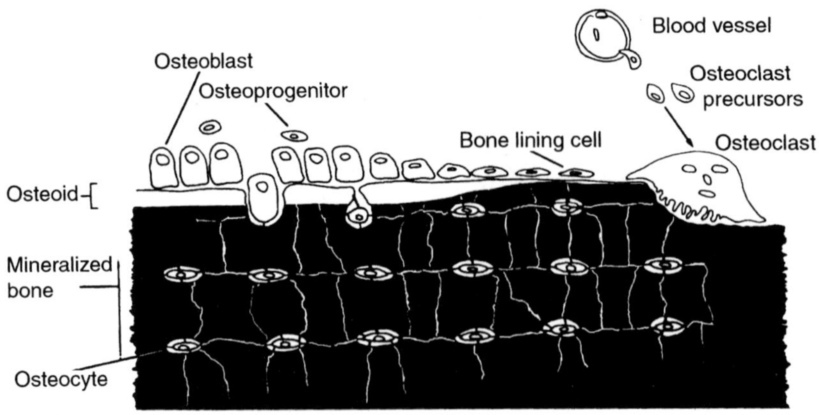

Figure 8.1 Origin and location of bone cells. Taken from Marks and Popoff (1988)

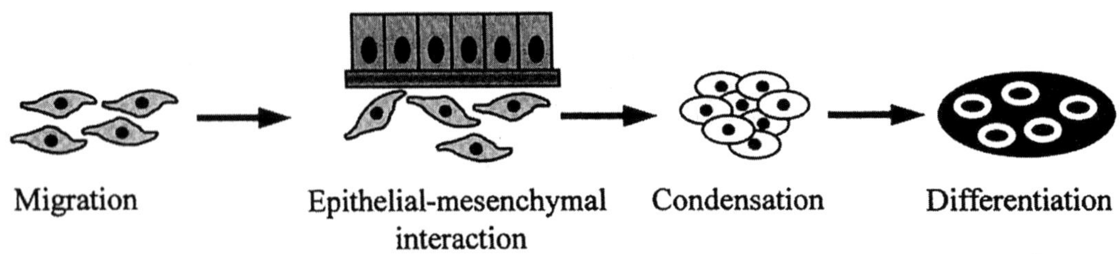

Figure 8.2 Phases of skeletal development. From J. P. Bilezikian, L. F. Raisz, and G. A. Rodan, editors. *Principles of Bone Biology,* Academic Press, San Diego.

exhibit osteolytic functions. Development and differentiation of osteoprogenitor cells to osteocytes is a very complex process involving many growth and transcription factors. Cell types are characterized by their proliferative capacity and extracellular matrix development.

- *Chondroblasts* (which differentiate to *chondrocytes*) form cartilage which is associated with bone and required for bone growth in some cases. Chondrocyte differentiation and cartilage formation requires a transcription factor named, *Sox9.*
- *Osteoclasts* are involved in bone *resorption* (breakdown) and remodeling. Osteoclasts are large multinucleated cells that form from fusion of hemopoietic cells of monocyte-macrophage lineage. The number of osteoclasts in adult bone is a very small relative to the number of osteoblasts.
- *Bone lining cells* are inactive cells found on the surfaces of bone and may serve as osteoblast precursors.
- Of course, associated closely with bone are tendons and ligaments. These structures are composed of connective tissue proper (fibers and ground substance) which is synthesized by fibroblasts.

Endochondral Ossification

Endochondral ossification occurs with the replacement of cartilage with bone (Figure 8.3). The axial and appendicular skeletons develop from a cartilagi-

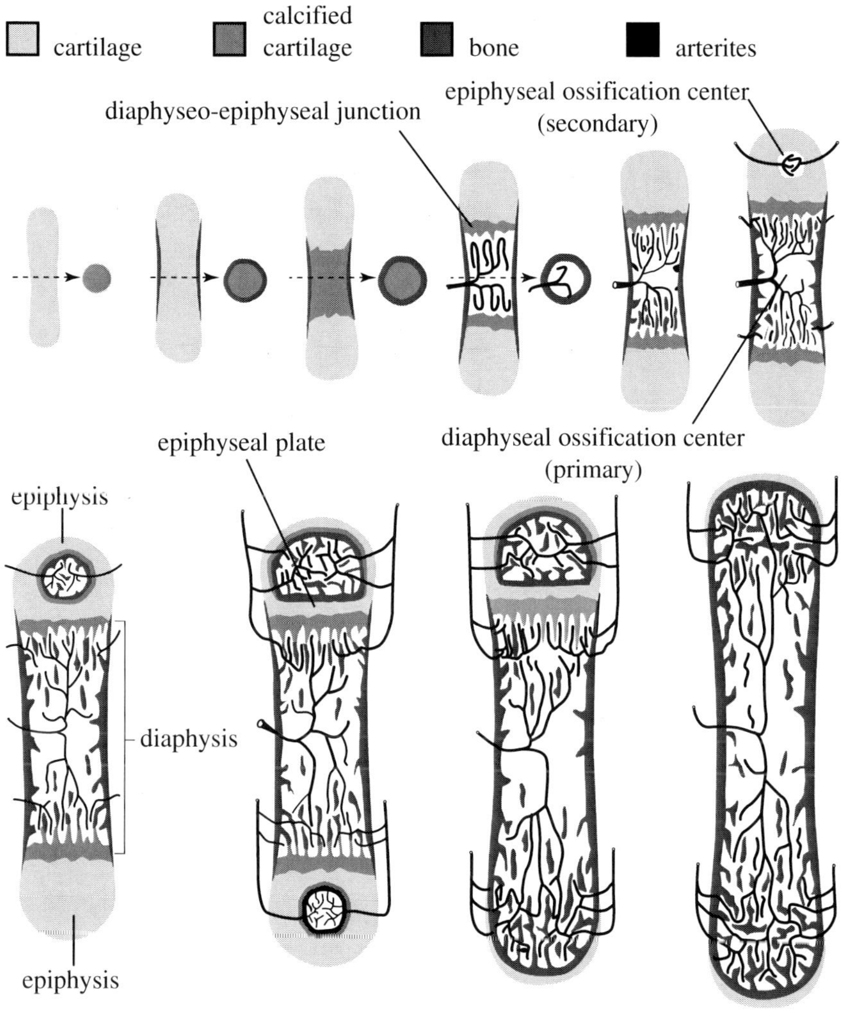

Figure 8.3 Long bone formation and growth by endochondral ossification. From J. P. Bilezikian, L. F. Raisz, and G. A. Rodan, editors. *Principles of Bone Biology,* Academic Press, San Diego.

nous structures which are established prenatally and serves as a template for ossification. Consequently, chondroblasts are major players in the early stages of endochondral ossification. As the cartilage model enlarges, the cartilagenous matrix near the center of the shaft (*diaphysis*) begins to calcify as chondrocytes mature and enlarge. This calcification also prevents nutrients from reaching the chondrocytes in the region and eventually the chondrocytes die and disintegrate. Along with these interior events, other events important for bone formation occur on the perimeter of the diaphysis. The *perichondrium* is converted into *periosteum* and osteoblasts become apparent in the inner layer of the periosteum. These osteoblasts eventually form a layer of bone around the diaphysis. A *primary ossification center* is established in the central portion of the diaphysis as a result of capillaries and osteoblasts that migrate into that region from the outer regions. The calcified cartilage matrix in the center is broken down and the action of the osteoblasts results in a replacement of the cartilage with new bone. Bone is formed from this center of ossification until the entire diaphysis is filled with spongy bone. As the bone grows, osteoclasts increase the inside diameter of the hollow central portion of the diaphysis so that a marrow cavity results.

As growth continues to the time of birth, a *secondary center of ossification* is established in the

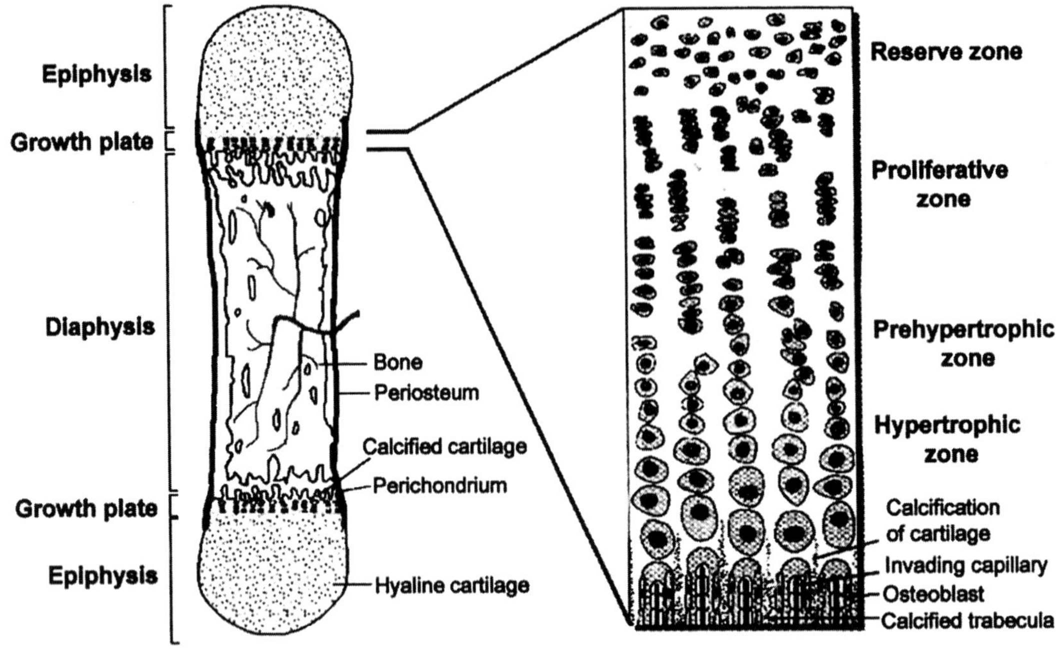

Figure 8.4 Organization of the growth plate. From J. P. Bilezikian, L. F. Raisz, and G. A. Rodan, editors. *Principles of Bone Biology,* Academic Press, San Diego.

center of each *epiphysis*. The osteoblasts in these centers produce spongy bone that replaces the cartilage of the epiphyses. A thin layer of *articular cartilage* remains on the surface of each epiphysis which is important in reducing the friction where ends of bones are in contact with one another in joints. A thin region of cartilage also remains between the epiphysis and the diaphysis. This cartilage is the *epiphyseal plate,* also named the *growth plate* because of the role it plays in increasing the length of long bones.

Increased length of long bones involves the action of both chondrocytes and osteoblasts at the growth plate (Figure 8.4). Growth of cartilage on the epiphyseal side of the growth plate and the accompanying ossification of cartilage on the diaphyseal side of the growth plate results in increased length in the long bone. Zones of the growth plate, perpendicular to the main axis of growth, are characterized by the type and shape of cells and their metabolic actions. The chondrocytes closest to the epiphyseal side of the growth plate are part of the *reserve zone* (also referred to as *resting zone* or *stem cell zone*) and are the source of the proliferating chondrocytes that form the *proliferative zone.* These proliferating chondrocytes produce extracellular matrix molecules, such as various types of collagens and proteoglycans. The proteoglycans produced include *aggrecan* (which contains keratin sulfate and chondroitin sulfate) and hyaluronic acid. The chondrocytes eventually differentiate and as they mature they begin to increase in size forming the *prehypertrophic zone* and finally the *hypertrophic zone.* The cells produce additional molecules for the extracellular matrix, such as *cartilage matrix protein, fibronectin,* and other types of collagen. The hypertrophying chondrocytes also produce proteins, such as *chondrocalcin, osteocalcin, osteonectin,* and *alkaline phosphatase* that contribute to the mineralization of the matrix. The mature chondrocytes eventually die as a result of apoptosis and leave behind this calcified matrix which is eventually replaced with bone through the action of the invading capillaries and osteoblasts. The region of growth between the cartilage and new bone is referred to as the *chondroosseous junction*.

Degradation of growth plate cartilage is important in allowing bone formation to occur at a normal rate. Insufficient cartilage degradation results in accumulation of growth plate cartilage at the expense of bone. This condition is known as *dyschondroplasia* and has been a serious problem in many poultry production systems. The incidences of tibial dyschondroplasia in poultry have tended to increase as broilers, turkeys, and ducks have been selected for increased breast yield and decreased days to market. *Osteochondrosis,* a type of dyschondroplasia, also occurs to some degree in domestic animals and is characterized by the lack of endochondral ossification of cartilage, including cartilage of the growth plate. *Epiphyseolysis* (separation of the epiphysis along the growth plate) may occur as a result of osteochondrosis. Osteochondrosis also may include the failure of endochondral ossification of articular cartilage. Disorders affecting the articular cartilage, such as *osteoarthritis* and osteochondrosis, often are referred to as *degenerative joint diseases*. All of these disorders may have serious consequences on animal productivity and well-being.

The regulation of lengthening of long bones involves a number of hormones and growth factors that have actions at the growth plate, specifically on its turnover. *Growth plate turnover* refers to the rate at which the chondrocytes progress through the various zones, and is dependent on species as well as stage of development. For example, in chickens turnover of chondrocytes in the growth plate is a day or less, whereas in some mammals it may be several days.

Termination of Long Bone Growth

The diaphysis of the long bone will continue to lengthen as long as the rate of cartilage growth remains faster than the rate of osteoblast invasion at the growth plate, or in other words, as long as the growth plate does not completely ossify. Termination of long bone growth in length occurs when cartilage growth stops, or slows to a rate that allows it to be overcome by ossification. This *epiphyseal closure* results in fusion of the epiphyses and diaphysis into a single bone.

The phenomenon of growth plate closure is used in determining the maturity of lamb carcasses (see Chapter 12). During processing of lamb, the feet of the lamb carcass are removed at the distal end of the metacarpal. In lambs less than 12 months of age, the epiphyseal cartilage is still present, and when the foot is broken off the limb, it separates at the distal epiphyseal plate of the metacarpal giving rise to a *break joint.* If carcasses are from older sheep e.g., greater than 12 months of age, the epiphyseal cartilage has ossified and the foot separates from the limb at the *metacarpophalangeal joint* rather than at the epiphyseal cartilage. This results in exposure of the articular cartilage at the distal end of the metacarpus which is referred to as a *spool* joint. Presence of a spool joint usually results in classification of the carcass as yearling mutton or mutton rather than lamb. Epiphyseal closure occurs at different times in different bones depending on the rate of development.

Just as long bone growth terminates at different times, growth of other bones terminates at different times as well. For example, sacral vertebrae reach maturity before lumbar vertebrae, and lumbar vertebrae mature before thoracic vertebrae. Thus, maturation of the vertebral column occurs in a posterior to anterior direction as noted by the ossification of the cartilage associated with the vertebral processes (see Chapter 12). This progression of bone hardening is used in estimating maturity of beef carcasses. An understanding of bone biology has allowed the meat industry to implement rather simple methods as markers of maturity of cattle and sheep.

Intramembranous Ossification

Intramembranous ossification is characterized by bone formation as a replacement of connective tissue, but in the absence of cartilage. Mesenchymal cells at the sites of bone formation differentiate directly to osteoblasts with no need for chondrocytes to form a cartilaginous template. Examples of this type of bone growth are exhibited in the bones of the skull. Bone growth begins when osteoblasts in connective tissue begin producing collagen fibers and bone matrix. The osteoblasts eventually differentiate into osteocytes and produce bone (Figure 8.5).

Increases in diameter of growing bone also occur by intramembranous ossification through *appositional growth* (Figure 8.6). In this situation, the osteoblasts that differentiate from the periosteum deposit new bone matrix within the periosteum. Osteoblast activity is highest for those cells that surround the periosteal blood vessels. New osteons are produced as a result of the concentric deposition of bone that results from the action of osteoblasts and the formation of osteocytes. These new osteons contribute to compact bone and increase the overall outside diameter of the bone.

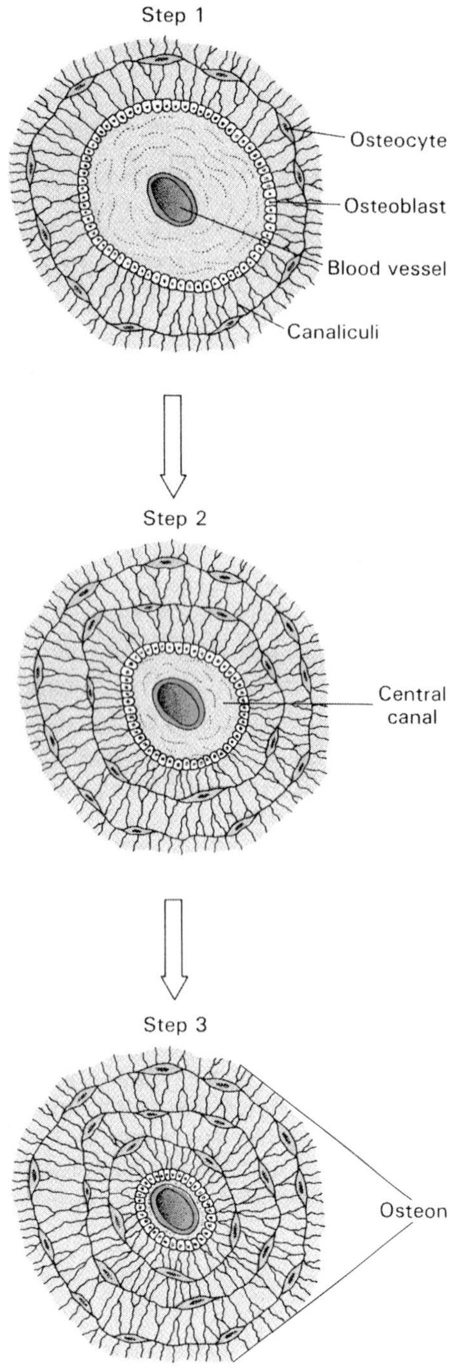

Figure 8.5 Intramembranous ossification. From *Fundamentals of Anatomy and Physiology, 3rd Edition* by Martini, © Reprinted by permission of Pearson Eduction, Inc., Upper Saddle River, NJ.

Bone Resorption

While osteoblasts and osteocytes are adding new bone to the outer surface of the bone, osteoclasts are removing bone at the inner surface and enlarging the marrow cavity. These osteoclasts are very large cells with up to 50 or more nuclei. Mononuclear osteoclast precursors proliferate in the bone marrow and their fusion into multinuclear cells appears to take place closer to their site of action in the bone. Much of the hormonal regulation of osteolysis is through modulation of osteoclast activity, but the interaction between osteocytes and osteoclasts also is important in regulating osteoclast activity. Osteoclasts make contact with the bone through a region referred to as *ruffled border* and subsequently have the ability to degrade both inorganic as well as organic matrices of bone (Figures 8.7 and 8.8). The degradation of the inorganic matrix is accomplished by solubilization of crystalline hydroxyapatite crystals. Osteoclasts appear to be capable of creating an acidic environment by pumping *carbonic anhydrase Type II* across the ruffled border. The resulting decrease in pH in the local region is key to the solubilization process. This process results in mobilization of minerals, and is important in regulating chronic concentrations of serum minerals in the animal. Degradation of the organic matrix is accomplished with various enzymes, including collagenase, and lyosomal and non-lysosomal proteinases.

Growth in bone diameter occurs by appositional growth and is a delicate balance between addition of matrix on the outside and its removal on the inside. Marrow cavities that are formed as a result of osteoclast activity during bone development are filled with marrow which is composed of a soft connective tissue. Two types of marrow may be present. *Red marrow* is the chief blood cell forming organ of the adult and is found in spongy bone of the epiphyses of long bones and the main bodies of the sternum, ribs, and vertebrae. *Yellow marrow* is composed mainly of adipose tissue and is found in the medullary cavity of bones.

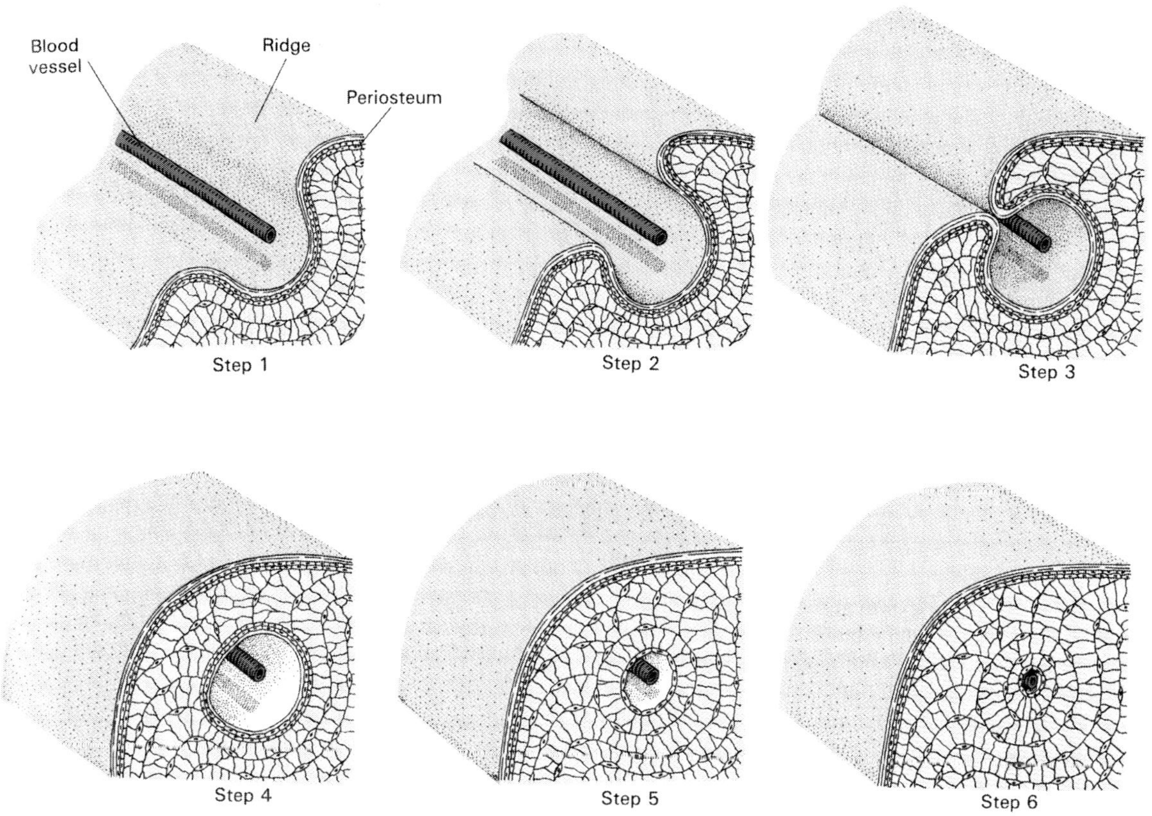

Figure 8.6 Appositional bone growth. From *Fundamentals of Anatomy and Physiology, 3rd Edition* by Martini, © Reprinted by permission of Pearson Eduction, Inc., Upper Saddle River, NJ.

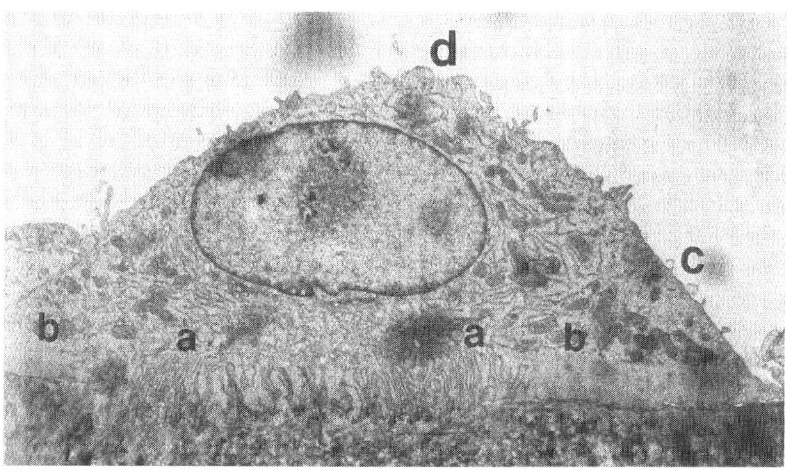

Figure 8.7 Micrograph (transmission microscopy) of a bone resorbing osteoclast. a, ruffled border; b, sealing zone; c, basement membrane; d, functional secretory domain. Figure from *Principle of Bone Biology* by John Bilezikian et al., copyright 2002, Elsevier Science (USA), reproduced with permission for the publisher.

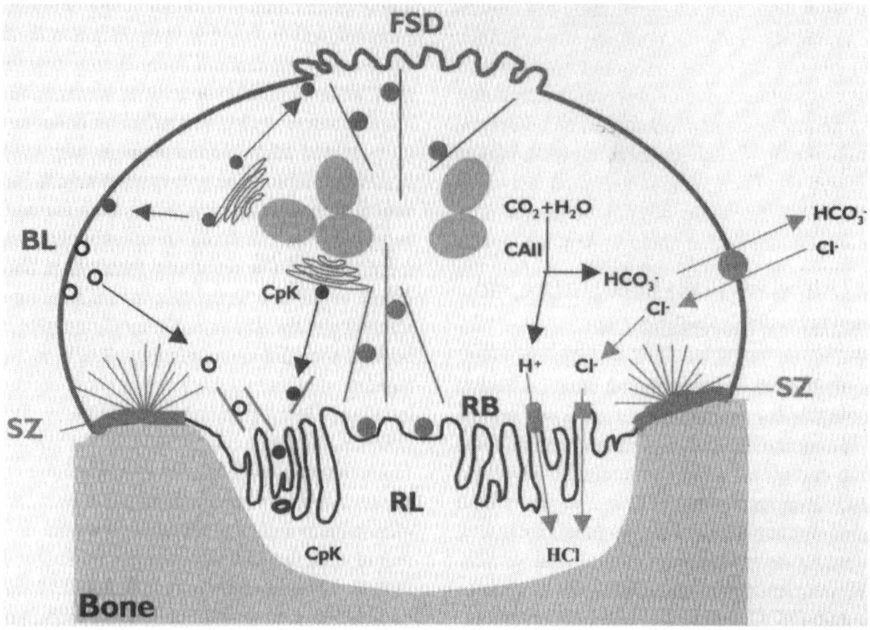

Figure 8.8 Action of a bone resorbing osteoclast. BL, basolateral domain; CAII, carbonic anhydrase II; CpK, cathepsin K; FSD, functional secretory domain; RB, ruffled border; RL, resorption lacunae; SZ, sealing zone. Figure from *Principle of Bone Biology* by John Bilezikian et al., copyright 2002, Elsevier Science (USA), reproduced with permission for the publisher.

Bone Remodeling

The deposition and resorption processes remodel bone so that the size and shape of the skeleton is changed during growth (Figure 8.9). Such changes are necessary for the skeleton to remain functionally competent. Osteoblasts form osteons by *osteogenesis,* whereas osteoclasts destroy osteons by resorption. The balance between these two processes determines net bone deposition. During growth processes described above, bone formation exceeds bone resorption. Once bone has reached the plateau of its accumulative growth curve, bone mass remains relatively constant and formation and resorption are in an appropriate balance. However, as animals age, bone resorption may exceed formation, and in such cases, bone mass begins to decrease. Throughout all of these stages of development, remodeling of bone can be quite extensive on location. For example, up to 18% of the mineral components of some bones are removed and replaced each year. The spongy bone of the femur head may be replaced two to three times per year, but compact bone of the bone shaft remains largely static.

Bone remodeling provides a mechanism for bone to adapt to new stresses, inducing heavily stressed bones to become thicker and stronger. Remodeling is usually initiated through osteoclast activation for removal of bone matrix and is followed by osteoblast formation for the production of new bone. Osteoblasts respond to the stress on bones and deposit additional bone in locations of stress to make bones stronger. In fact, osteocytes may serve as *mechano-receptors* that mediate stress-induced changes in bone formation. For example, increased force of muscles will result in increased bone growth at the *protuberances* where tendons connect to bones. The amount of force or stress on the bone appears to be closely related to the degree of mineralization and porosity, and also affects the orientation at which collagen fibers are deposited.

Various methods of measuring bone modeling are available and are classified as either *biochemical* or *histomorphometric techniques.* Biochemical markers, such as serum alkaline phosphatase and

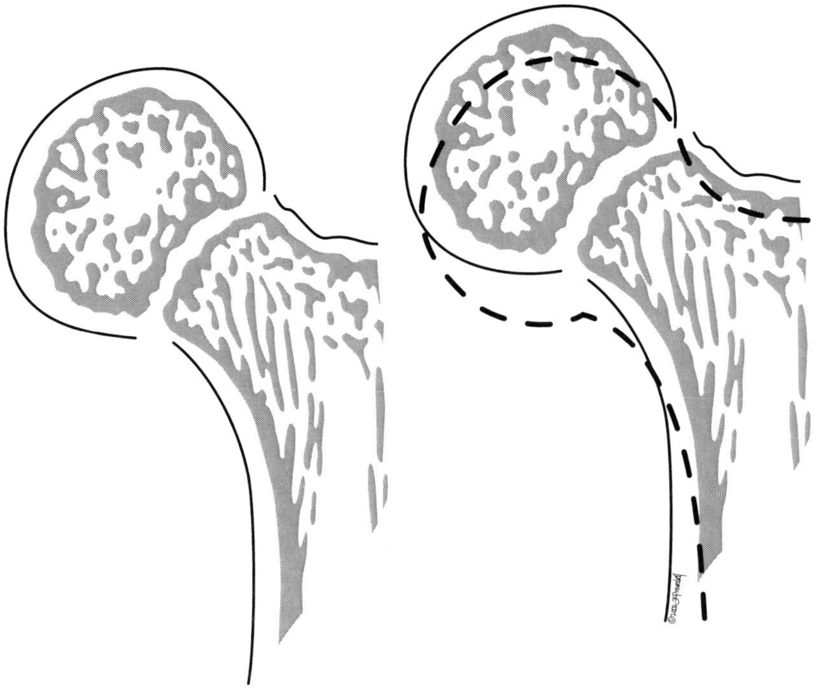

Figure 8.9 Regional changes in cartilage and bone that create increased length and width in long bones. Courtesy of Faye C. Prevedell.

osteocalcin, often are used to assess overall bone activity. Bone histomorphometric assays provide additional information on changes in bone structure and bone cell function. Fluorochrome compounds, such as tetracycline and calcein green, bind to calcium at sites of mineralization. Administering these compounds to animals and subjecting bone sections to histological analyses allow measurements of mineral apposition rate and new bone formation rate and may be used in conjunction with measurements of osteoid volume and trabecular bone volume, thickness, number, etc. These techniques allow investigation of developmental changes in bone formation, as well dietary and hormonal effects on formation.

Factors That Affect Bone Growth and Remodeling

Much of bone remodeling is the result of hormone action. For example, the coordinated actions of *calcitonin* and *parathyroid hormone* have major roles in remodeling (Figure 8.10). Calcitonin is a hormone secreted by the thyroid gland that decreases osteoclast activity resulting in decreased mobilization of calcium from bone and consequently decreased blood calcium concentrations. Parathyroid hormone (PTH) is secreted by the parathyroid gland and acts on bone, kidney, and the gastrointestinal tract to increase blood calcium concentrations. Its action on bone includes an increase in osteoclast activity. Many of these factors are mediated through other signals from osteoblasts and osteocytes, and often mobilization of calcium out of bone via osteocytes is called *osteocytic osteolysis*. PTH also participates in the regulation of phosphorus metabolism. Thus, bone plays a significant role in homeostasis of calcium and phosphorus and is an integral component of the endocrine system.

Other hormones, especially *estrogen* and *testosterone,* have dramatic affects on the rate of epiphyseal closure. Estrogen and testosterone hasten the ossification of epiphyseal cartilage (the mechanisms are described in a later chapter). As stated above, once epiphyseal plates ossify, additional growth of

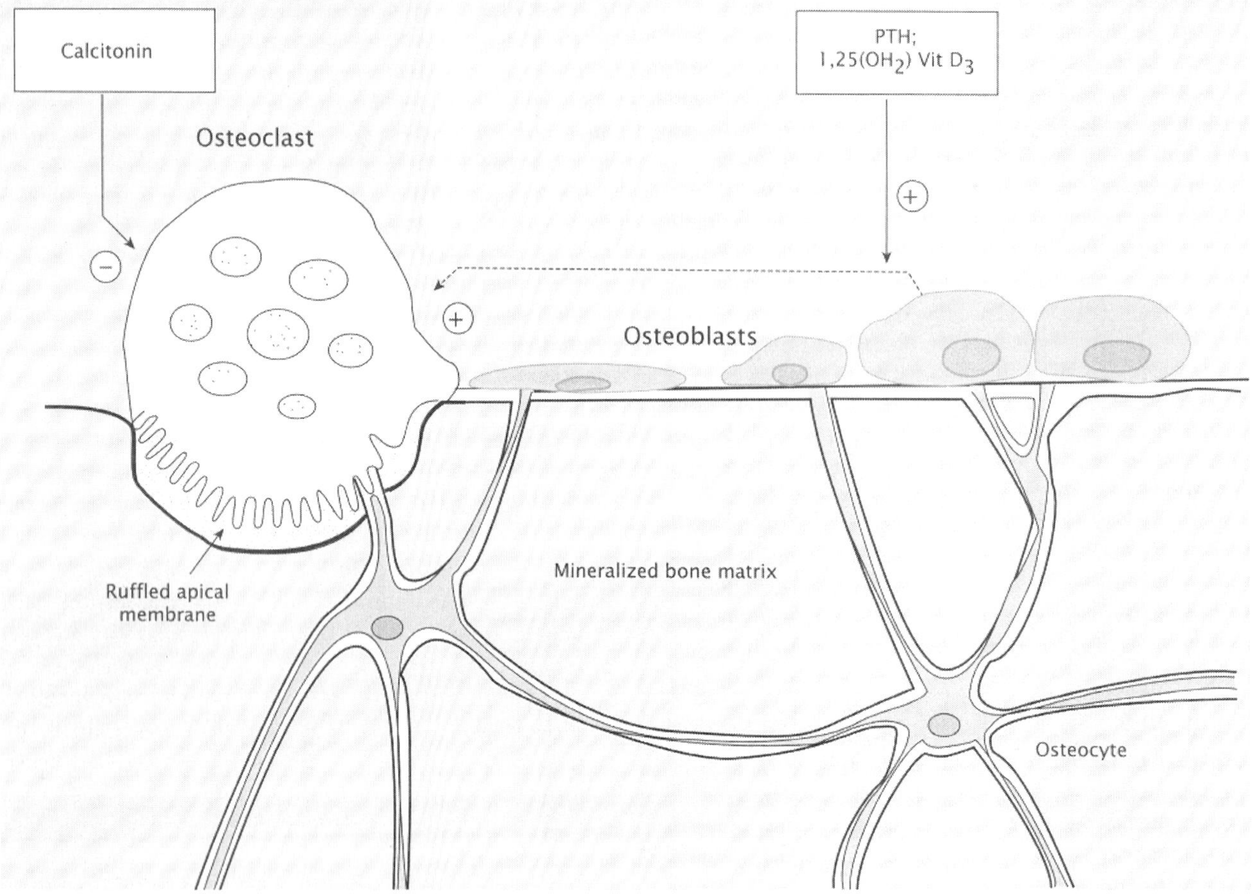

Figure 8.10 Bone remodeling by osteoblastic and osteoclastic cell activity. *Osteoclasts* are responsible for bone breakdown and de-mineralization (resorption) of mature bone, whereas *osteoblasts* form new bone and mineralize it. Parathyroid hormone (PTH) and 1,25-$(OH)_2$ vitamin D_3 stimulate osteoclastic bone resorption indirectly by facilitating release of osteoblastic cytokines (IL-6; GM-CSF), whereas calcitonin has a direct inhibitory effect on osteoclast activity. From *Basic Endocrinology* by Constanti, Bartke and Khardori. (Gordon and Breach, 1998) Reprinted by permission of Taylor and Francis Books.

long bones can only occur in diameter. Because growth plate closure is modulated by sex hormones, castration leads to continued growth in the length of long bones and results in larger, taller animals. Certainly, this retained skeletal growth ability of castrate males or females does not continue indefinitely and growth plate closure will ultimately occur from the presence of other hormones. Estrogens are more effective in mediating the closure of the growth plate than testosterone and as a result, intact females are generally smaller statured than their intact male counterparts. On the average, growth in the length of long bones is greatest in castrates, intermediate in intact males, and least in intact females. Other factors, such as nutrition and management practices that affect the mature size of farm animals, may do so through delaying the onset of sexual maturity.

Obviously, nutrient requirements for bone development are considered when formulating diets of livestock and poultry. Vitamin and mineral (even trace mineral) deficiencies have resulted in abnormal bone growth. The relationship between plane of nutrition and abnormal bone development may involve the effect of nutrition on overall body growth rate and the proportional accretion of bone, fat, and muscle. Management strategies that uncouple the balance among bone, fat, and muscle to a point where an immature skelton must bear excess muscle and (or) fat mass, exert detrimental force on the skeleton, including articular and growth plate cartilage. In some cases, restriction of feed intake, and consequently average daily gain, has reduced the incidence of osteochondrosis in mammals and birds. Potential economical losses and well-being issues associated with skeletal problems must be considered when developing strategies that enhance growth rates and carcass yields of meat-animals.

Many of the effects of nutrition on bone growth are likely mediated by hormones and growth factors. For example, altering the ratio of n-6 and n-3 polyunsaturated fatty acids in diets and feeding of fatty acid supplements, such as conjugated linoleic acid, may have significant effects on bone metabolism, and potentially bone formation, through changes in the insulin-like growth factor-I axis and prostaglandin synthesis pathways. Regulation of long bone growth by hormones and growth factors, including IGF-I modulation (one of the most abundant growth factors in bone), will be discussed in a later chapter.

Chapter 9

Whole Body Animal Growth

Introduction

At the most elementary levels, growth is an increase in size and is usually expressed relative to a given period of time. We may relate to growth in these terms because we have all experienced an increase in body size during our lives. However, the growth process is complex because animals grow at different rates, different adult sizes are common, and body parts grow at different relative rates. Rate of body growth often has an impact on the composition of the body. Composition of the weight gain is particularly important in domestic livestock that are used as a source of high quality protein food. Factors such as nutrition, genetics, environment and health status may affect the rate of total body growth as well as the composition of weight gain. In order to assess growth, an understanding of the various methods of expressing growth is necessary.

Growth Curves

Sigmoidal Curve

A typical growth curve is shown in Figure 9.1 and exhibits a characteristic *sigmoidal* shape. The x-axis is usually represented by units of time, whereas the units of the y-axis may be quite variable. The y-axis often is used to illustrate weight e.g., grams, kilograms, pounds; however, other measures of size may be used as well e.g., volume and length.

The sigmoidal curve is *accumulative* or *cumulative,* meaning that the weight of the animal is expressed as the total at any given time. Growth of all animals is sigmoidal when time and growth units are used as variables. The shape of the curve is different among species, breeds, sexes, etc. For example, those animals with large mature weights require more time to attain a mature weight. Furthermore, the curve may not be as smooth as shown in Figure 9.1, but may be stair-stepped in shape (Figure 9.2) as a result of the growth "spurts" that may occur at sporadic intervals.

Sigmoidal curves are useful for studying growth and can reveal much information about the dynamics of animal growth. During prenatal growth, it is clear that early stages are characterized by much slower growth rates than later stages (Figure 9.3). Water, as a percent of the total fetal weight, decreases because there is a concomitant increase in amount of other tissue components rapidly accumulating in the fetus. These changes accompany many cellular changes that are necessary for facilitating early tissue growth. A graph of both prenatal and postnatal growth (weight vs. time) illustrates different phases that occur from conception through maturity (Figure 9.4). Key features of a cumulative growth curve are:

- Conception
- Birth
- *Self-Accelerating Phase*
- *Inflection Point*
- *Self-Retarding Phase*
- *Maturity* (Figure 9.4).

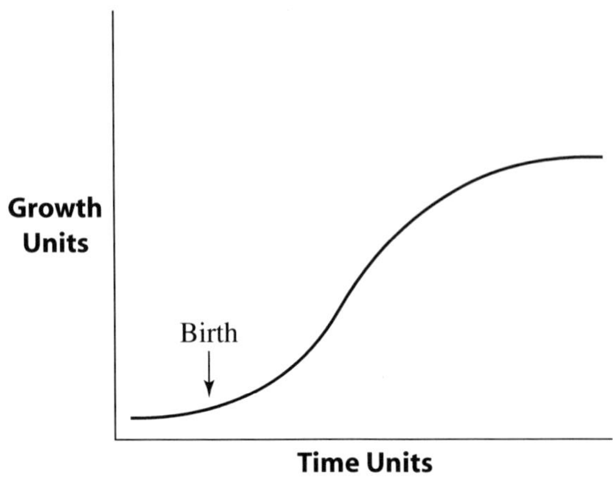

Figure 9.1 Typical growth curve. From *Livestock & Carcasses,* by Boggs et al. Copyright © 1998 by Kendall/Hunt Publishing Company. Reprinted by permission.

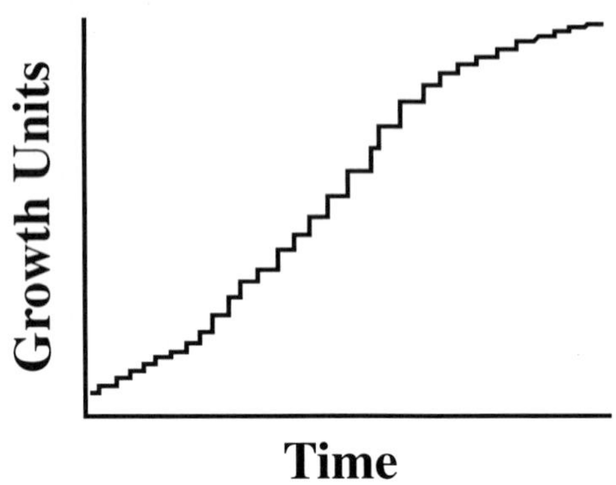

Figure 9.2 Non-smoothed growth curve illustrating the actual "stair-stepped" growth occurring in most animals.

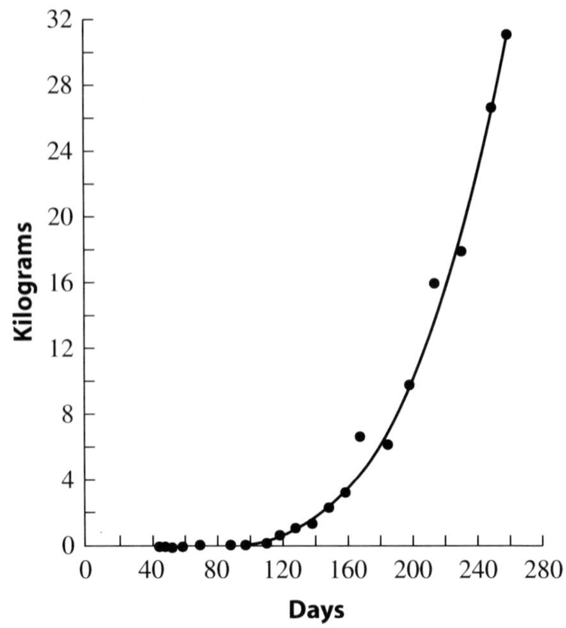

Figure 9.3 Prenatal weight growth of the bovine fetus.

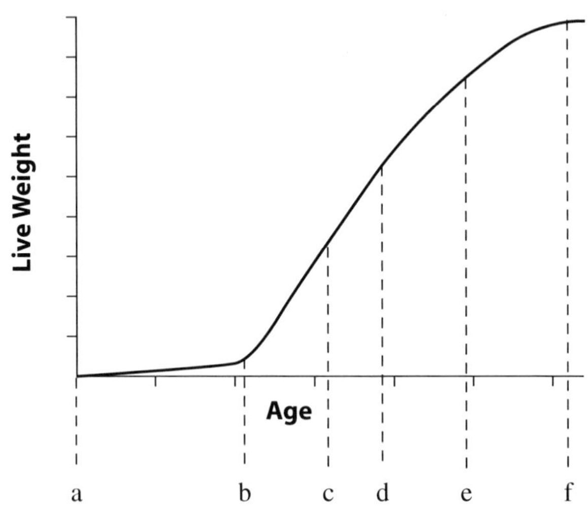

Figure 9.4 Identification of major points along a typical growth curve for growth in sheep: conception (a), birth (b), self-accelerating (c), inflection point or puberty (d), self-retarding phase (e) and maturity (f). From F. N. Owens, P. Dubeski, and C. F. Hanson, Factors that alter the growth and development of ruminants, *Journal of Animal Science* (1993), vol 71, pg. 3138.

The self-accelerating phase is the period of time in which an animal has reached a level of complexity that enables it to grow rapidly. For example, in ruminants this occurs after the digestive system develops to handle greater nutrient intake, digestion, and utilization. At some point after initiation of this phase, an inflection point in the exponential portion of the growth curve is attained. This point often coincides with puberty. At puberty, nutrients are directed

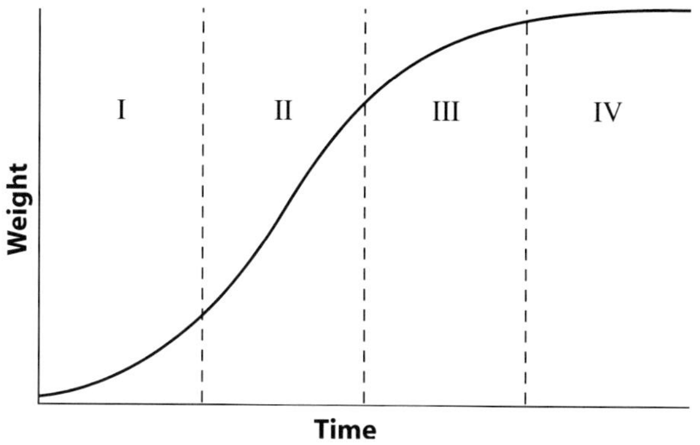

Figure 9.5 Alternative means of segmenting a typical growth curve.

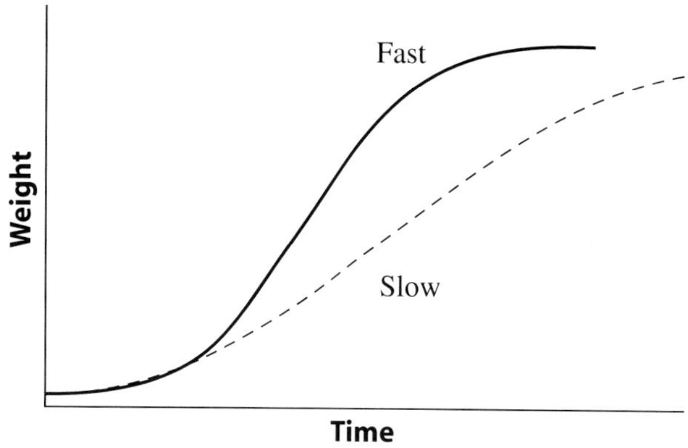

Figure 9.6 Growth curves of fast and slow growing animals.

toward maturation of reproductive tissues and other tissues not intimately associated with body growth *per se,* which tend to slow or change the slope of a growth curve. The self-retarding phase, which occurs shortly after, is the period during which growth begins to slow. Reasons for this are not readily known, but are likely related to a reduction in ability to deposit body protein.

Another way to illustrate phases of growth is to divide the postnatal growth curve into four distinct phases (Figure 9.5). General characteristics of the phases are as follow:

- **Phase I** represents 15 to 20% of total growth; slow growth of all tissues; tissue growth order is organs > bone > muscle.
- **Phase II** represents ~75% of total growth; organs reach mature size; bone growth completed; muscle growth maximal; fat accumulates slowly.
- **Phase III** 80 to 90% of growth attained; 80 to 90% of muscle deposited; rapid accumulation of fat begins.
- **Phase IV** 90 to 95% of additional growth is fat, 5 to 10% of gain is muscle.

Age and Maturity

The *chronological age* of an animal refers to the age in absolute time units e.g., days, weeks, months, or years. Used for plotting growth data, chronological age leads to some misunderstanding of growth processes of different animals. All animals within a species, breed, or sex do not grow, develop, or fatten at the same rate, or mature at the same chronological age. Many factors can alter the growth process causing the resulting growth curve to assume a much different shape. For example, if two pigs from the same litter are isolated and one animal is fed a diet that restricts maximal growth while the other is fed to achieve maximal growth, it is obvious that the latter would not reach the same weight as the former at the same chronological age (Figure 9.6). However, the animals may have the same physical and chemical characteristics when they finally reach the same weight, as illustrated by the same ending point on the two curves. The only difference is that the growth-restricted animals would take more time to reach the same endpoint i.e., the growth curve is shifted to the right. Therefore, great caution must be taken when studying data and interpreting data collected over time.

Physiological age is used to overcome some of the drawbacks of using chronological age as a point to compare different animals or populations. Physiological age refers to a stage of development that may be discerned by specific physical or chemical stages of development or function, such as height, weight, composition, or puberty. Animals may reach their definable stages at vastly different chronological ages. Physiological age is sometimes used as a means to establish *maturity.* Maturity is defined as the point in time when an animal reaches its highest level of complexity or development. This is a rather vague definition and is subject to interpretation, and it is necessary to clarify the type of maturity being studied. Animals consist of many different systems that collectively function to maintain the productivity of the organism. Each of these systems

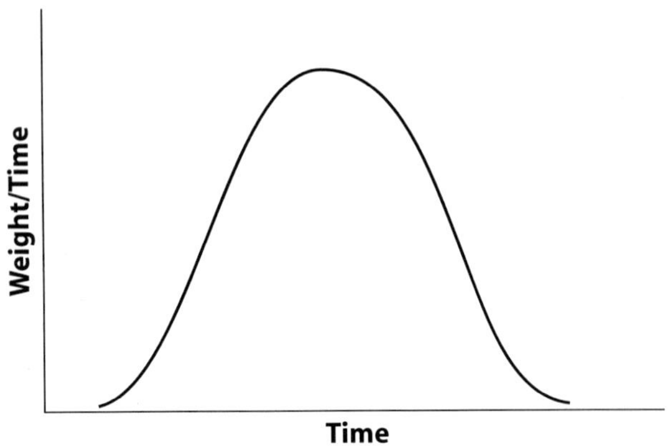

Figure 9.7 Absolute growth rate (weight/unit time).

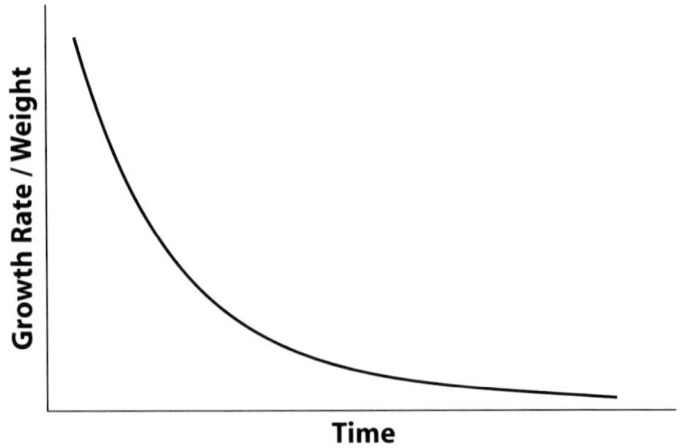

Figure 9.8 Relative growth rate (growth rate/unit weight).

or characteristics mature or reach their highest level of complexity at different chronological ages. Often scientists use the time at which a given system matures i.e., *physiological maturity* to make observations rather than using an absolute or chronological age. One example is *sexual maturity,* the age at which an animal reaches the ability to successfully reproduce.

Another means of comparing animals in growth studies is by using *chemical maturity,* the time at which animals reach the same *compositional maturity* or *endpoint.* In other words, comparisons of growth rates or days to market are made among animals having the same body composition. This will be developed further in subsequent sections in this chapter, but it is important to understand that use of chronological age has limitations and that physiological age and compositional maturity may be more appropriate at times for evaluating growth and development of animals.

Absolute Rate Curves

Another way of expressing animal growth is by plotting *absolute growth rate* i.e., gain per unit of time, against time (Figure 9.7). The most common example of this in the animal industry is *average daily gain* or weight gain per unit time. The shape of this curve is very different from that of the accumulative growth curve. In the case of absolute growth, one can readily observe that growth rate is very low early, increases to a maximum (corresponding to the time at which the slope of the accumulative growth curve is at its steepest point), and declines thereafter.

Relative Growth Rate Curves

Relative growth rate growth curves, as the name implies, are used to show growth in relation to total weight i.e., growth rate per unit weight plotted against time (Figure 9.8). These types of curves can be used to assess whole body growth or individual tissue growth. Relative whole body growth rate is very high during the initial phases of an animal's life primarily because the animal is so small. However, as the animal grows, the relative growth rate decreases rapidly.

All of the above means of depicting growth are related, as weight and age are the primary inputs. To appreciate the relationships among these attributes, it is helpful to visualize these curves on the same graph (Figure 9.9). The most striking result from this manipulation is that relative growth rate is, by and large, the greatest during the earliest phases of growth when the animal is the smallest and the absolute rate of growth is lowest. Conversely, when the animal has reached its maximal growth units, minimal relative and absolute growth is occurring. Both of these events are easily understood because changes in growth units are minimal. However, during the steepest portion of the accumulative growth phase, maximal absolute growth rate is observed. Relative growth rate during this phase decreases dramatically because there

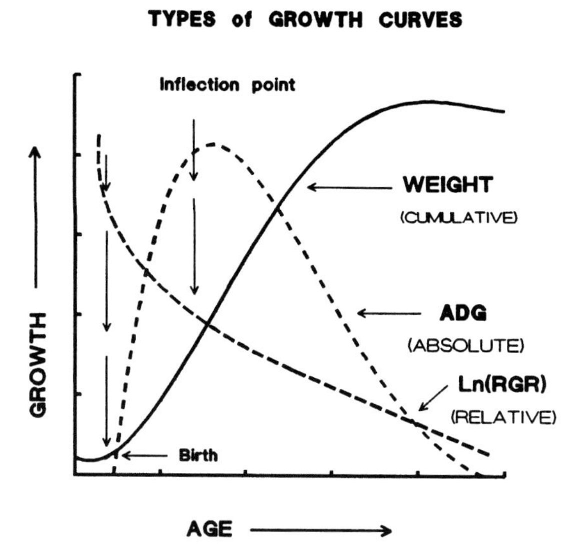

Figure 9.9 Relative changes in animal weight (cumulative growth), absolute growth rate and relative growth rate over the lifetime of an animal. From *Proceedings: 21st Century Concepts Important to Meat-Animal Evaluation* by R. G. Kauffman, editor. Copyright © 2000 by University of Wisconsin, Dept. of Meat & Animal Science. Reprinted by permission.

is a significant accumulation of growth units (weight) during this phase.

Changes in Body Form

Dramatic physical changes can easily be observed during growth of any animal. Figure 9.10 shows that the size of the head in relation to the rest of the body changes during the lifetime of a human. These same types of changes occur in all animals. Such changes manifest themselves differently depending upon the way the body and its various components are studied. Those components of the animal that are more important physiologically tend to grow and develop earlier (Figure 9.11a). From a gross anatomical standpoint, for example, the head of an animal is more fully developed at birth compared with other aspects of the body. Development of the head is followed by the neck, then the thorax, and finally the loin to their maximal stage of development. Clearly, an animal can survive with less total mass in the area of the loin, as long as its head attains the appropriate normal size for vital functions. This same type of priority occurs for other components. Nervous tissue of the brain, for example, is more critical than bone tissue, so the nervous system develops earlier than the skeletal system. Similarly,

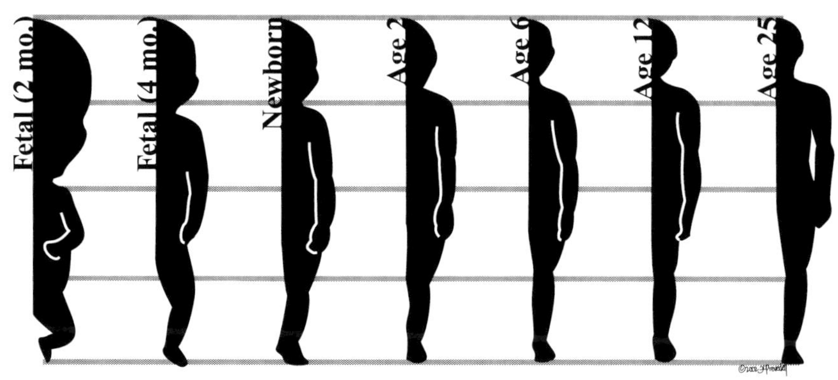

Figure 9.10 Relative changes in body proportions during growth. Courtesy of Faye C. Prevedell.

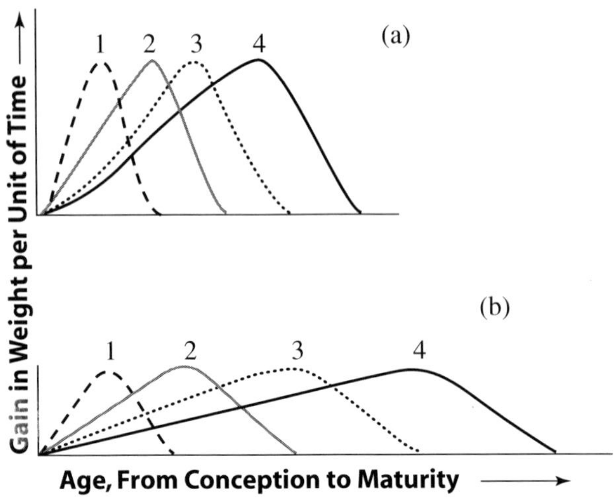

1	2
Head	Neck
Brain	Bone
Metatarsus	Tibia-fibula
Kidney Fat	Intermuscular Fat

3	4
Thorax	Loin
Muscle	Fat
Femur	Pelvis
Subcutaneous Fat	Intramuscular Fat

Figure 9.11 Growth rates of various tissues and proportions of the body of animals fed for rapid (A) or slow growth (B). From F. N. Owens, P. Dubeski, and C. F. Hanson, Factors that alter the growth and development of ruminants, *Journal of Animal Science* (1993), vol 71, pg. 3138.

bone tissue is earlier-maturing than skeletal muscle reflecting its physiological importance earlier in life. White adipose tissue or fat is less important to survival of the animal and is less developed during early stages of life.

To reinforce the idea that portions or parts of animals develop and grow differently, it is useful to understand that there is priority of development, even within tissues of the body. For example, individual bones or groups of bones grow at different rates during growth. In particular, the cannon or distal leg bones of most animals develop earlier than bones closer or more proximal to the body. This distal to proximal order of development of bones is physiologically relevant as young animals need greater development of the lower leg to attain immediate and necessary mobility after birth. Various depots within adipose tissue also exhibit different rates of development. Those depots surrounding vital organs (visceral fat deposits) develop earlier than depots less crucial for survival (intermuscular, subcutaneous and intramuscular). Time required for various tissues or body parts to develop fully varies, but because of their inherent priorities, the order of development never changes (Figure 9.11).

Changes in Body Measurements

Just as body form changes, so do measurements used to quantify different components of growing animals (Table 9.1). These measurements are particularly useful because many are predictors of composition and are useful for understanding how different tissues change during the growth process.

Carcass Length

Carcass length has been used in some pork carcass merit programs because longer carcasses tend to produce higher percentages of lean tissues. Although carcass length increases with age, if it is expressed relative to live or carcass weight, this value decreases with age (Table 9.1). Carcass length in pigs is the distance between the most anterior portion of the *aitch bone* (portion of the pubis of the pelvis that is exposed when the carcass is split along the median plane) and the most anterior portion of the first rib (Figure 9.12). This measurement is indicative of bone growth and reflects the overall size of the skeleton. An animal will attain a large proportion of its skeletal size before reaching maximal weight because the majority of bone growth occurs earlier than that of other tissues. As a result, any measurement that directly or indirectly evaluates skeletal growth will likely decrease with time when normalized to (or divided by) the weight of the animal.

Longissimus Muscle Area

The *longissimus dorsi* muscle is among the largest muscles of the carcass from livestock species and its cross-sectional area (loin eye area or ribeye area) or its diameter (loin depth) is often used as a predictor of muscling or muscularity. Absolute cross-sectional area of the longissimus muscle increases during the lifetime of an animal. If, however, the longissimus muscle area is normalized to carcass or

TABLE 9.1 COMPARISON OF LINEAR MEASUREMENTS, LEAN CUTS, AND CARCASS FAT OF PIGS AT VARIOUS LIVEWEIGHTS

Live weight. kg (lbs)	Carcass weight kg (lbs)	Carcass length		Backfat thickness		Longissimus muscle area		% of Carcass	
		cm (in)	per 100 lb.	cm (in)	per 100 lb.	10th rib cm² (in²)	per 100 lb.	Yield of four lean cuts	Chemical fat
55.8 (122.8)	38.8 (85.4)	66.5 (26.12)	30.8	1.75 (.69)	.81	22.0 (3.42)	4.0	63.8	20.8
75.8 (166.8)	54.8 (120.6)	73.0 (28.7)	23.7	2.34 (.92)	.76	26.3 (4.07)	3.4	60.4	26.7
90.1 (198.2)	68.2 (150.0)	75.9 (29.9)	19.9	2.81 (1.11)	.74	30.4 (4.71)	3.1	60.1	31.7
106.8 (235.0)	81.2 (178.6)	80.1 (31.5)	17.6	3.23 (1.27)	.71	35.5 (5.50)	3.0	59.5	34.3
127.4 (280.3)	99.7 (219.3)	84.8 (33.4)	15.3	3.91 (1.54)	.70	33.4 (5.17)	2.4	53.8	44.5
146.0 (321.2)	115.5 (254.1)	87.0 (34.3)	13.5	4.19 (1.65)	.65	38.3 (5.94)	2.3	52.4	45.2

Shields et al. (1983) J. Anim. Sci. 57:43.

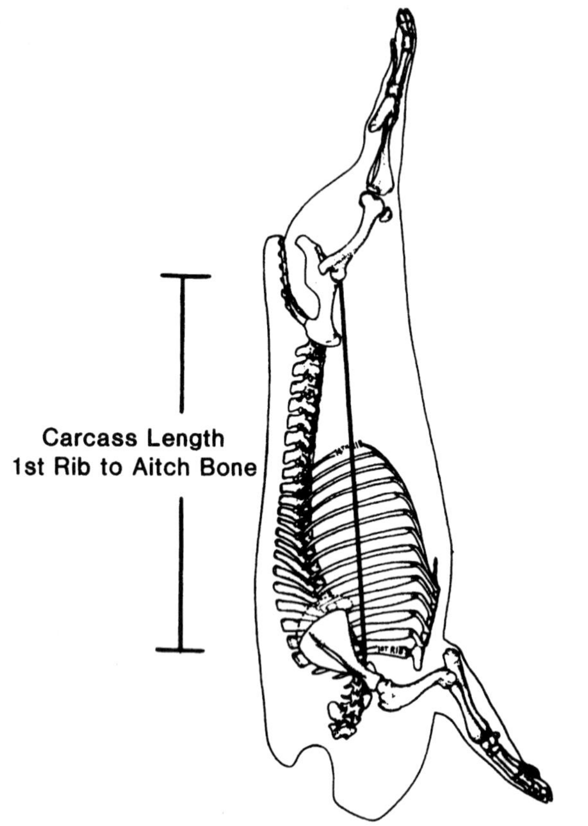

Figure 9.12 A skeletal diagram showing the anatomical locations used to evaluate carcass length.

live weight, this value decreases over time (Table 9.1) demonstrating that deposition of muscle during the latter phases of animal growth is not proportional to that of other tissues and that muscle is being diluted by deposition of other non-muscle tissues.

Subcutaneous Fat Thickness

Another measurement used as an index of composition in most meat animals is subcutaneous fat thickness. In the aforementioned indicators of carcass or animal muscularity, measurements to predict carcass adiposity usually increase with age (Table 6.1). When normalized or adjusted for carcass or live weight, backfat thickness decreases only slightly with age. Again, this is because no single tissue is capable of keeping pace with the combined growth of all other tissues of the body. However, because fat is a later developing tissue than muscle, fat deposition increases at a greater rate and thus, results in less change when normalization to weight.

Changes in Body Components

Taking into account the proportions of different body compartments at the time of slaughter provides much information regarding the growth processes. Time related changes in these compartments may aid in understanding growth processes.

Body Weight

Evaluating weight gain was probably the first method used to evaluate an animal's ability to produce edible products but sole utilization of this variable is not without problems (Table 9.2). First, weight *per se* is not necessarily a good indicator of the economic value of animals. Animals, especially ruminants, have an extensive digestive system that varies in content depending on the state of feeding. The sheer volume of the rumen represents a large variable in the weight of the animal depending on the amount of rumen content or *fill* remaining at the time of slaughter. For example, cattle can each consume over 50 kg of roughages and water if allowed *ad libitum* or *free choice* access. Therefore it is standard practice to withhold feed and water for 24 hours prior to slaughter of cattle to secure shrunken weights.

Empty Body Weight

One means of accounting for digestive system content in animal research is by calculating *empty body weight*. This takes into account the weight of digestive tract contents at the time of harvesting. This is accomplished by flushing the contents of the digestive tract and weighing the empty tissues. This clearly is a mandatory practice if changes in live weight are to be used as end points in evaluation. From Table 9.2, it is obvious that empty body weight increases with live weight.

Dressing Percentage

During slaughter, portions of the animal are removed in a process called *dressing* the animal, which means removal of all internal organs, hide, feet and head (depending on the processing method). The endpoint of this process yields many

TABLE 9.2 DISTRIBUTION OF BODY COMPONENTS OF CATTLE DURING GROWTH AND DEVELOPMENT*

	Liveweight (kg)					
	111	**204**	**313**	**517**	**853**	**888**
	Age (months)					
Component	**3**	**5**	**10**	**19**	**44**	**47**
Percentage of empty weight ** to live weight	88.0	84.3	87.6	88.8	90.4	92
Percentage of carcass weight to live weight	54.2	53.7	56.5	60.5	65.3	69
Percentage of carcass weight to empty weight	61.6	63.7	64.5	68.1	72.6	74
Percentage of hide and hair weight to empty weight	10.5	8.2	8.4	7.4	5.9	6.2
Percentage of blood weight to empty weight	6.2	5.2	5.0	4.1	3.3	3.5
Percentage of stomach and intestines weight to empty weight	4.8	4.6	5.0	4.7	2.5	2.3
Percentage of liver weight to empty weight	1.8	1.8	1.5	1.2	0.8	0.8
Percentage of offal fat weight to empty weight	1.4	3.9	4.7	5.2	7.0	4.7

Moulton, C.R., P.F. Towbridge and L.D. Haigh. "Studies in Animal Nutrition, II. Changes in Proportions of Caracass and Offal on Different Planes of Nutrition." Mo. Agr. Exp. Sta. Res. Bul. 54 (1922).
* Cattle were fed *ad libitum.*
** Empty weight is the liveweight minus the intestinal tract contents.

products, the most important of which is the carcass consisting largely of muscle, bone, and fat. *Dressing percentage* is calculated by dividing the carcass weight by the shrunken live weight of the animal (Table 9.3).

Over time and with growth of an animal, dressing percentage increases, meaning that the amount of carcass (predominantly muscle, bone and fat) increases as a percentage of live weight. During later stages of growth, increased dressing percentage is due largely to increased fat accumulation. The majority of fat deposited near market weight occurs in depots associated with the carcass rather than in tissues that are removed during slaughter. This does not mean that fat is not deposited in and around the internal organs, but that the relative percentage of total fat being deposited during the later stages is directed to depots associated with the carcass (Table 9.2).

Obviously, the absolute weight of the carcass increases with increasing live weight. Similarly, the proportion of the animal that yields a carcass (dressing percentage) increases with live weight. Consequently, the non-carcass components, such as the hide, blood, stomach, intestines, liver, etc., decrease in proportion to live weight or age. Therefore, younger animals have a greater proportion of their body as slaughter by-products. The observation that non-carcass components decrease as a percentage of live weight over time implies that most all of these tissues reach their mature size earlier during growth than tissues of the carcass. In other words, after some point in the growth of an animal, growth of those structures removed during the dressing process (internal organs, head, feet, etc.) is nearly complete and additional gain in weight is far less than that occurring during earlier stages of the growth process.

CARCASS WEIGHT

From a meat-animal standpoint, carcass weight as a growth endpoint is highly desirable and is the cornerstone of most progressive marketing programs. Carcass weight is routinely used as primary endpoint along with other predictors of carcass composition. Even though animal slaughter practices around the world differ and may yield carcasses with differing anatomical specifications e.g., head on vs. head removed, the resulting carcass contains primarily muscle, bone and fat, which are the most economically important components of the live animal. As a result, variations in hide weights, which may vary from 25 to 50 kilograms, as well as the other non-carcass components may be removed as a source of biological variation. Carcass weight over

Table 9.3 Dressing Percentages of Various Kinds of Livestock by Grades*

Grade	Range	Average
Cattle		
Prime	62–67	64
Choice	59–65	62
Good	58–62	60
Standard	55–60	57
Commercial	54–62	57
Utiltiy	49–57	53
Cutter	45–54	49
Canner	40–48	45
Calves and Vealers**		
Prime	62–67	64
Choice	58–64	60
Good	56–60	58
Standard	52–57	55
Utility	47–54	51
Cull	40–48	46
Lambs (Wooled)		
Prime	49–55	52
Choice	47–52	50
Good	45–49	47
Utility	43–47	45
Cull	40–45	42
Barrows and Gilts***		
U.S. No. 1	68–72	70
U.S. No. 2	69–73	71
U.S. No. 3	70–74	72
U.S. No. 4	71–75	73
Utiltity	67–71	69

United States Department of Agriculture, Consumer and Marketing Service, Livestock Division.
* All percentages are based on hot weights.
** Based on hide-off carcass weights.
*** Based on packer style dressing (ham facings, leaf fat, kidneys, and head removed).

time also follows a classical growth curve demonstrating that a growth curve can be used for describing the growth of not only an animal but also a component of an animal.

Changes in Carcass Components

Although carcass weight is useful to express the growth of animals, it should be recognized that the composition of the carcass has a major effect on eco-

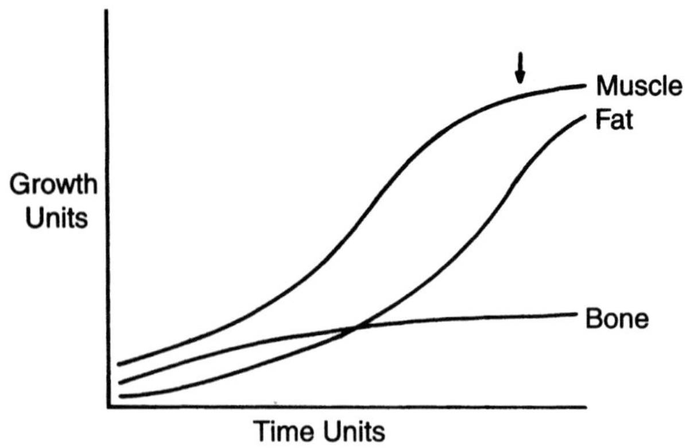

Figure 9.13 Postnatal growth curves of bone, muscle and fat. From *Livestock & Carcasses,* by Boggs et al. Copyright © 1998 by Kendall/Hunt Publishing Company. Reprinted by permission.

nomic value of the animal. This may best be illustrated by considering animals that have the same growth rate and ultimate live and carcass weights, but have significant differences in carcass composition. On one extreme, a carcass may be quite lean and heavily muscled while another may be extremely fat and thinly muscled. If these carcasses originate from animals with the same growth rate, one must conclude the animal possessing the leanest, heaviest muscled carcass is genetically superior because it deposited less fat and more muscle. Growth of individual carcass components (muscle, bone and fat) may be plotted against time to better characterize these differences. Collection of this type of data is laborious because it is necessary to dissect carcasses into muscle, fat and bone. However, resulting data are more valuable because they provide a detailed picture of how economically-relevant tissues are growing over time.

Figure 9.13 shows the accumulative growth of muscle, bone and fat in animals over time. As apparent from this figure, the curves for muscle and fat resemble classical sigmoidal growth curves. Of course, evidence of the true shape of these curves is somewhat elusive given the scale of the y-axis. Growth of bone is particularly difficult to visualize because it changes relatively little compared to the other tissues of the carcass. Nonetheless, bone weight increases at a decreasing rate. The skeletal system must be sufficiently developed at birth to accommodate the weight of the animal. However, after the skeletal system is developed, further skele-

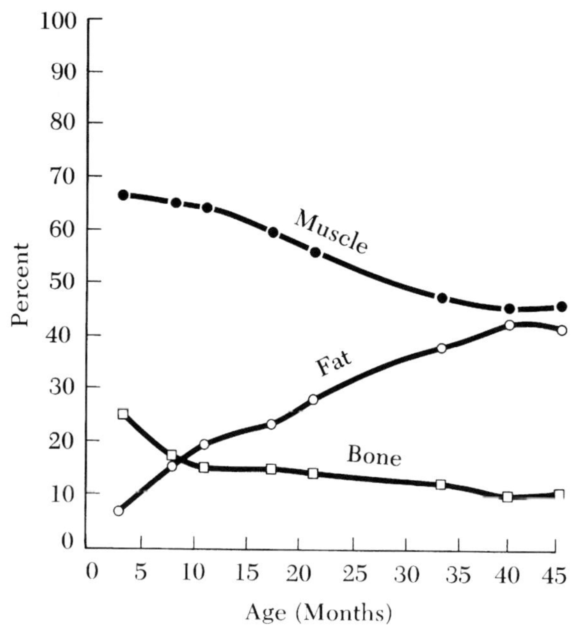

Figure 9.14 Changes in the percentage of bone, muscle, and fat in beef carcasses during growth. Data from Moulton, C. R., P. F. Trowbridge and L. D. Haigh, *Studies in Animal Nutrition;* "III. Changes in Chemical Composition on Different Planes of Nutrition," *Mo. Agr. Exp. Sta. Res. Bul.* 55 (1922). Courtesy of W. H. Freeman and Company.

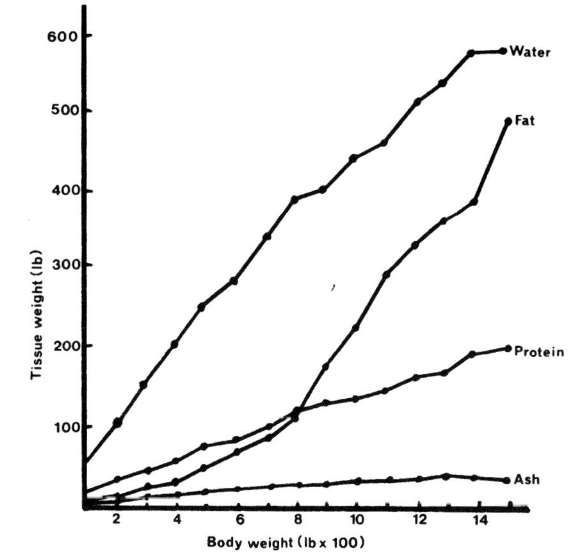

Figure 9.15 Chemical component changes with empty body weight of growing steers. From R. T. Berg and R. M. Butterfield, *New Concepts of Cattle Growth,* Sydney University Press, Sydney.

tal growth is minimal. In contrast, muscle growth increases dramatically and quickly becomes a large carcass component. Fat, on the other hand, is quite low at birth and remains low until muscle growth slows or plateaus. Whether muscle growth begins to slow because fat deposition increases or vice versa remains an intense area of research among animal scientists today. Nonetheless, maximal rates of bone, muscle, and fat deposition occur at different times during animal growth.

Another way to plot the growth of bone, muscle, and fat is by expressing the weight of each tissue as a percentage of total carcass weight. As is evident from Figure 9.14, there is a substantial difference in the shape of each curve when compared to those of absolute tissue weights. The shape of these curves ultimately determines the compositional maturity of the animal and its carcass. The percentage of the carcass represented by muscle tissue is quite high at birth and slowly decreases with time. At the same time, bone represents a relatively low percentage of the carcass and remains so throughout the course of the animal's life. However, fat tissue represents a very small proportion of the carcass initially, but becomes greater with age. The decrease in percent muscle and bone is due to this significant increase in fat percentage. Remember that absolute fat accretion (Figure 9.14) occurs even during the earliest phases of animal growth, but because total muscle weight as well as deposition rates are greater for muscle than for fat, the percentage of muscle far exceeds that of fat in early animal growth. Of course, during the later phases of growth when animals continue to consume feed and muscle growth slows, excess nutrients are deposited in the form of fat.

Changes in the Chemical Composition

Because bone, muscle, and fat are deposited at different rates, the chemical composition of an animal also changes with time and weight. Figure 9.15 shows the cumulative increase in water, fat, protein, and ash with body weight. Although the absolute amount of each chemical component increases with body weight, lipid increases and water and protein decrease as a percent of body weight (Figure 9.16).

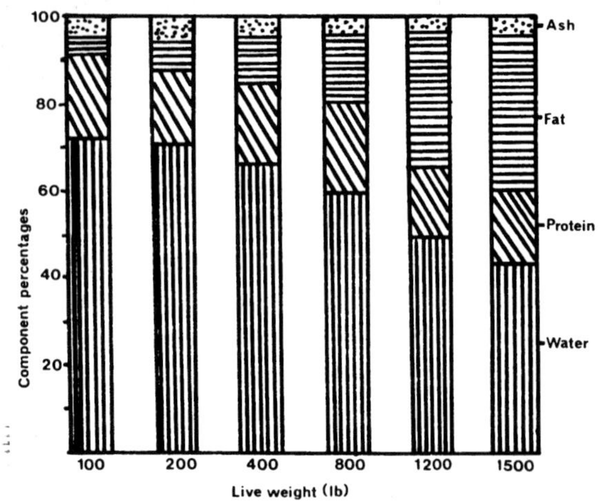

Figure 9.16 Changes in empty body composition in growing steers. From R. T. Berg and R. M. Butterfield, *New Concepts of Cattle Growth,* Sydney University Press, Sydney.

Water and Protein

Muscle ranges from 68 to 85% water on a weight basis. During the bulk of animal growth, the water to protein ratio in muscle stays near 4:1. This decreases somewhat with age because of increased percent protein and decreased percent water. Percent water is greatest at the earliest stages of muscle development, but decreases with the accumulation of other muscle components. Changes in myofibrillar and sarcoplasmic protein concentrations occur during the growth of muscle (Table 9.4). For example, *myoglobin,* a sarcoplasmic protein and the major pigment of muscle, increases in percentage with age and is responsible for the darker appearing muscle observed in older animals.

Adipose tissue, as a modified connective tissue, possesses much the same characteristics during the earliest stages of its development. The water content of connective tissue is quite high early in life and decreases with age. In contrast to muscle, the protein content of adipose tissue is the greatest early and decreases later in life. This difference is, of course, related to the impact of triglyceride deposition within the tissue which dilutes water and protein so that they decrease as a percentage of the total weight of the tissue (Table 9.5). Earlier developing adipose tissue depots, such as kidney depots, are more mature than later developing depots, such as intermuscular and subcutaneous depots; thus, kidney fat which may be > 90% fat has much lower protein and water percentages than the subcutaneous and intermuscular depots.

The same general changes in water content and protein content would be observed for bone as that of adipose tissue, because inorganic salts deposited in modified connective tissue to increase bone strength also dilute water and protein.

Connective Tissue

The absolute amount of connective tissue in animals increases with age. Accretion of collagen and elastin fibers must be sufficiently increased to form the myriad of fascicles of muscle bundles in adult muscle. Furthermore, the reinforcement of the myotendonal junctions requires substantial increases in connective tissue. However, the percent of connective tissue in muscle decreases with age because there is a dilution of connective tissue by muscle fibers and their associated proteins. Individual muscle fibers grow in diameter and length. As a result, greater and

Table 9.4 Changes in Concentration of Proteins in Mammalian Skeletal Muscle During Growth

	Immature*		Adult*	
Species	**Sarcoplasmic**	**Myofibrillar**	**Sarcoplasmic**	**Myofibrillar**
Pig	3.5	10.6	5.2	12.8
Cattle	5.0	8.4	5.3	13.0
Human	3.2	11.0	4.3	12.8
Rat	4.1	10.0	5.2	11.2

Bailey and Zobrisky, (1968).
*Grams/100 grams of tissue.

TABLE 9.5 PERCENTAGE CHEMICAL COMPOSITION OF FAT FROM DEPOTS OF STEERS

Depot	Water	Protein	Fat
Subcutaneous	17.23	6.10	76.69
Intermuscular	22.71	6.57	70.94
Kidney	5.48	1.11	93.39
Mesenteric	14.71	2.75	82.68

From R.T. Berg and R.M. Butterfield, New Concepts of Cattle Growth, Sydney University Press, Sydney.

greater proportions of the muscle are contractile proteins. As pointed out earlier, this dilution effect of muscle on connective tissue occurs only when muscle is still growing. When muscle begins to experience atrophy during senescence, the percentage of connective tissue in muscle may begin to increase.

The relative amount of connective tissue in bone and adipose tissue also is affected by the deposition of inorganic salts and lipid.

NUCLEOPROTEIN

Nucleoproteins are those proteins that are responsible for coordinating the structure and function of the nucleus. These proteins often are overlooked because they represent such a small portion of the total protein in skeletal muscle. The absolute amount of nucleoprotein increases with age, but there is a slight decrease in the percentage of these proteins in muscle because of the dramatic increase in muscle fiber size. The magnitude of the change is not as dramatic as one might expect because additional nuclei are continually added to the muscle during the growth phase through satellite cell recruitment.

In contrast to bone and muscle tissues, the contribution and maintenance of nuclei concentration in adipose tissue is influenced by DNA and nucleoprotein content which decreases with the influx of triglyceride deposition. This, of course, could be masked by recruitment of additional adipocytes, which occurs in the later stages of adipose tissue depot development.

FAT

Fat is one of the more variable constituents of many tissues and increases dramatically during growth. As an animal matures, fat accumulates in the various depots, including those of the muscle. Fat accumulation increases very little in compact bone, but substantial changes occur in the *intraosseus fat* which impacts the resulting fat content of the skeleton.

ASH (MINERALS)

The total amount of minerals in the body of animals increases with age. These changes are not, however, universal to all minerals. For example, the concentration of sodium decreases during muscle growth, while the concentration of potassium increases. Furthermore, increases in calcium and phosphorous are observed, while other minerals decrease.

Differential Allometric Growth

Another way of studying animal growth is by studying the *allometric growth* of various tissues or components. *Allometry* is growth of one component in relation to the whole. As pointed out earlier, tissues grow at different rates depending on when the bulk of growth occurs in relation to the whole body. Deriving these data is accomplished by dissecting and weighing individual tissues or components and plotting the log weight of the tissue or component against the log of the total carcass or live weight. Collectively, these data may be used to derive the following equation: $y = ax^b$, where y equals the weight of the specific tissue or component, "a" equals a constant, "x" equals the log weight of the total tissue carcass, or body and "b" equals the *growth coefficient* of the component. In mathematical terms, "b" dictates the slope of the line and whether it increases (positive) or decreases (negative) with the weight of the whole.

- If b is greater than 1.0, the tissue or component is growing at a faster rate than the whole and this is considered to have a *high growth impetus* (Figure 9.17).
- If b is less than 1.0, the tissue or component is growing at a slower rate than the whole and has a *low growth impetus.*
- If b = 1.0, the tissue or component is growing at the same rate as the whole and has an *average growth impetus.*

Generally, those tissues or components that have growth coefficients greater than one are considered *late developing* tissues. Fat is one component that is rather late developing because significant accumulation of body fat does not occur until muscle growth begins to slow. Fat has a high growth impetus when considered over the entire life of an animal because fat develops at a much faster rate in the later stages of animal growth. Total muscle has a growth impetus of about 1.0 and therefore, grows uniformly over the entire growth period.

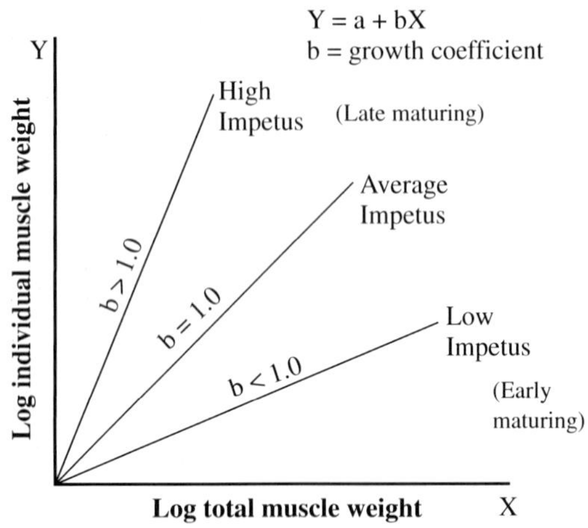

Figure 9.17 Theoretical representation of monophasic growth patterns. From R. T. Berg and R. M. Butterfield, *New Concepts of Cattle Growth*, Sydney University Press, Sydney.

Muscles may be loosely grouped according to their location within the body (Table 9.6) and this is useful to study the allometric growth of various muscle groups. Examples of muscles or groups of muscles that are late developing in cattle are those muscles associated with the abdominal wall, neck to thoracic limb, and those located in the proximal pelvic limb (Table 9.7). The abdominal wall muscles likely develop in response to functional demand created by the rumen while increases in the neck and forelimb are likely associated with increased weight-bearing duties as well as potential secondary sex characteristics. The proximal pelvic limb muscle development is simply a reflection of the increased growth experienced by the larger muscles of the hind leg, which are located more proximal in the leg.

In contrast to late developing muscles or tissue components, there are several *early developing* muscles or tissue components (Table 9.7). Bone is an example of a tissue component that is early developing or has a growth coefficient less than

TABLE 9.6 MAJOR MUSCLE GROUPINGS

Muscle Groups
Proximal pelvic limb
Distal pelvic limb
Proximal thoracic limb
Spinal column
Abdominal
Neck-thorax-thoracic limb

TABLE 9.7 GROWTH PATTERNS OF STANDARD MUSCLE GROUPS OF STEERS USING GROWTH COEFFICIENTS

Muscle Group	Growth Coefficients "b"	Relative Time of Development
1. Proximal pelvic limb	High-average or high	Late
2. Distal pelvic limb	Low	Early
3. Surrounding spinal column	Average	Average
4. Abdominal Wall	High-average or high	Late
5. Proximal thoracic limb	Low-average	Early
6. Distal thoracic limb	Low-average or low	Early
7. Thorax to thoracic limb	High	Very late
8. Neck to thoracic limb	Average-high	Very late
9. Neck and thorax	Low-average	Very late

From R.T. Berg and R.M. Butterfield, New Concepts of Cattle Growth, Sydney University Press, Sydney.

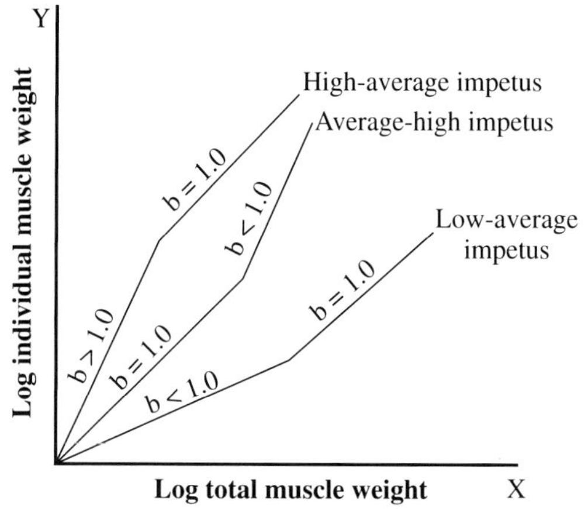

Figure 9.18 Theoretical representation of diphasic growth patterns. From R. T. Berg and R. M. Butterfield, *New Concepts of Cattle Growth,* Sydney University Press, Sydney.

one. Remember, this does not mean that bone is not growing. Of course, the weight of the skeleton increases several-fold from birth to maturity. However, incremental increases in the weight of bone are surpassed or dwarfed by the dramatic increases in other components such as muscle tissue and ultimately fat. Individual muscles or groups of muscles that are early developing are those located in the distal aspects of the thoracic and hind limbs. Other tissues having a growth coefficient approximately equal to 1.0 grow at a similar rate to the rest of the body. Muscles surrounding the spinal column grow similarly to the total muscle of an animal (Table 9.7).

Muscles or tissues that have growth coefficients that are constant over time are considered to be *monophasic* (Figure 9.17). However, growth coefficients may be generated for tissues over a longer period of time and therefore, may yield dramatically different and sometimes contrasting coefficients. These tissues are considered *diphasic* (Figure 9.18) and develop at two different rates. For example, the growth coefficient of fat during early postnatal growth is less than one but increases to greater than one as fat deposition dramatically increases. Expressing tissue growth in this manner is very useful and provides additional insight into how and when muscles or various components grow in relation to the total body.

Factors Affecting Carcass Composition

Genetics

Genetics is the science of *heredity,* which is the process of passing along genes from one generation to another. Genetic differences have come about through evolution, which can produce mutations in genetic material, or through intensive selection influences that favor a particular set of genes. The *genotype* of an animal refers to the unique set of genes that an animal possesses in its genome. Often, the biological connotation of this term is used to identify the specific and absolute DNA sequence contained in one allele. In this text the former definition reflects more accurately the broader nature of the term. The ability of an animal to fully express outwardly its genetic ability, or *genetic potential,* is often not attained. The main reason for this is that the physical surroundings, or *environment,* immediately surrounding the animal may limit its ability to reach its full potential. Environment does not necessarily impact the animal in a negative manner, as some environments can actually boost productivity and thus improve the outward expression of an animal's genetic potential. Ultimately, therefore, the outward expression of any trait of an animal results from the interaction between the genotype of the animal and the environment in which it resides. This outward expression of genotype is termed *phenotype.*

The contribution of the genetic component of an animal's phenotype can be quantified. These *heritability estimates,* which reflect the ability of an animal to pass a particular gene or set of genes to the next generation, are used by geneticists to develop selection programs that help seedstock producers improve productivity of their livestock. Traits associated with compositional performance are moderately heritable (Table 9.8), meaning that breeders can make moderate gains in composition if these traits are used in genetic selection systems.

Species

There is little question that carcass composition differs with species (Table 9.9). Species vary in:

- Composition
- Their ability to deposit different tissues
- The relative proportionality of each body part.

Table 9.8 Heritability Estimates of Carcass Characteristics of Cattle, Poultry, Sheep and Swine

Characteristics	Species	Heritabiltiy estimate* approximate average
Fat thickness	Cattle	38
	Sheep	20
	Pigs	49
Longissimus muscle cross-sectional area	Cattle	70
	Sheep	48
	Pigs	48
Percent carcass fat	Broilers	30

*Estimates were complied from various sources.
From *Principles of Meat Science, Fourth Edition* by Aberle et al. Copyright © 2001 by Kendall/Hunt Publishing Company. Reprinted with permission.

Table 9.9 Empty Body Composition

Species	N	Age	Water	Fat	Protein	Ash
Pigs	714	1–923	30.7–80.8	1.1–61.5	8.3–19.6	1.3–5.6
Sheep	221	90–805	39.6–73.8	4.9–46.6	10.7–19.5	1.7–5.8
Cattle	256	1–4,860	39.8–77.6	1.8–44.6	12.4–20.6	3.0–6.1

From Reid, et al. (1968), NAS Publ. 1598, pg. 20.

Evolution undoubtedly has played an important role in the development of species or *speciation.* Intermingled with evolution, however, are increased selection pressures, which are ultimately responsible for most of the species differences. On a more practical and useful note, as genetic companies and seedstock producers attempt to improve productivity further, it becomes increasingly evident that improvement of some economically valuable characteristics is not possible whereas improvement of others is rapid and quite fruitful.

Evolution, Selection and Domestication

Since early times, meat processors have known that some parts of carcass are more valuable than others. During animal domestication, management became more and more intensive. Animals were sheltered from the environment, fed special diets and raised in controlled, non-wild environments. Obviously, this process required thousands of years, but gradually the *selection pressure* for survival in the wilderness was attenuated and replaced by man-made pressures in which controlled matings led to improvements in the overall meat-producing ability of these domesticated animals. Therefore, the ability of the wild animal to run for extended periods of time to escape a predator was reduced, where as the animals capable of depositing greater and greater amounts of bulky muscle ultimately flourished. Furthermore, dietary changes allowed those animals capable of ingesting great volumes of feedstuffs to grow faster and these genotypes found favor in the eyes of breeders. As time passed, various species and breeds evolved into producers of protein. This process resulted in many economically important changes in the composition of meat producing animals. The results of these changes are helpful in providing insight into how much change may be expected through genetic selection.

Muscle Distribution

The effects of selection and domestication on muscle distribution are illustrated by comparing an uns-

TABLE 9.10 COMPARISON OF MUSCLE DISTRIBUTION (%) BETWEEN ADULT CATTLE AND DEER

	Weight of muscle group as % of total muscle weight			
Muscle Group	**Heifers**[1]	**Female deer**[2]	**Deer relative to heifers**	**Index***
1. Proximal pelvic limb	31.5	35.7	+4.7	113
2. Distal pelvic limb	4.3	5.8	+1.5	135
3. Around spinal column	12.1	14.7	+2.6	121
4. Abdominal wall	11.5	6.8	–4.7	59
5. Proximal thoracic limb	12.4	11.4	–1.0	92
6. Distal thoracic limb	2.3	3.1	+0.8	135
7. Thorax to thoracic limb	10.3	7.8	–2.5	76
8. Neck to thoracic limb	5.2	3.3	–1.9	63
9. Neck and thorax	9.0	8.9	–0.1	99

*Index = Deer value relative to cattle = 100
[1]Heifers mean total side muscle weight 51.0 kg N = 22
Source: Berg and Mukhoty, 1970.
[2]Female deer mean total side muscle weight 9.9 kg N = 3
[3]Angus × Charolais bull total side muscle weight 193.9 kg N = 1
[4]Mature male deer total side muscle weight 28.3 kg N = 1
Source: Berg, unpublished.
From R. T. Berg and R. M. Butterfield, *New Concepts of Cattle Growth,* Sydney University Press, Sydney.

TABLE 9.11 COMPARISON OF MUSCLE DISTRIBUTION (%) AMONG INTACT MALE BISON, DOMESTIC CATTLE AND SHEEP

	Bos taurus bull	**Bison bull**	**Merino Ram**
Proximal pelvic limb	28.4	30.7	24.9
Distal pelvic limb	4.3	4.2	4.4
Spinal column	12.4	12.7	15.5
Abdominal wall	9.7	6.9	10.6
Proximal thoracic limb	12.5	13.0	10.8
Distal thoracic limb	2.3	2.3	2.8
Thorax to thoracic limb	10.5	8.8	8.6
Neck to thoracic limb	5.5	8.7	7.6
Neck to thorax	12.5	11.4	12.4

Comparison of muscle distribution (%) among intact male Bison, domestic cattle and sheep. From R. T. Berg and R. M. Butterfield, *New Concepts of Cattle Growth,* Sydney University Press, Sydney.

elected white-tailed deer or bison with selected beef heifers. Data from detailed dissection studies reveal that the relative distribution of muscle has been altered with domestication (Tables 9.10 and 9.11). Unselected animals have a greater percentage of muscle located in the proximal pelvic limb i.e., those muscles located in the upper portions of the hindlegs. In the wild, there is likely a selective advantage for having a higher proportion of the total muscle mass in the proximal limb, perhaps by enhancing animal speed and agility. Proximal pelvic limb muscles are especially important in increasing speed of limb movement, a primary determinant in controlling animal speed. In most animals, especially those domesticated farm animals, the rear legs function to thrust the body forward, whereas the front legs merely act to oppose gravity and keep cranial aspects of the body raised. As a result, those animals most adept for living in the wild, are those with greater advantages in running and jumping. Therefore, even though bison and white-tailed deer are referred to as unselected in this example, wild

TABLE 9.12 MUSCLE WEIGHT DISTRIBUTION OF SEVERAL SPECIES EXPRESSED RELATIVE TO *BOS TAURUS*

Muscle groups	Bos taurus	Pig	Sheep	Water buffalo	Moose	Deer	Bison
1. Proximal pelvic limb	100	97	94	103	117	111	108
2. Distal pelvic limb	100	91	109	105	155	132	98
3. Around spinal column	100	141	139	79	96	122	102
4. Abdominal wall	100	103	97	79	58	56	71
5. Proximal thoracic limb	100	94	90	117	133	89	104
6. Distal Thoracic limb	100	79	130	107	158	115	100
7. Thorax to thoracic limb	100	70	72	96	81	74	84
8. Neck to thoracic limb	100	93	109	120	89	82	158
9. Neck and thorax	100	97	90	116	64	105	91

From R. T. Berg and R. M. Butterfield, *New Concepts of Cattle Growth,* Sydney University Press, Sydney.

TABLE 9.13 MUSCLE WEIGHT DISTRIBUTION OF STEERS AND BARROWS

Muscle group	Weight of muscle group as % of total muscle weight			
	Steers[1]	Barrows[2]	Barrows relative to steers	Index*
1. Proximal pelvic limb	29.6	28.7	–0.9	97
2. Distal pelvic limb	4.3	3.9	–0.4	91
3. Around spinal column	12.3	17.4	+4.9	141
4. Abdominal wall	10.9	11.2	+0.3	103
5. Proximal thoracic limb	12.6	11.8	–0.8	94
6. Distal thoracic limb	2.4	1.9	–0.5	79
7. Thorax to thoracic limb	10.6	7.4	–3.2	70
8. Neck to thoracic limb	5.4	5.0	–0.4	93
9. Neck and thorax	1.8	3.1	+1.2	97

*Index = pig value relative to cattle = 100
[1]Steers mean total side muscle weight 69.3 kg N = 85
Source: Berg and Mukhoty, 1970.
[2]Barrows at total side muscle weight 17.3 kg N = 12
Source: Richmond and Berg, 1971.
From R. T. Berg and R. M. Butterfield, *New Concepts of Cattle Growth,* Sydney University Press, Sydney.

animals are highly selected, too, for the ability to escape predators. Consistent with this idea, is the observation that domesticated and selected animals have more of their weight in the thorax and thoracic limb. Additionally, domesticated cattle have a greater proportion of their weight in muscles associated with the abdominal wall. This difference is likely related to the fact that domesticated ruminants have extensive rumens and demand additional muscularity in the abdomen compared to their unselected or wild counterparts. Alternatively, it may also be related to the shear volume of diet consumed by domesticated animals.

Species differences in muscle distributions are pronounced and discernible (Table 9.12). Muscles in the barrow and steer (Table 9.13) show that differences in distribution of muscle in pigs and cattle primarily involves muscles around the spinal col-

TABLE 9.14 FAT DISTRIBUTION AMONG CARCASS DEPOTS (% TOTAL FAT)

Depot	Cattle	Sheep	Swine	Poultry*
Kidney	16	14	5	32[1]
Intermuscular	48	41	15	32[2]
Subcutaneous	35	45	79	36

[1]Abdominal fat pad
[2]Total muscle lipids
*Ricard et al., British J. Poul, Sci. (1983) 24:511

umn and those associated with the thorax and thoracic limb. Muscles surrounding the spinal column, especially the longissimus muscle, function to extend the spine or backbone, a function rudimentarily important to the rooting action of pigs. As far as the disparity between thoracic limb muscles, with the exception of poultry, domestic meat animals have no bony connection between the thoracic limb and the axial skeleton. As a result, soft tissues are the only structures attaching the front leg to the thorax. Because cattle are ruminants, and therefore must forage extensively to meet their nutritional needs, it is necessary for their front legs to bear weight for extended periods of time. It is important to understand that no cause and effect relationship has been shown for these phenomena and therefore, it is equally possible that pigs root because they are best suited for such a task.

The examples show that muscle distribution is different between domesticated and non-domesticated animals as well as across species. The differences are best explained as responses to the functional demands placed on the muscles.

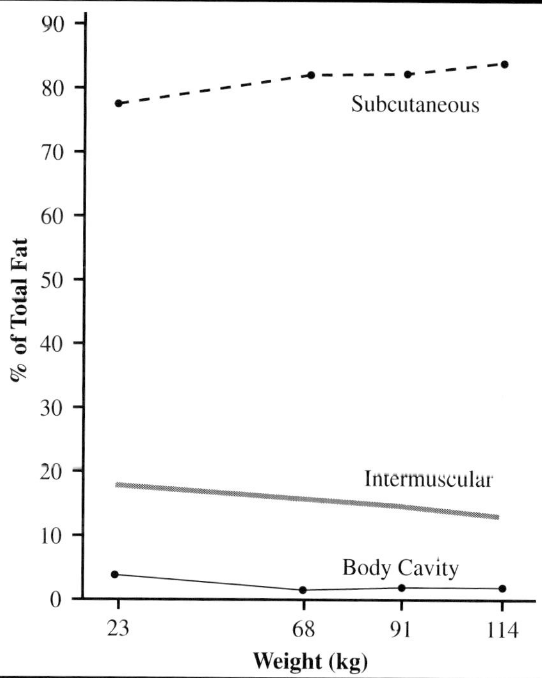

Figure 9.19 Growth and distribution of fat in growing pigs. From P. Walstra, *Growth and Carcass Composition from Birth to Maturity in Relation of Feeding Level and Sex in Dutch Landrace Pigs,* H. Veenman & Zonnen B.V., Wageningen.

FAT DISTRIBUTION

Species is a major determinant in the deposition of fat in specific anatomical locations (Table 9.14) and at different times during growth (Figures 9.19 and 9.20). During growth, subcutaneous fat deposition increases rapidly and represents the bulk of fat in pig carcasses. This is particularly evident by the fact that some pigs can have in excess of 2.5 cm of fat located just beneath the skin at slaughter weight.

In contrast to pigs, adult cattle and sheep yield carcasses with fat that is more evenly distributed across the subcutaneous and intermuscular depots (Figure 9.20). Initially, carcass fat is equally distributed across perirenal and intermuscular depots. However, a rapid increase in percentage of fat located in the intermuscular depot occurs, while perirenal fat percentage decreases. Progressively, subcutaneous fat begins to accumulate and reaches a percentage approximately equal to fat associated with the intermuscular depot.

Even more diverse, chickens contain a high percentage of fat in the abdominal cavity and under the skin (Table 9.15). These differences are most likely due to earlier unintended selection pressure.

It is important to note that the deposition of fat within a depot can vary at different locations.

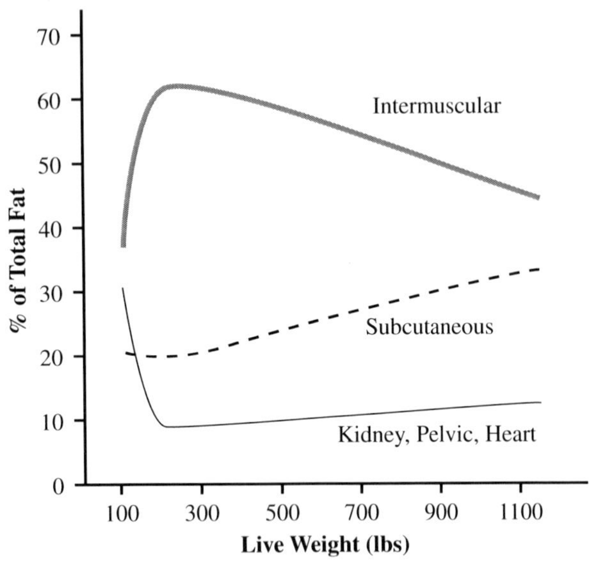

Figure 9.20 Growth and distribution of fat in growing cattle. From R. T. Berg and R. M. Butterfield, *New Concepts of Cattle Growth*, Sydney University Press, Sydney.

TABLE 9.15 SEPARATION OF ABDOMINAL LEAF FAT (ALF) IN THE BROILER CARCASS

Variable	Female	Male
ALF (% Live Weight)	3.9	3.2
Fat Pad (% ALF)	58.7	60.4
Gizzard Fat (% ALF)	20.7	19.6
Viscera Fat (% ALF)	17.2	16.8
Heart Fat (% ALF)	3.4	3.2

Health et al. (1980) Poultry Sci. 59:2456.

Figure 9.21 is a topographical illustration of subcutaneous fat deposition in a growing steer and illustrates the different sites within the subcutaneous adipose tissue depot that have increased propensities for fat deposition.

BREED OR STRAIN

Different breeds of livestock differ in their ability to grow, even though, in some cases, the differences are quite subtle. Within a species, animal breeders have placed various degrees of selection pressure on different traits, which have led to development of many different breeds or strains of livestock in all species. In fact, there are literally hundreds of different cattle, swine, sheep and poultry breeds throughout the world. All of these breeds or strains were developed based on some *traits* or set of attributes that breeders felt were beneficial to their clientele. One of the earliest criteria for selecting animals was genetic potential to produce milk (*dairy type*) or meat (*meat type*). Another trait that was likely viewed as important during early selection programs was overall body size, mainly a function of skeletal size and often simply referred to as *frame size.* Cattle, more so than pigs and sheep, originated not only as a source of food but also as beasts of burden. Pulling implements in the field or wagons to market were tasks well suited for oxen. As a result, cattle breeds have been developed that possess very large skeletons, often referred to as *large-framed* animals. Other selection efforts have resulted in *small-framed* cattle or such cattle may have been the basis from which large-framed cattle were selected. Regardless, there are vast differences among all breeds of cattle today in overall mature size and in ability to produce milk or meat (Figure 9.22). Holstein cattle represent a large-framed, dairy-type breed, whereas Charolais represent a large-framed, beef-type breed. Within a given production criterion (milk versus meat type) animals also may be selected for frame size as is the case with dairy breeds in which Jersey are substantially smaller-framed than Holstein cattle. Likewise, the same can be stated for Angus and Charolais cattle that are traditionally considered to be small-framed and large-framed cattle, respectively. Some Angus cattle today are as big as Charolais cattle, but a difference in the mean frame size across both breeds still exists.

Frame Size

As introduced above, *frame size* is a function of skeletal growth and maturation. As is the case with most biologically controlled characteristics, frame size is a continuum, whereas *frame scores* within a chronological age range from those considered *large-framed* to those *small-framed* animals (Figure 9.23). Large-framed animals are those individuals

Figure 9.21 Topographical maps illustrating the location of fat deposition (top) and relative growth impetus (bottom) of various muscle groups in cattle. From *Proceedings: 21st Century Concepts Important to Meat-Animal Evaluation* by R. G. Kauffman, editor. Copyright © 2000 by university of Wisconsin, Dept. of Meat & Animal Science. Reprinted by permission.

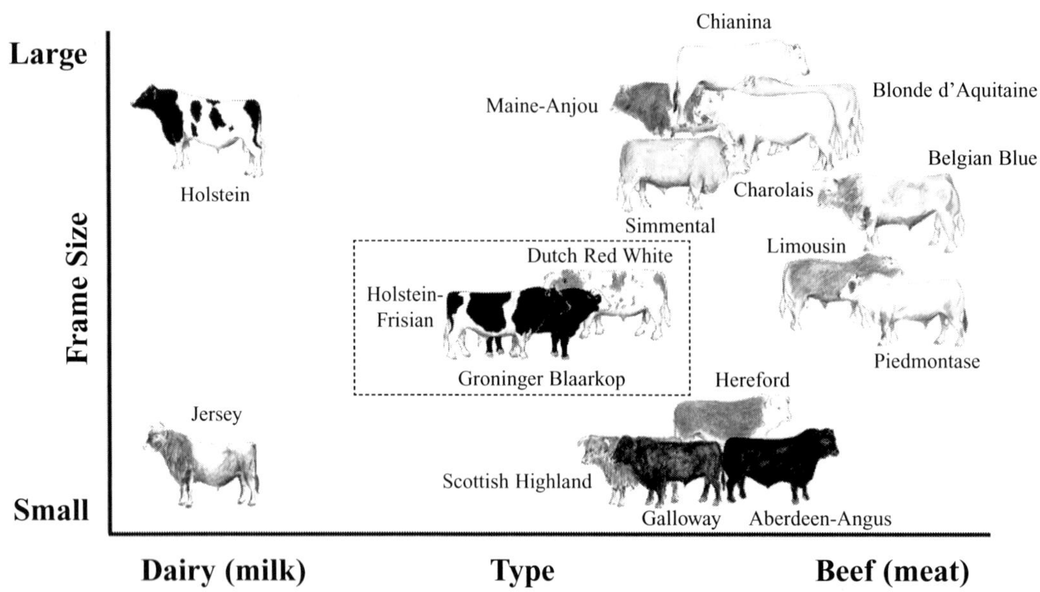

Figure 9.22 Schematic illustrating the difference in frame size and type among cattle breeds. From P. L. Borgstrom I.V.O. Zeist. Dashed box, 'dual purpose' breeds.

that reach heavier weights at a given compositional maturity or *fat endpoint*. Because it requires greater time for large-framed animals to attain a given compositional maturity than small-framed animals, large-framed animals are considered *later maturing* animals (Figure 9.24) and accordingly, *small-framed* animals are considered *early maturing*. At a constant chronological age, later maturing, large-framed animals are physiologically younger than early maturing, small-framed animals (Figure 9.25). At a constant chronological age, early maturing animals contain a higher percentage of carcass fat than late maturing animals (Table 9.16). If comparisons are made at the same weight (y-axis), later maturing animals are leaner while earlier maturing animals are more mature, fatter and "further along on the growth curve" (Table 9.17). Perspectives of physiological age help explain the difference in growth and fattening rates among different breeds of livestock.

Another way to compare the difference between large- and small-framed cattle is to compare them at the same compositional endpoint or maturity (Table 9.18). For example, Chianina (large-framed, late maturing breed) and Hereford/Angus crossbred (small-framed, early maturing breed) cattle fed to the same compositional endpoint (19% fat trim) have the same percentage of muscle on a carcass weight basis.

Cattle may be compared at the same quality grade endpoint i.e., 5% intramuscular fat in the longissimus muscle (Table 9.19). In this case, large-framed, later maturing cattle are compositionally equivalent to small-framed cattle. Large-framed cattle, however, have a slight advantage in leanness, which means that they have the propensity of depositing marbling at a slightly lower level of overall fatness than small-framed animals.

Selection in dairy-type cattle has yielded rather dramatic changes when compared with traditional beef cattle. One of the biggest differences between dairy and beef cattle is the difference in dressing percentage. Normal dressing percentages of beef cattle range between 60 and 67% depending on fill, muscularity and fatness. In contrast, dairy cattle usually yield carcasses weighing less than 60% of their live weight. The reason for this lowered dressing percentage is the reduced muscle mass of dairy cattle and a concomitant increase in the weight of the digestive system as a percent of live weight. The practical significance of this difference in body type is illustrated by the differential in live market price between dairy and beef cattle.

Visual Body Types as Feeders	Ht at Should. at Slau. (in.)			Slaughter Wt (lbs)			Ave. Daily Gain (lbs)			Fat Cover (in.)		
	A	CX	H	A	CX	H	A	CX	H	A	CX	H
	42.1	—	42.4	877	—	889	2.7	—	2.7	0.57	—	0.53
	43.8	—	43.4	984	—	958	2.8	—	2.8	0.65	—	0.50
	45.5	45.5	45.3	1024	1029	1055	2.9	2.9	3.0	0.73	0.35	0.66
	45.9	46.8	47.0	1035	1093	1107	2.9	3.0	3.0	0.74	0.34	0.56
	47.1	47.9	48.4	1106	1133	1175	2.9	3.1	3.2	0.80	0.30	0.60
	—	49.2	—	—	1137	—	—	3.0	—	—	0.33	—
	—	49.5	—	—	1211	—	—	3.2	—	—	0.34	—

Figure 9.23 Relationship between frame size, mature weight, growth rate and fat thickness in cattle (A, Angus; H, Hereford; CX, Charolais cross). From D. R. Mulvaney, How growth affects composition, in Proceedings: 21st Century Concepts Important to Meat-Animal Evaluation, Publication number 285, University of Wisconsin Extension, Madison, Wisconsin.

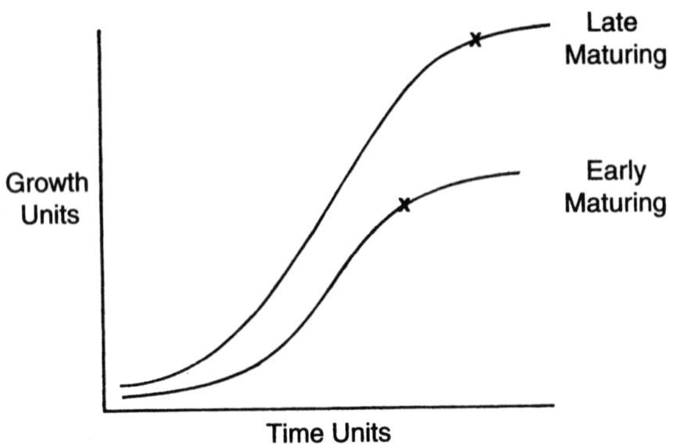

Figure 9.24 Growth curves of early and late maturing animals. The X at the upper inflection of the curves indicates the onset of chemical maturity. From *Livestock & Carcasses,* by Boggs et al. Copyright © 1998 by Kendall/Hunt Publishing Company. Reprinted by permission.

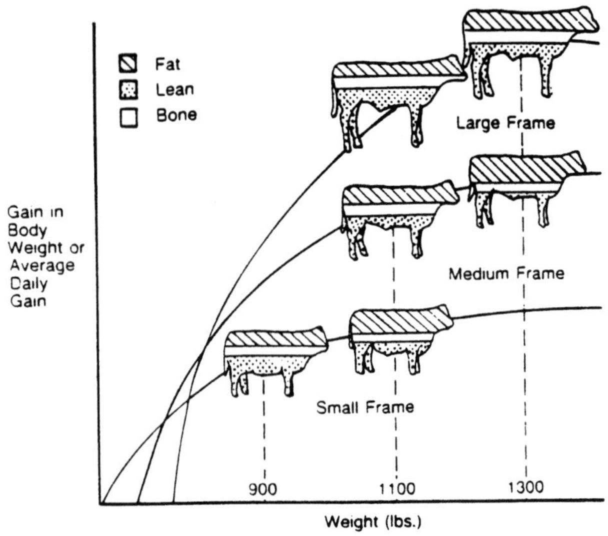

Figure 9.25 Differences in frame scores, cumulative growth and composition of growing cattle. From D. R. Mulvaney, How growth affects composition, in Proceedings: 21st Century Concepts Important to Meat-Animal Evaluation, Publication number 285, University of Wisconsin Extension, Madison, Wisconsin.

TABLE 9.16 EFFECT OF FRAME SIZE AND BREED ON BEEF CARCASS COMPOSITION AT A CONSTANT AGE (473 DAYS)

	Hereford Angus X	Red Poll X	Maine Anjou X	Chianina X
% Lean	67.2	67.4	71.0	73.8
% Fat trim	20.9	20.2	15.7	12.3
% Bone	11.9	12.4	13.2	14.0

From Koch et al. JAS 49:448 (1979)

TABLE 9.17 EFFECT OF FRAME SIZE AND BREED ON BEEF CARCASS COMPOSITION AT A CONSTANT WEIGHT (634 LB. CARCASS WT.)

	Hereford Angus X	Red Poll X	Maine Anjou X	Chianina X
% Lean	66.6	66.2	72.0	74.4
% Fat trim	21.6	21.7	14.3	11.4
% Bone	11.8	12.1	13.6	14.2

From Koch et al. JAS 49:448 (1979)

Table 9.18 Effect of Frame Size and Breed on Beef Carcass Composition at a Constant Maturity (19% Fat Trim)

	Hereford Angus X	Red Poll X	Maine Anjou X	Chianina X
% Lean	68.8	68.4	68.6	69.0
% Fat trim	—	—	—	—
% Bone	12.3	12.7	12.5	12.1

From Koch et al. JAS 49:448 (1979)

Table 9.19 Effect of Frame Size and Breed on Beef Carcass Compostion at a Constant Quality Grade (5% Fat in Longissimus)

	Hereford Angus X	Limousin X	Charolais X	Simmental X
% Lean	67.1	69.6	70.9	69.7
% Fat trim	20.3	18.5	16.2	17.2
% Bone	12.6	11.9	12.9	13.1

From Koch et al. JAS 49:448 (1979)

Selection of muscle seems to be inversely related to milk production and thus most animals used for milk production are inherently less muscular. At a constant weight, dairy cattle are leaner than beef cattle (Table 9.20). This is supported by the fact that beef cattle begin to deposit fat at a younger age, and consequently, a higher percentage of the carcasses of dairy cattle is muscle and bone. The amount of muscle is closely related to amount of bone in animals and may be a useful method to evaluate superior genotype. In the case of cattle, mature muscle to bone ratios range from about 4 to 1 to nearly 7 to 1 for dairy cattle and double-muscled cattle, respectively. In all species, changes in muscle to bone ratios are primarily responsible for muscling differences across breeds. Obviously muscle to bone ratios change with time (Figure 9.26). During the earliest periods of life, the muscle to bone ratio is relatively close, 2.5:1. In cattle, as stated earlier, muscle grows at a much faster rate than bone and therefore mature muscle to bone ratios reach a plateau of around 5:1 in typical beef cattle. Even though selection for milk production has reduced the muscle to bone ratio (Figure 9.26), this selection pressure has not changed muscle distribution in these animals (Table 9.21). Recall from Chapter 6 that muscle fiber size is not different between dairy and beef breeds. Therefore, the difference in the amount of muscle developed is due to a reduction in muscle fiber numbers, which occurs during prenatal development. Essentially, dairy type cattle develop less muscle uniformly throughout the body during the prenatal stages of development.

Table 9.20 Effect of Cattle Breed on Carcass Composition

Breed	% Bone	% Muscle	% Fat
British beef	13.5	53.6	32.8
Zebu	15.4	58.5	26.1
Dairy	16.9	58.5	24.2

From Cole et al. JAS 23:71 (1964)

Table 9.21 Muscle Distribution among Breed Types

Muscle Group	% of Total Muscle Weight		
	Hereford	Crossbred	Holstein
Proximal hind	29.5	29.5	29.6
Spinal	12.3	12.3	12.3
Abdominal	10.8	10.4	11.3
Proximal front	12.5	12.6	12.5
Thorax-leg	10.0	10.8	10.4
Neck-thorax	10.4	10.4	9.4

From R.T. Berg and R.M. Butterfield, New Concepts of Cattle Growth, Sydney University Press, Sydney.

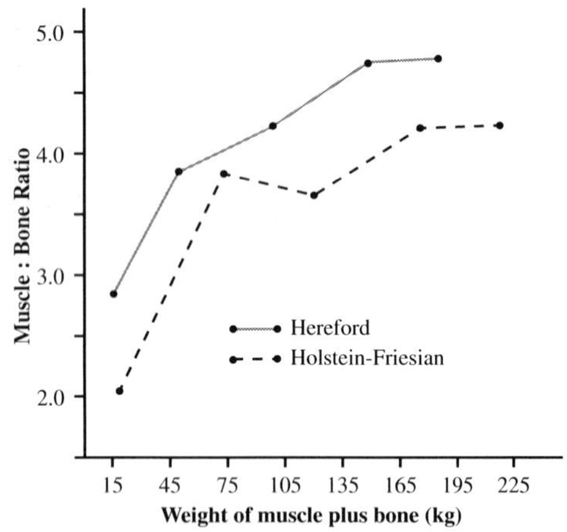

Figure 9.26 Differences between muscle to bone ratios of Holstein-Friesian and Hereford steers. From R. T. Berg and R. M. Butterfield, *New Concepts of Cattle Growth,* Sydney University Press, Sydney.

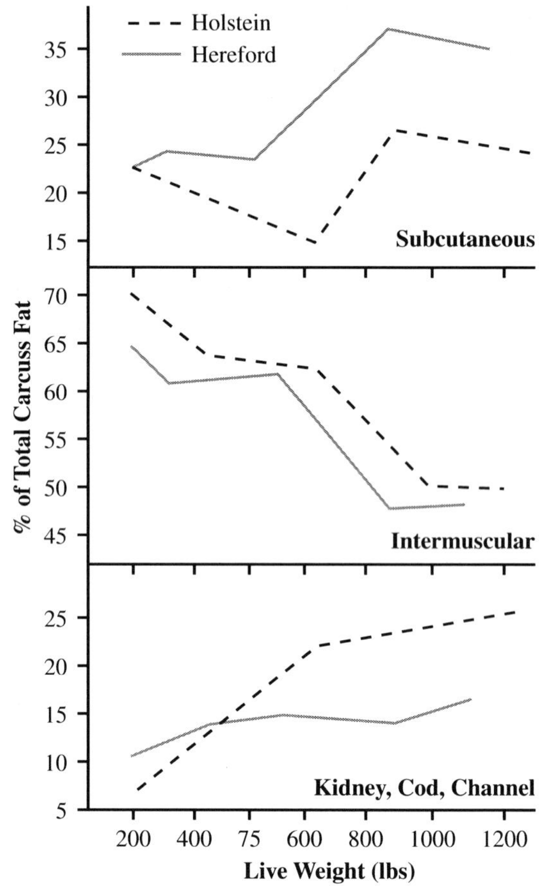

Figure 9.27 Effect of breed type on fat distribution. From R. T. Berg and R. M. Butterfield, *New Concepts of Cattle Growth,* Sydney University Press, Sydney.

Selection for the dairy type cattle has not only affected the time of fat deposition but also has shifted the distribution of fat on the carcass (Figure 9.27). As stated earlier, carcass fat is generally evenly distributed between the intermuscular and subcutaneous depots in mature cattle. However, dairy cattle have much less fat (on a percentage basis) located in the subcutaneous region but more in the perirenal area than beef cattle.

Table 9.22 Muscle to Bone Ratios and Distribution of Muscle Groups in Meat and Mothering Type Pigs (100 kg Live Wt.)

	Duroc & Yorkshire	Pietrain	Duroc × Landrace
Muscle: bone	5.6	7.8	5.2
% muscle	44.0	54.6	43.2
% muscle in:	29.0	30.0	30.8
Prox. Pelvic limb			
Prox. Thoracic limb	10.9	10.8	11.2
Spinal Column	15.5	15.6	15.8

From Walstra, (1991)

Wool versus Meat Type

As outlined for cattle, breeds of sheep have been developed in response to selection pressure for two economically divergent commodities, meat and wool. As is the case in cattle, there is a strong inverse relationship between production of fine wool and meat. As a result, fine wool breeds differ dramatically in body form and ultimate composition. Differences observed between these two diverse breed types mirror those observed for beef and dairy type cattle. Fine wool breeds are larger framed and fatten at heavier weights. Meat type breeds of sheep have superior conformation as measured by muscle to bone ratios.

Table 9.23 Effect of Genetic Strain on Carcass and Abdominal Fat in Broilers

	Fat line	Lean line
Live wt., lbs.	3.14	3.39
Abdominal fat, % body wt.	2.54	.64
Subcutaneous fat, % body wt.	2.6	.93
Total lipids in thigh muscle, % fresh tissue	1.79	1.80

From Richard et al. Brit. J. Poul. Sci. 24:511 (1983)

Mothering versus Meat Type

One of the most important traits in livestock production is the ability of animals to raise young; of course, predicated on this is the ability of females to conceive and produce sufficient milk. Two breeds of pigs, Landrace and Yorkshire, are well known for their ability to raise young. Conversely, Pietrain pigs are extremely muscular and are thought to be ideal for meat. Consistent with that shown for dairy and meat type cattle, Table 9.22 shows that Pietrain pigs have greater muscle to bone ratios and a higher proportion of muscle in their carcasses. However, as with cattle and sheep, selection for "type" has not altered the distribution of muscles in pigs.

Layer versus Broiler Type

Selection for leaner lines of poultry has mainly affected abdominal and subcutaneous fat depots (Table 9.23). Little difference has been observed in content of fat within the muscle (intramuscular fat). In contrast to livestock species, genetic selection for muscle growth in poultry has resulted in some change in the distribution of muscle. The poultry industry has been successful in increasing the amount of breast muscle as a proportion of total carcass muscle.

Conformation

Conformation is defined as the proportionate shape of an animal or carcass. Conformation has and continues to be one of the major forms of assessing the desirability of animals and their carcass by-product. In fact, many European carcass grading systems consider conformation as a crucial ingredient for establishing the value of the carcass and subsequent payment to the producer. In order to eliminate the subjective nature of this measurement, many companies use computer-aided topography technology to assess conformation. Although the term conformation may not be used explicitly, everyone wants an animal or carcass that is "well-proportioned." Selection for a more "ideal conformation," however, has not improved the percentage of muscle located in regions of the carcasses that demand a higher price (Table 9.24).

Because most meat type animals have a "more ideal conformation" than dairy, mothering, and fine wool type animals, meat type animals have greater muscle to bone ratios. The differences in muscle to bone ratios are evident throughout the entire lifetime of the animals. Unfortunately, the amount of fat also increases with perceived improvements in conformation. One of the more disappointing, yet quite revealing results is that the distribution of muscle in animals with superior conformations does not change. Thus, animals with better conformation may indeed be more muscular but will likely result in a carcass that is fatter as well.

TABLE 9.24 DISTRIBUTION OF MUSCLE AMONG CATTLE BREEDS SELECTED FOR DIFFERENT CONFORMATIONS

Breed	% "Expensive" muscles of total muscle weight
Polled Hereford	55.92
Hereford	55.59
Angus	55.61
3/4 Brahman	56.38
1/2 Brahman	56.84
Unimproved	56.11

Butterfield, 1963
From R. T. Berg and R. M. Butterfield, *New Concepts of Cattle Growth,* Sydney University Press, Sydney.

SEX

Growth and composition are dependent on the sex of the animal. A comparison of growth curves characteristic of four species across three sex classes (intact male, intact female and castrate male) is shown in Figure 9.28. Without exception in farm animals, intact males are capable of reaching greater body weights and attain compositional maturity (not necessarily skeletal maturity) at a later chronological age than do male castrates and intact females. Furthermore, intact males are generally heavier at any given chronologicial age. With the exception of pigs, intact females generally mature the earliest of the sex classes, whereas castrates are intermediate to the two intact sexes. For reasons yet unknown, intact gilts mature later and reach heavier weights than barrows. Consequently, gilts are leaner than barrows at the same chronological age. The relative

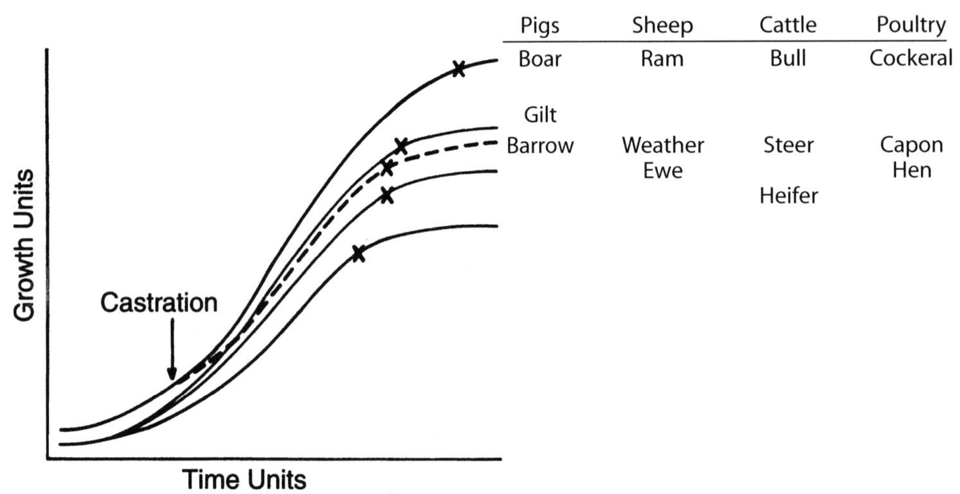

Figure 9.28 Growth curves of the various sexes of pigs, sheep, cattle and poultry. The X at the upper inflection of the curve indicates the onset of maturity. The approximate point of castration is indicated by an arrow on the lower left of the figure. From *Livestock & Carcasses,* by Boggs et al. Copyright © 1998 by Kendall/Hunt Publishing Company. Reprinted by permission.

TABLE 9.25 DISTRIBUTION OF MUSCLE, FAT AND BONE IN BULL AND STEER CARCASSES

	% of Carcass Weight			
	Forequarter		Hindquarter	
	Bulls	Steers	Bulls	Steers
Muscle	36.4	30.8	33.4	33.2
Fat	5.8	8.9	4.8	7.5
Bone	9.8	9.8	7.9	7.9

From R. T. Berg and R. M. Butterfield, *New Concepts of Cattle Growth,* Sydney University Press, Sydney.

differences in carcass composition among bulls and steers are shown in Table 9.25. Intact males are the later maturing gender of the species, and consequently are leaner than castrates or females at the same chronological age. Muscle to bone ratios favor intact males. Muscle to bone ratios for bulls, steers and heifers are 5:1, 4.8:1 and 4.9:1, respectively.

In order to interpret some growth curves, it is necessary to consider the effect of sex on the growth of muscle, bone and fat. The ability of intact males to produce muscle is high, yet fat deposition is much reduced, supporting the concept that later maturing animals are leaner and heavier muscled. Heifers begin to deposit fat tissue earlier and to a greater extent than bulls or steers. This likely is a consequence of the hormonal changes necessary for reproductive competence at the time of sexual maturity.

Distribution of various tissues is also dependent on sex:

- Intact males (bulls) have a greater proportion of their muscle located in the forequarter of their carcasses, or more specifically, in the neck to thorax region (Table 9.25 and Figure 9.29). Enlarged neck muscles are considered secondary sex characteristics in males of most species and results from increased androgen binding receptors in these tissues (Chapter 10).
- A greater proportion of the total muscle of heifers is located in the proximal pelvic limb and in the abdominal wall (Table 9.26).
- Castrate males are intermediate to either of these extremes.
- Similar trends are observed for pigs and sheep.

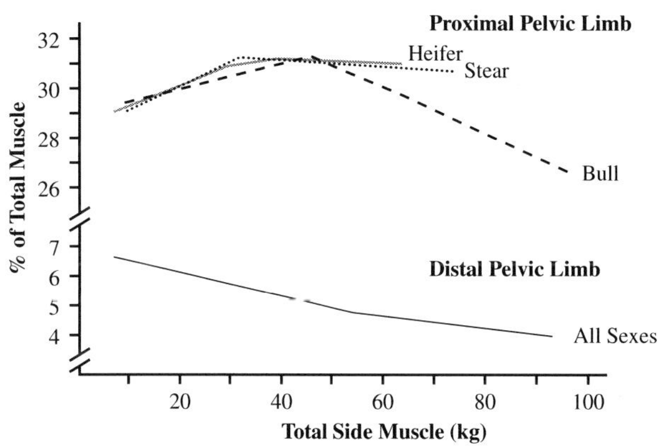

Figure 9.29 Effect of sex on muscle distribution in cattle. From R. T. Berg and R. M. Butterfield, *New Concepts of Cattle Growth,* Sydney University Press, Sydney.

From a practical standpoint, those individuals involved in marketing livestock must have a good working knowledge of interpreting growth curves and knowing how resultant changes in body composition may alter carcass composition and value. For example, market weight heifers usually are valued less than steers because heifers will be physiologically more mature and will yield a carcass with a higher percentage of fat. This excess fat will require additional labor for trimming and greater trimming losses. On the other hand, because heifers are earlier maturing, they are more likely to have greater marbling in the lean (Chapter 12). Likewise, a hog buyer is likely to pay more for gilts than barrows given the same weight and genotype because gilts tend to be later maturing and therefore leaner. On the other hand, feeding intact males

Table 9.26 Muscle Weight Distribution in Bull, Steer and Heifer Carcass (% of Total Carcass Muscle)

Muscle group	Bull	Steer	Heifer
1. Proximal pelvic limb	28.4	29.6	31.5
2. Distal pelvic limb	4.3	4.3	4.3
3. Spinal column	12.1	12.3	12.1
4. Abdominal wall	9.7	10.9	11.5
5. Proximal front limb	12.5	12.6	12.4
6. Distal front limb	2.3	2.4	2.3
7. Thorax to front limb	10.5	10.6	10.3
8. Neck to front limb	5.5	5.4	5.2
9. Neck to thorax	12.5	10.1	9.0
Expensive Muscles			
Group 1,3,5	53.2	54.3	56.1
Group 1,3	40.8	41.8	43.6

From R. T. Berg and R. M. Butterfield, *New Concepts of Cattle Growth,* Sydney University Press, Sydney.

is rarely practiced in the US because they are generally very aggressive and require additional management precautions because of their need to establish an order of dominance. More importantly, meat quality of intact males reduces carcass value as will be discussed in Chapter 12.

Chapter 10

Hormonal Regulation of Growth

INTRODUCTION

Hormones have a variety of effects on growth, reproduction, homeostasis, and integration of metabolism. Many of the growth processes described in previous chapters are regulated and influenced by hormones and growth factors, and consequently, these chemicals are significant to tissue development and carcass composition. The nervous system also plays a major role in hormonal regulation of growth. Furthermore, many of the growth factors involved in tissue development are produced in a manner that targets them to specific cell types.

Hormone Actions

The definition of a hormone has changed over the years as more is learned about the endocrine system. Hormones were originally defined as chemicals secreted into body fluids by ductless endocrine glands. More recently, many define hormones as chemical messengers secreted by various tissues, not necessarily secreted by ductless glands. Some chemical messengers may not be referred to as hormones, but rather as hormone-like materials or *growth factors*. There are many mechanisms by which chemical messengers can exert actions on the body (Figure 10.1). Hormones act in an *endocrine* manner when secreted by cells and then transmitted via the bloodstream, usually at low concentrations e.g., 10^{-9} to 10^{-15} M, to act on distant target cells, i.e., by *bloodstream conveyance*. In a *neurocrine* action, a hormone is synthesized in a cell body of a neuron, e.g., hypothalamic neuron, and stored in axons such as neurotransmitters, but secreted into the bloodstream to act on relatively distant target cells. In this case, the messenger is referred to as a *neurohormone* and it acts in a combined axonal and bloodstream conveyance manner. This action is an integral part of the *neuroendocrine system* because it relies on close coordination between the nervous and endocrine systems. The neuroendocrine system is a key regulator of animal growth and development because the *hypothalamic-pituitary-peripheral* gland axes allow a system of checks and balances on hormone synthesis and secretion for control and regulation of nutrient partitioning and tissue growth and development.

Hormones may also act near to their site of secretion. This *local conveyance* can be accomplished in two ways. First is *paracrine* action in which a hormone from one cell is conveyed to an adjacent cell of different type over short distances via interstitial fluid. Second is *autocrine* action, in which a hormone from one cell acts on itself or on neighboring cells of the same type. A final type of hormonal action denoted *intracrine* action has been proposed for a limited number of growth factors in which a growth factor is able to act intracellularly and does not require secretion to alter an intracellular process. Some hormones and growth factors elicit their actions using a combination of these mechanisms.

Hormones may also exhibit *tissue-specificity*. It is this tissue-specificity that allows hormones to act on target tissues without affecting other tissues or organs. Tissue-specificity is usually accomplished through *receptors* that are only present in target tissues, or whose affinity for specific hormones is tissue-dependent, or through intracellular signaling pathways that are tissue-specific (Figure 10.2). Receptors for hormones may be located at the cell

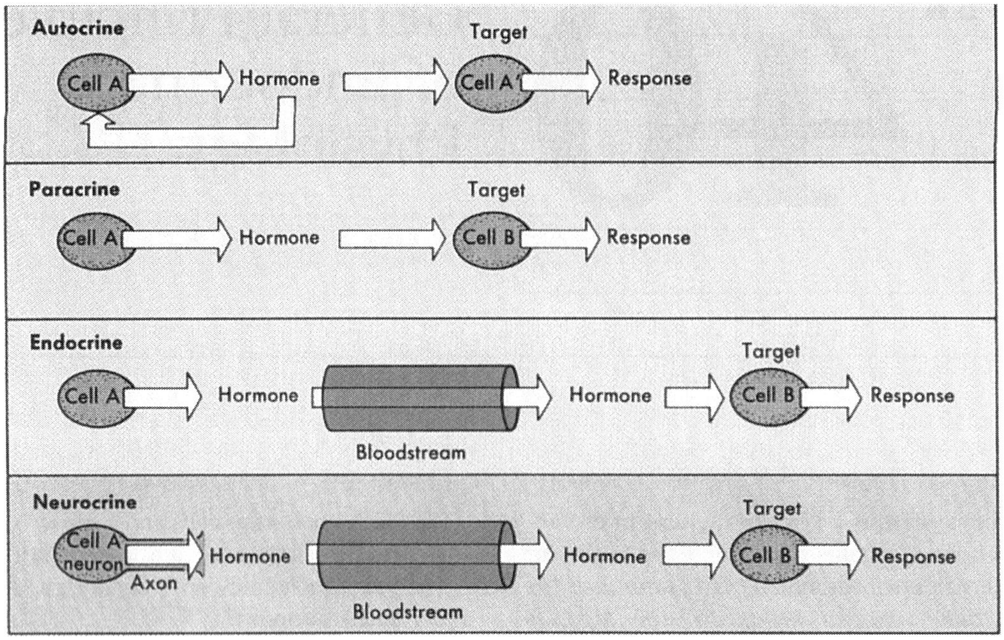

Figure 10.1 Different mechanisms of hormonal signaling. From *Physiology, Second Edition* by Robert Berne & Matthew Levey, Copyright © 1988 by C.V. Mosby Company. Reprinted by permission.

surface or can be intracellular and may trigger different intracellular signaling pathways. Hormones that bind *cell surface receptors* e.g., protein and catecholamine hormones, often act through various *protein kinases* and *protein phosphorylation* to alter *enzyme activity, nutrient* or *ion transport,* or *gene regulation.* Hormones that bind *intracellular receptors* located in the cytosol or nucleus, e.g., steroid and thyroid hormones, usually act as hormone-regulated transcription factors to directly alter gene expression. *Negative feedback* loops also may regulate hormonal action by controlling the secretion and ultimate concentration of the hormone in the circulation or extracellular fluid (Figure 10.3). This is accomplished through a change in synthesis and (or) secretion of releasing or inhibiting factors or tropic hormones.

Chemical Nature of Hormones

A common method for classifying hormones is by their chemical nature. Some hormones are classified as *peptides* or *amino acid derivatives.* This class includes *proteins,* e.g., growth hormone and insulin, *glycoproteins,* e.g., leuteinizing hormone and follicle stimulating hormone, *dipeptides,* e.g., 3,5,3′-triiodothyronine and thyroxine, and *single amino acid derivatives,* e.g., catecholamines such as *epinephrine.* These are classified as water-soluble hormones. Other hormones are classified as *steroid hormones* and are derivatives of cholesterol. Examples of steroid hormones are *testosterone, estrogen, progesterone,* and *glucocorticoids.* These are lipid-soluble, not water-soluble. Later discussion will review the relationship between the chemical nature of hormones and their action and provide a basis for strategies for administering hormones to animals.

Hypothalamic-Pituitary-Peripheral Gland Axes

The central organ of the neuroendocrine system is the *hypothalamus.* Secretions from the hypothalamus regulate secretions of hormones from the *pitu-*

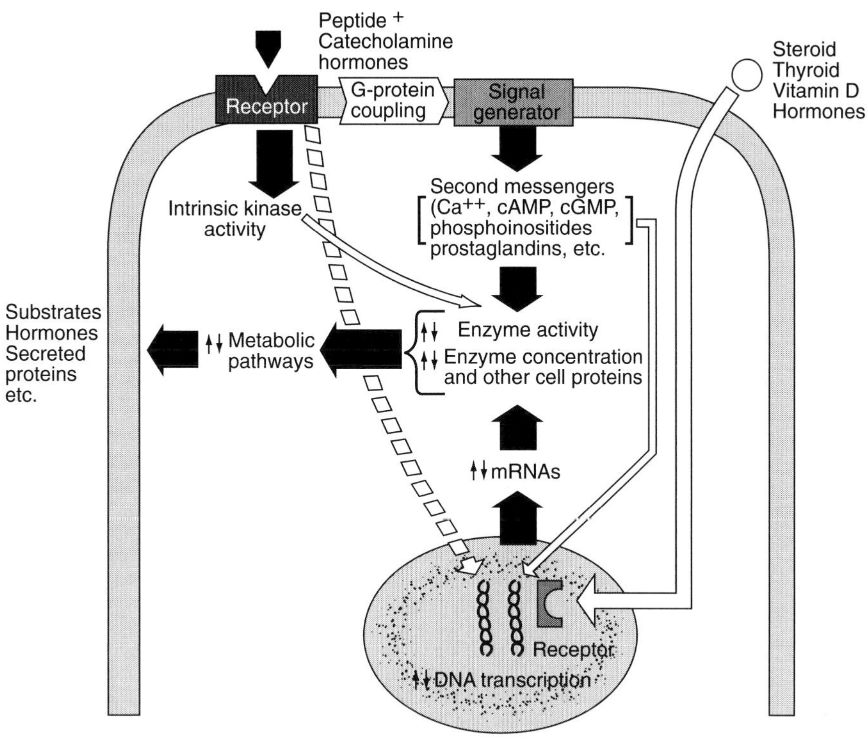

Figure 10.2 Hormone action on cells. From *Physiology, Second Edition* by Robert Berne & Matthew Levey, Copyright © 1988 by C.V. Mosby Company. Reprinted by permission.

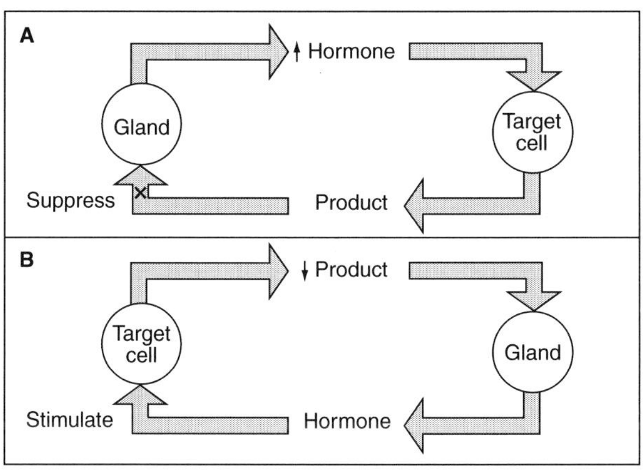

Figure 10.3 Potential mechanisms of negative feedback. From R. M. Berne and M. N. Levy, *Physiology* 2nd edition, Mosby-Year Book, Inc, St. Louis, Missouri.

itary gland (Figure 10.4). The pituitary gland (or hypophysis) is located at the base of the brain with a *neural stalk* (pituitary stalk) that ties the pituitary gland to the brain. The upper portion of this stalk is referred to as the *median eminence* and is adjacent to the hypothalamus, an area of grey matter in the basal part of the forebrain.

The *pituitary gland* is divided into two sections; the *adenohypophysis* and the *neurohypophysis.* The adenohypophysis in mammals includes the *pars tuberalis, pars intermedia,* and *pars distalis,* but there is no pars intermedia in birds. The pars distalis is referred to as the *anterior pituitary* in both mammals and birds. The neurohypophysis includes the *pars nervosa* and *pars eminens* (neural stalk and median eminence). The posterior lobe includes the pars nervosa and intermedia in mammals, but only the pars nervosa in birds.

The posterior pituitary consists of axons whose cell bodies (referred to as *principal nuclei*) are located in the hypothalamus. Hormones initially

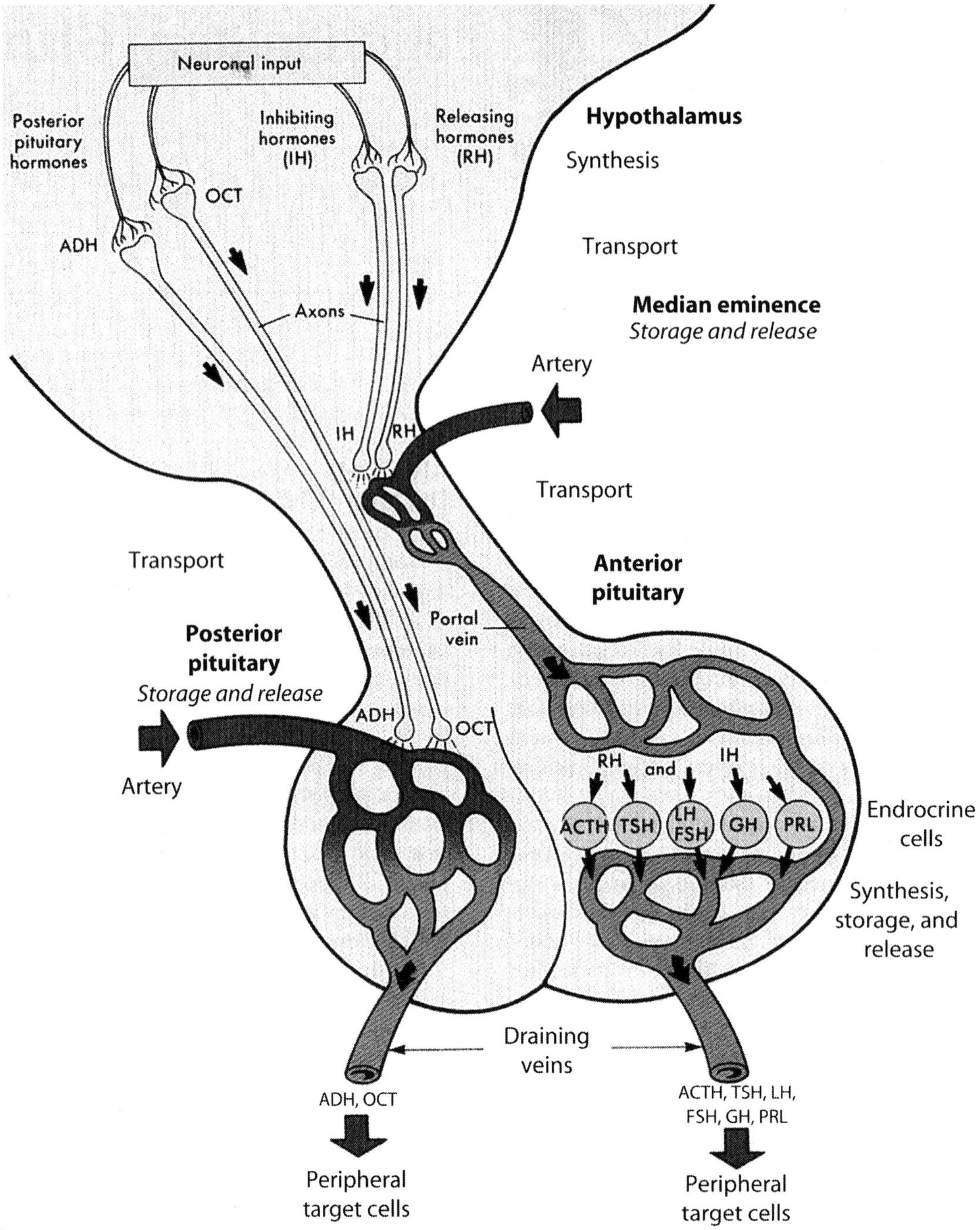

Figure 10.4 Relationship between the hypothalamus and the pituitary gland. *ADH,* Antidiuretic hormone; *OCT,* oxytocin; *ACTH,* adrenocorticotropic hormone; *TSH,* thyroid-stimulating hormone; *LH,* leutinizing hormone; *FSH,* follicle-stimulating hormone; *GH,* growth hormone; *PRL,* prolactin. From *Physiology, Second Edition* by Robert Berne & Matthew Levey, Copyright © 1988 by C. V. Mosby Company. Reprinted by permission.

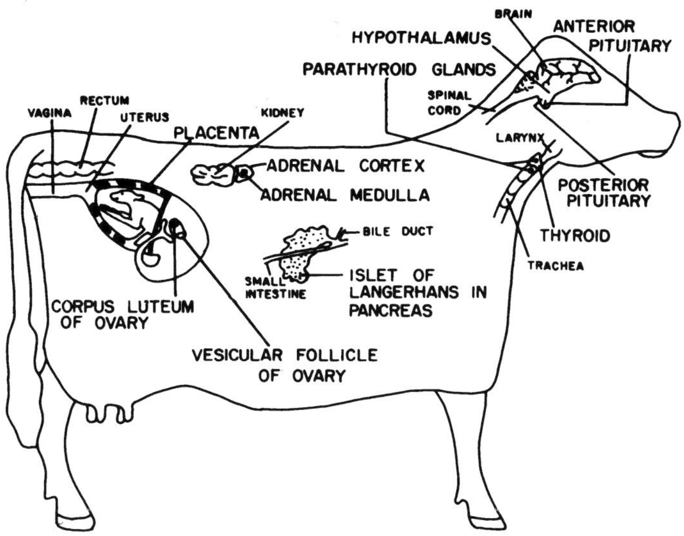

Figure 10.5 Gross location of the major endocrine glands in a cow. From R. R. Anderson, R. J. Collier, A. J. Guidry, C. W. Heald, R. Jenness, B. L. Larson, and H. A. Tucker, *Lactation,* The Iowa State University Press, Ames, Iowa.

synthesized in the cell bodies are transported via axons to nerve terminals within the posterior pituitary gland for storage. Antidiuretic hormone (ADH or vasopressin) and *oxytocin* are two hormones synthesized in the hypothalamus, but are stored and secreted by the posterior pituitary gland. ADH is involved in regulating water balance and oxytocin is responsible for regulating uterine contractions and myoepithelial cell contraction in the mammary gland. The intermediate lobe (pars intermedia) is present in most mammals and contains cells responsible for synthesizing *melanocyte-stimulating hormone* (MSH).

The anterior pituitary consists of various types of hormone-producing glandular cells. Synthesis and secretion of these *tropic hormones* i.e., hormones that act on endocrine glands, are regulated by *releasing and inhibiting hormones* that are synthesized in the hypothalamus neuron cell bodies and stored in nerve terminals *(synaptosomes)* of the median eminence. When stimulated, synaptosomes release hormones into the hypothalamic-hypophyseal portal system for transport to the anterior pituitary via the long portal veins. *Growth hormone-releasing hormone* (GHRH) from the hypothalamus stimulates synthesis and secretion of growth hormone *(somatotropin)* from the *somatotrophs,* whereas *somatostatin* inhibits synthesis and secretion of somatotropin. *Gonadotropin-releasing hormone* (GnRH) induces *gonadotrophs* to produce *follicle-stimulating hormone* (FSH) and *luteinizing hormone* (LH). *Corticotropin-releasing hormone* (CRH) induces *corticotrophs* to produce *adrenocorticotropic hormone* (ACTH). *Thyrotropin-releasing hormone* (TRH) induces *thyrotrophs* to produce *thyroid stimulating hormone* (TSH). *Prolactin* is synthesized by the *lactotrophs.* These tropic hormones that are synthesized in the anterior pituitary are secreted into the peripheral circulation to elicit a response in target endocrine glands. Classical endocrine glands include the *thyroid, parathyroid, pancreas, adrenal, gonads* (*ovaries* and *testes*), and *placenta* (Figure 10.5). FSH and LH act on gonads, ACTH acts on the adrenal gland, TSH acts on the thyroid gland, and prolactin acts on the mammary gland and the gonads. However, other organs and tissues, such as *pineal body, thymus, liver, kidneys,* and *gastrointestines,* also are targets of the tropic hormones. For example, growth hormone induces the synthesis of *insulin-like growth factor-I* in several different tissues of the body. A number of negative feedback loops exist to regulate hormone

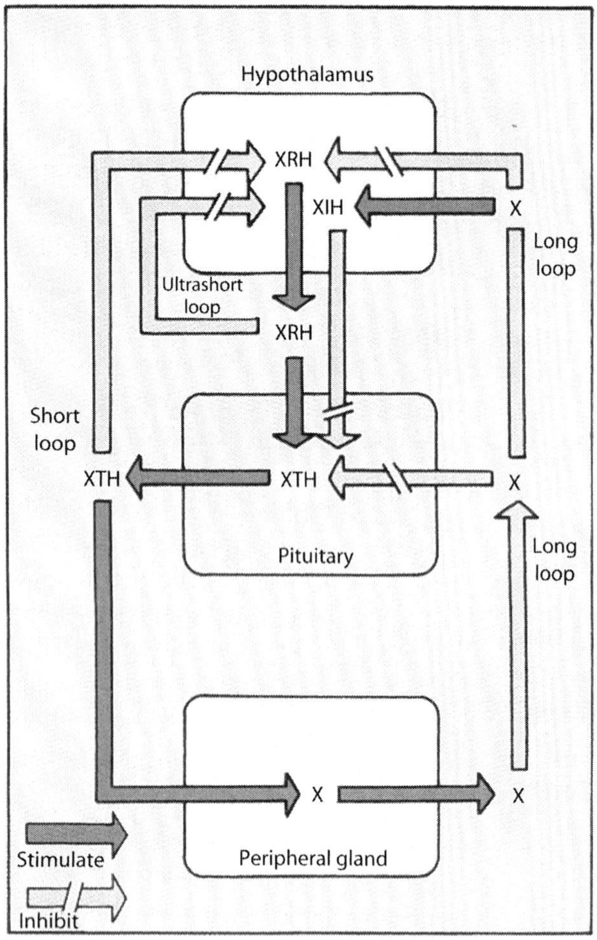

Figure 10.6 Potential negative feedback loops of a classical hypothalamus-pituitary-peripheral tissue axis. *X,* Peripheral gland hormone; *XTH,* pituitary tropic hormone; *XRH,* hypothalamic releasing hormone; *XIH,* hypothalamic inhibiting hormone. From *Physiology, Second Edition* by Robert Berne & Matthew Levey, Copyright © 1988 by C.V. Mosby Company. Reprinted by permission.

secretion in the various hypothalamus-pituitary-peripheral gland axes (Figure 10.6). These are important in the coordination of releasing hormone, tropic hormone and peripheral gland hormone secretions. Perturbation of this feedback system can result in a change in peripheral gland hormone secretion which may have beneficial effects on animal growth and composition (discussed in more detail later in this chapter). The remainder of this chapter is devoted to hormones that have major effects on growth and composition of meat animals.

Androgens

There are two types of androgens, *testicular androgens* and *adrenal androgens. Testosterone* and *androstenone* are the two testicular androgens. Testosterone, the major androgenic hormone, is synthesized by Leydig cells of the testes. Testosterone is responsible for maturation of the male reproductive tract, male sex behavior, spermatogenesis, and body growth. Androstenone is a pheromone and has received considerable attention in swine production because of its contribution to *boar taint* in pork. Androstenone is stored in the salivary gland and when secreted it is ingested by the boar and eventually accumulates in fat depots. It is often detected by consumers when pork from boar carcasses is cooked. Adrenal androgens are *17 keto steroids* and are synthesized by the *zona reticularis* (inner layer) of the cortex of the adrenal gland.

Major growth related physiological effects of testosterone involve bone and muscle. Testosterone increases deposition of bone salts and tends to increase linear growth, but the adult levels of testosterone stimulate epiphyseal closure. The actions of androgens on periosteal osteoblasts result in increased bone mass, specifically cortical thickness. Consequently, bone mass is usually greater in males than in females. The actions of androgens on bone likely involve both osteoblasts and osteoclasts and there is likely interaction between the two cell types so that bone formation and resorption are closely coupled (Figure 10.7).

Androgens stimulate muscle development in a variety of ways. Androgen secretion is:

- High *in utero*
- Declines after birth
- Then increases at puberty.

Thus, it is possible that prenatal androgen influences myogenesis which affects ultimate muscle development. Testosterone shortens the G1 phase of the cell cycle in muscle cell cultures. If a similar effect exists *in vivo,* then the increased number of myoblasts that are formed as a result of *in utero* androgens may be responsible for the increased number of muscle fibers often observed in males as compared to females. Moreover, the higher concentrations of androgens later in postnatal life contribute to larger fibers from induction of muscle protein accretion.

Multiple mechanisms exist by which androgens alter muscle protein metabolism:

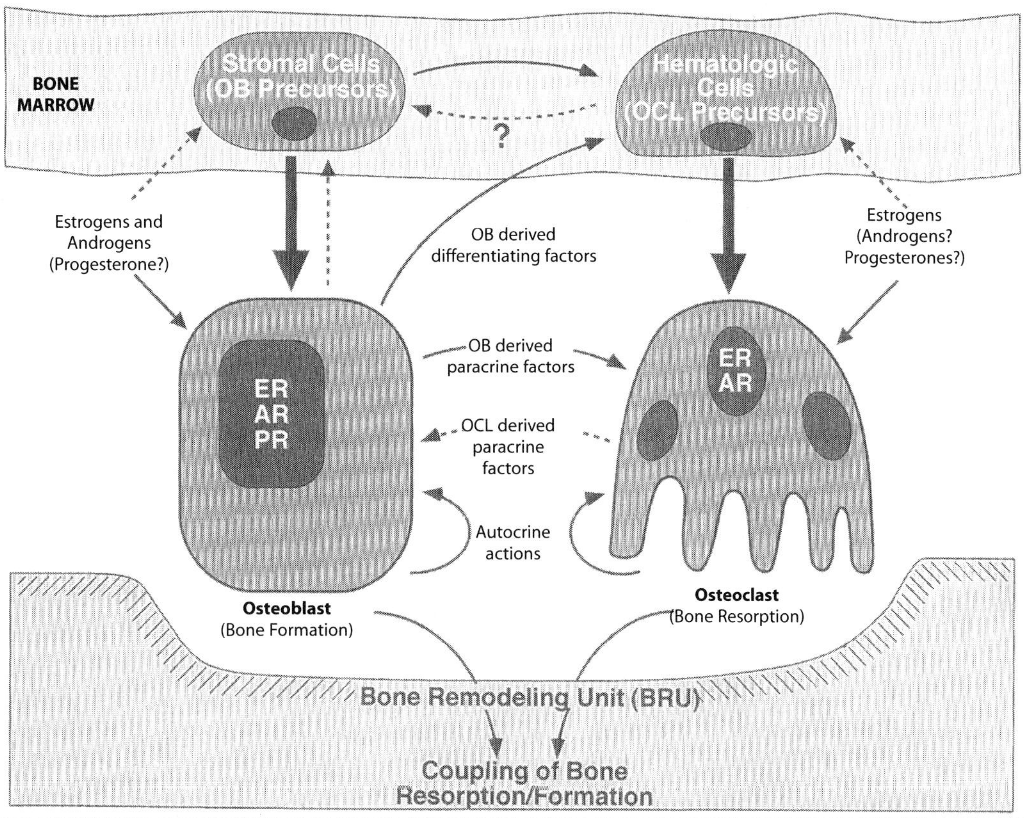

Figure 10.7 Effects of various hormones and growth factors on bone resorption and formation. *OB,* osteoblast; *OCL,* osteoclast; *ER,* estrogen receptor; *AR,* androgen receptor; *PR,* progesterone receptor. From Spelsberg et al (1999).

- First, androgens bind to specific intracellular androgen receptors in muscle to increase protein synthesis.
- Second, testosterone is also aromatized to *estradiol,* so some of the effects of testosterone are, therefore mediated through its estrogenic metabolites.
- Third, androgens interfere with the anti-anabolic effects of cortisol (cortisol actions are discussed later), allowing androgens to elicit their effect more easily.
- Fourth, many of the androgen effects are mediated through growth hormone.

In fact, castrated males have lower circulating growth hormone concentrations than intact males, whereas exogenous androgens increase growth hormone concentrations. Other hormones also may mediate the effect of androgens on tissue growth.

Androgens tend to increase both protein synthesis and protein degradation, but protein synthesis is increased to a greater extent; thus, protein accretion is increased. In some cases e.g., ewes and heifers, protein synthesis and degradation are decreased, but the degradation is decreased to a greater extent than the synthesis; thus, the end result is still a net increased accretion of protein in the muscle. Although administration of androgens to chickens has little effect on growth, they have been shown to improve growth of turkeys.

In mammals, there are some differences in muscle distribution as a result of androgens. As covered briefly in Chapter 9, in cattle, the muscles of the forequarter of the male, especially those in the neck and crest region, are more sensitive to androgens.

Consequently, the muscles of this region become more developed and represent a larger proportion of the total body muscle mass in bulls than in steers and heifers.

Castration of domestic animals diverts energy from growth of muscle to deposition of fat. Because fat deposition requires more energy per unit of weight than required for muscle, castrates have lower feed conversion ratios than intact animals. Therefore, to achieve leanness, and if palatability and tenderness are unrelated to fat deposition and marbling, the intact male is preferable. Producers frequently castrate male animals to modify behavior and improve meat quality e.g., to eliminate boar taint in pigs, but this causes deposition of more fat and less muscle during growth (See Chapter 12 for more details on meat quality).

Estrogens

The ovary is the source of three estrogens:

- *Estrone*
- *Estriol*
- *Beta-estradiol*

The latter is the major hormone synthesized by follicles of ovaries. These three hormones are involved in body growth, maturation of the female reproductive tract, female sex behavior, and mammary duct growth. Physiological effects of estrogens impact bone, adipose tissue, and muscle. Adult levels of estrogen stimulate epiphyseal closure via influences on chondrocyte proliferation and bone formation. Estrogens appear to be highly effective at inducing epiphyseal closure; thus, females tend to have shorter skeletons than males. Similar to androgens, estrogens also influence the coupling of bone formation and resorption through effects on osteoblast and osteoclast activities (Figure 10.7).

Estrogens facilitate deposition of fat and stimulate muscle growth. As a result of these effects, estrogens represent a major class of anabolic growth promotants for some meat-animals. Estrogens are anabolic for ruminants, but are less effective in pigs, and cause fattening in poultry. Estrogens are most effective in castrated male ruminants in which they improve gain and feed conversion by 10 to 20%, but the improvement is less pronounced in females. Furthermore, estrogens have little effect on growth in intact males, except for increased carcass fat. Major response to estrogens in steers is increased carcass protein. Administration of estrogenic anabolic steroids (such as *zeranol; Synovex S;* and *diethylstilbestrol,* DES) increases growth hormone secretion rate and thyroxine concentrations in steers. These interact to increase muscle protein deposition and growth. Effects on protein synthesis and degradation are not clear, but involve decreased rates of degradation with variable effects on synthesis. Diethylstilbestrol is a very effective growth promotant, especially in steers, but was withdrawn from use in the 1970's because it was found to be carcinogenic in tests using laboratory animals. However, several other naturally occurring estrogenic growth promotants have been approved for use and are very effective in altering growth and composition of cattle.

Progestins

The *progestins* are a class of steroid hormones of which progesterone is a member. *Progesterone,* which is involved in maintenance of pregnancy and mammary growth, is synthesized by the corpus luteum of mammals. Progestins also have been shown to improve growth and feed conversion efficiency of cyclic heifers. Melengestrol acetate (MGA) is a synthetic progestin that is 100-times more potent than progesterone in stimulating growth. Feeding MGA to heifers improves gain and feed conversion by 10% without changing carcass composition. The mechanism by which this compound exerts its effect is probably through suppression of estrus and its related problems i.e., reduced feed intake and hyperactivity.

Synthetic Hormone Combinations

Use of anabolic steroids (i.e., steroids that result in increased tissue accretion) for improving production is most effective in cattle (Table 10.1). Androgens improve growth, feed conversion, carcass protein content of cattle and pigs; however, they are most effective in cattle, especially heifers. In steers and heifers, testosterone alone is anabolic, but if combined with estrogens, the combination is more effective than estrogens alone. *Trenbolone acetate* is a synthetic steroid that is a weak androgen, but a very potent anabolic steroid, especially if combined with estrogen (Table 10.2). Steers respond best to implants containing such a combination of androgen

TABLE 10.1 HISTORY OF THE FDA APPROVAL DATES AND STATUS FOR ANABOLIC AGENTS[a,b]

Trade name	Anabolic agent	Date approved	Approved for use in
Synovex-S	200 mg progesterone/ 20 mg E_2 benzoate	20/2/56	Steers
Synovex-H	200 mg testosterone propionate/20 mg E_2 benzoate	16/7/58	Heifers
Ralgro	36 mg zeranol	5/11/69	Calves Steers Heifers
	or 12 mg zeranol		Lambs
MGA[c]	0·25–0·50 mg/day MGA	3/6/77	Heifers
Compudose	24 mg E_2	12/3/82	Calves Steers Heifers
Steer-oid	200 mg progesterone/ 20 mg E_2 benzoate	12/11/82	Steers
Synovex-C	100 mg progesterone/ 10 mg E_2	9/4/84	Calves
Heifer-oid	200 mg testosterone propionate/20 mg E_2 benzoate	24/7/84	Heifers
Finaplix-S	140 mg TBA	17/6/87	Steers
Finaplix-H	200 mg TBA	17/6/87	Heifers

[a]Sources: Taylor (1984) and McEvoy *et al.* (1987).
[b]Adapted from Collins *et al.* (1989).
[c]48-hour withdrawal period.
Composition of various hormone implants, date of approval and target species. From D. L. Hancock, J. F. Wagner and D. B. Anderson. Protein accretion, in *Growth Regulation in Farm Animals, Advances in Meat Research* vol 7, A. M. Pearson and T. R. Dutson, editors, Elsevier Applied Science, London.

TABLE 10.2 PERFORMANCE AND CARCASS CHARACTERISTICS OF IMPLANTED CATTLE WITH ESTRADIOL (E_2) AND TRENBOLONE ACETALE (TBA) ALONE OR IN COMBINATION[a]

Treatment	ADG (kg)	% increase	% F/G	% increase	Quality grade	Yield grade	Cutability
Control	1.39	0	6.79	0	11.39	2.21	49.96
E_2	1.55	11.5	6.29	7.4	11.42	2.43	50.15
TBA	1.49	7.2	6.29	7.4	11.09	2.25	50.54
E_2 + TBA	1.62	16.5	5.95	12.4	10.78	2.29	50.38

[a]Adapted from Trenkle (1987). Taken from pooled data in five-trial summary.
From W. G. Bergen and R. A. Merkel, Protein accretion, in Growth Regulation in Farm Animals, Advances in Meat Research vol 7, A. M. Pearson and T. R. Dutson, editors, Elsevier Applied Science, London.

Table 10.3 Concentrations of Estradiol (E_2) and Estrone (E_1) in Tissues of Steers Following Estrogen Implantation[a,b]

Tissue estrogen	Treatment	Tissue concentration (pg/g)			
		Lean	Kidney fat	Kidney	Liver
E_2	Estrogen	3.4	15.0	21.0	10.0
	Control	5.8	6.8	6.7	4.0
	Difference	–2.4	8.2	14.3	6.0
E_1	Estrogen	10.4	35.0	19.0	9.0
	Control	4.8	10.5	7.9	6.5
	Difference	5.6	24.5	11.1	2.5

[a]Adapted from Wagner and Pankhurst, (1981).
[b]Implants were not removed.
From Growth Regulation in Farm Animals, Advances in Meat Research vol 7, A. M. Pearson and T. R. Dutson, editors, Elsevier Applied Science, London.

Table 10.4 Daily Human 17-β Estradiol (E_2) and Estrone (E_1) Production[a]

Sex, age	Serum conc.[b]		Total production[c]	
	E_2 (pg/ml)	E_1 (pg/ml)	E_2 (pg)	E_1 (pg)
Children, 5–60 days	12.5	12.0	16 875 000	26 520 000
Males, adult	37.2	38.0	70 308 000	97 660 000
Females,[d] 41 years	400.1	—	540 000 000	—

[a]Adapted from Wagner and Pankhurst, (1981).
[b]Adapted from Winter *et al.* (1978). VanTheir and Loriaux (1979), and Korenman *et al.* (1978) for children, adult males and females, respectively.
[c]Based on serum clearance of estrogens (Longcope *et al.*, 1968):
E_1 adult females = 2 210 000 ml/day;
E_1 adult males = 2 570 000 ml/day;
E_2 adult females = 1 350 000 ml/day;
E_2 adult males = 1 890 000 ml/day.
[d]Regularly cycling females at peak estrogen production.
From Growth Regulation in Farm Animals, Advances in Meat Research vol 7, A. M. Pearson and T. R. Dutson, editors, Elsevier Applied Science, London.

and estrogen. This combination may exert its actions on muscle by binding to testosterone and estrogen receptors in skeletal muscle. A significant decrease in protein degradation and a slight decrease in protein synthesis result in increased protein accretion. There is some evidence that suggests an increased recruitment of satellite cells accompanies the induced myofiber hypertrophy. A combination of TBA and estrogenic compounds also increase growth hormone secretion which contributes to the effectiveness of the combination implant on growth and composition.

The use of hormonal implants for improving the efficiency of beef production has led some to question whether hormone residues remain in the beef at concentrations that elicit biological activity. However, little difference exists in the amounts of estrogen in beef from non-implanted steers compared with beef from steers implanted with estrogen (Table 10.3). In fact, levels in a serving of beef are much lower than levels in servings of other foods. Furthermore, the amount of estrogen consumed from food products is negligible compared with the amount of estrogen synthesized daily in the human body (Table 10.4).

TABLE 10.5 MAJOR EFFECTS OF GROWTH HORMONE

Direct	Indirect
Reduced glucose transport and metabolism	Promotion of growth and endocrine effects
Increased lipolysis	Reduction in insulin receptors, e.g. in the liver
Increased amino acid transport	Localized decrease in adipose tissue; free fatty acids released, which then provide an energy source for muscles
Increased protein synthesis	In muscle, liver and adipose cells
	Increases in both transcription and translation
Increased Insulin-like growth factor-I (IGF-I) production	From liver and other tissues
	Chondrocyte, osteoblast and adipocyte formation

Growth Hormone (GH)

As described earlier in this chapter, GH or somatotrophin (ST) is produced by the anterior pituitary and acts in an endocrine manner to affect a number of different tissues in the body (Table 10.5). The liver can synthesize growth factors that act as hormones to regulate growth; thus, the liver plays a major role in mediating growth and development of various systems, not only by playing a central role in metabolism, but also by acting as an endocrine gland. One of the major actions of GH is the production of IGF-I (insulin-like growth factor-I; also a somatomedin) by various tissues, predominantly the liver. IGF-I mediates many of the effects of GH and may act in endocrine, paracrine, and autocrine manners, and can elicit feedback inhibition to regulate GHRH and GH secretion (Figure 10.8). High concentrations of IGF-I inhibit GHRH and GH release and subsequently lead to reduced IGF-I production. Conversely, if IGF-I concentrations are too low, GHRH and GH release is not inhibited, so they act to increase IGF-I production. *Somatostatin,* which is produced by the hypothalamus and the *delta cells* of the pancreas, is also a component of the feedback inhibition loops. Somatostatin action decreases GH secretion resulting in decreased IGF-I production. Thus, GHRH, GH, IGF-I, and somatostatin all participate to maintain IGF-I concentrations at the appropriate concentrations. Strategies to increase IGF-I production have involved the development of methods to perturb this feedback system.

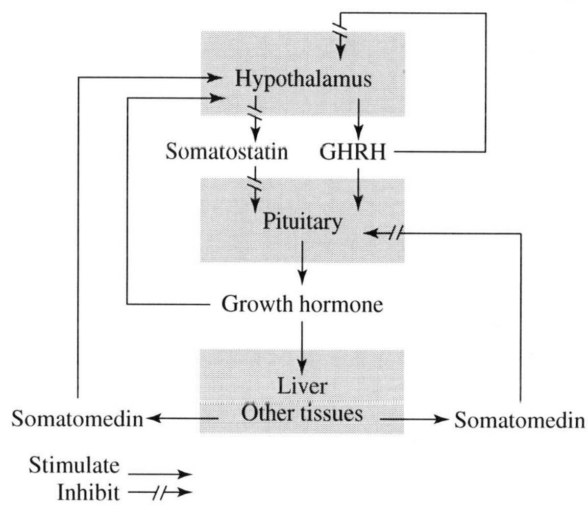

Figure 10.8 Regulation of growth hormone secretion. From *Physiology, Second Edition* by Robert Berne & Matthew Levey, Copyright © 1988 by C.V. Mosby Company. Reprinted by permission.

GH has a variety of effects on different tissues, including bone, muscle, and fat. Its effect on bone is mediated through IGF-I which increases chondrocyte proliferation and osteoblast activity. Consequently, GH will lead to increased length of bones prior to epiphyseal closure. GH also will increase periosteal growth contributing to increased bone mass.

GH increases lean growth by increasing rates of muscle protein synthesis and decreasing protein degradation. Increases in RNA and DNA often accompany the increased protein accretion. The increased DNA is likely a result of increased satellite

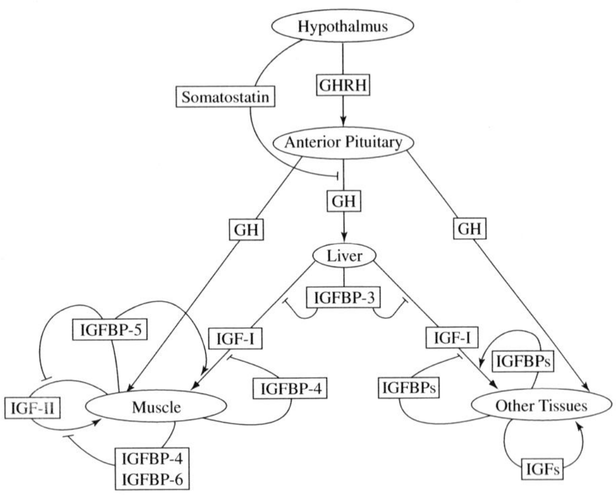

Figure 10.9 Growth hormone-insulin-like growth factor axes. Arrows indicate stimulatory effects, and parallel lines indicate inhibitions. From *Endocrine Reviews* (1996), Vol. 17, No. 5, pp. 481–517, by James R. Florini et al. Copyright 1996 by The Endocrine Society. Reprinted by permission of The Endocrine Society.

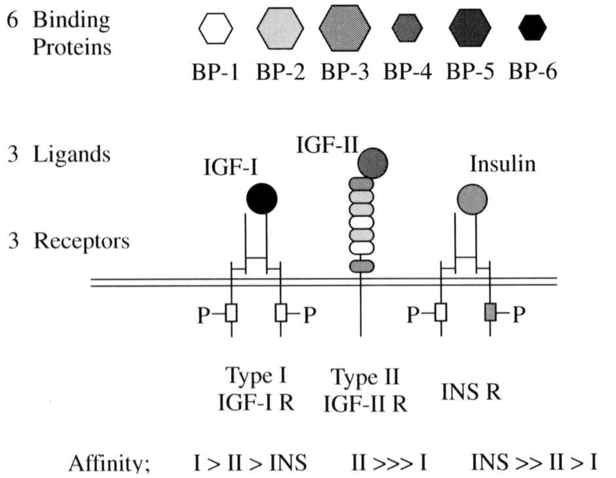

Figure 10.10 Receptors, ligands and binding proteins of the insulin-like growth factor signaling axes. From *Endocrine Reviews* (1996), Vol. 17, No. 5, pp. 481–517, by James R. Florini et al. Copyright 1996 by The Endocrine Society. Reprinted by permission of The Endocrine Society.

cell activity. Growth hormone-induced muscle growth is mediated by IGF-I and is accomplished by endocrine or paracrine/autocrine manners (Figure 10.9). GH acts on the liver to induce IGF-I secretion which is then transported to the muscle where it acts to increase muscle growth. Alternatively, GH can act directly on the target tissue (muscle). Either myogenic cells, or other cell types in close proximity to myogenic cells, produce IGF-I which act in a paracrine or autocrine manner to induce protein accretion. IGF-I actions on muscle include increased uptake of glucose and amino acids. Various IGF-binding proteins (IGFBP) are involved in transporting the IGF-I to the target tissue, increasing the half-life of IGF-I, and modulate the interaction of IGF-I with its cell receptor. The *IGF family* is quite complex, but IGF-I and the type I IGF receptor mediates most of the IGF-mediated GH actions (Figure 10.10).

GH increases lipolysis by stimulating mobilization of fatty acids from adipocytes. In some cases, for example in poultry and pigs, GH decreases lipogenesis. GH decreases insulin-stimulated glucose metabolism (i.e., decreases insulin sensitivity, and increases catecholamine-stimulated lipolysis.

Because of the effects on adipose and muscle metabolism, GH has been researched extensively as a metabolism modifier for some meat animals. GH is a protein hormone, so it is not orally active and therefore is usually administered systemically as an injection. Treatment of pigs results in increased body weight gain, improved feed conversion, and decreased feed intake. In addition, carcasses have more muscle and less fat (Table 10.6). GH is also effective in other species, but its response is often dependent on dosage level, frequency of administration, and animal age. For example, response is variable and greater in older broilers than younger broilers. As discussed in the next chapter, treatment of animals with GH may alter the dietary protein and energy requirements because of increased muscle protein deposition. However, GH also may increase the efficiency at which nutrients are used for growth contributing to the decreased feed intake that accompanies GH treatment.

Alternatives to GH administration have been investigated to improve animal growth and composition. Growth hormone releasing hormone (also referred to as *somatotropin releasing factor*) has been administered to animals to obtain effects similar to GH administration (Table 10.7). Increases in performance and composition require appropriate dosages and treatment periods. Immunizing animals

TABLE 10.6 EFFECTS OF PORCINE SOMATOTROPIN (pST) ON GROWTH AND CARCASS PERFORMANCE IN GROWING PIGS

Observation	pST Dose (μg/kg BW)[a]				
	0	50	100	150	200
Number of animals	10	10	9	8	10
Growth performance:			*Percentage difference from controls* (%)		
Average daily gain (kg)	0.904	+8.0	+13.1	+13.3	+13.3
Feed/gain	3.22	–21.1	–28.0	–31.4	–35.1
Daily feed intake (kg/day)	2.90	–14.8	–18.6	–22.1	–26.9
Carcass yield and composition:					
Dressing percentage (%)	76.6	–1.4	–3.5	–4.6	–4.2
Muscle mass (kg)	30.92	+27.9	+30.3	+32.7	+37.6
Adipose mass (kg)	22.78	–35.7	–54.6	–65.7	–74.1
Bone mass (kg)	7.28	+10.2	+14.3	+16.2	+17.3
Skin mass (kg)	3.66	+14.8	+36.1	+35.0	+38.3

[a]Somatotropin was administered by daily injections and adjusted biweekly to increased live weight. Growth performance data were analyzed by analysis of variance using live weight as the covariate (B. Krick & R. D. Boyd, unpublished data). Carcass data are summarized from Thiel *et al.* (1990). Tissue mass was determined by physical dissection of the half carcass. From Growth Regulation in Farm Animals, Advances in Meat Research vol 7, A. M. Pearson and T. R. Dutson, editors, Elsevier Applied Science, London.

TABLE 10.7 EFFECTS OF HUMAN SOMATOTROPIN RELEASING FACTOR (hSRF) ON GROWTH AND CARCASS PERFORMANCE IN GROWING LAMBS

Response	Treatment[a,b]		
	Control	5 μg hSRF	10 μg hSRF
Growth performance:		*% Difference from controls*	
Number of animals	18	20	20
Average daily gain	304	+13	+1.6
Feed/gain	4.99	–18	–19
Composition of carcass gain:			
Number of animals	9[b]	10	10
Protein accretion rate (g/day)	17.2	+30.8	+34.9
Water accretion rate (g/day)	55.6	+19.6	+28.8
Lipid accretion rate (g/day)	79.9	–21.2	–28.4
Ash accretion rate (g/day)	5.0	+32	+42
Semitendinosus muscle weight (g)	91.6	+10.5	+15

[a]Lambs received saline or the indicated dose of hSRF four times per day (0100, 0700, 1300 and 1900 hrs) for 42 days. Data are summarized from Beermann *et al.* (1990).
[b]Half of the lambs were withdrawn from treatment after 42 days. Data shown are for lambs treated 96 days. Carcass composition was determined by chemical analysis of the total ground carcass using the comparative slaughter technique (initial slaughter group included).
[c]($P < 0{\cdot}05$) vs control.
[d]($P < 0{\cdot}01$) vs control.
From Growth Regulation in Farm Animals, Advances in Meat Research vol 7, A. M. Pearson and T. R. Dutson, editors, Elsevier Applied Science, London.

against somatostatin also has some promise in increasing GH actions through the relief of the inhibition of somatostatin on GH release. The significant impact of GH on animal growth and composition has led to research efforts to over-express the GH gene in animals. Production of transgenic animals over-expressing GH has been accomplished with limited success.

GH, along with insulin and glucagon (especially in poultry), are major hormones important in the coordination of metabolism, especially in controlling the distribution of nutrients among various tissues. This coordination, controlled by several endocrine hormones, is referred to as *homeorhesis.* The liver is an important organ in homeorhesis. When nutrients are limiting, the increased GH increases lipolysis, but decreases growth because IGF-I becomes uncoupled from GH i.e., IGF-I decreases. These changes facilitate the transfer of calories from adipose to vital functions.

Insulin and Glucagon

Islets of Langerhans contain four types of cells that comprise the endocrine portion of the pancreas:

- *Alpha cells* synthesize glucagon
- *Beta cells* synthesize insulin
- *Delta cells* synthesize somatostatin
- A fourth cell type synthesizes pancreatic polypeptide.

Insulin increases the storage of glucose, fatty acids, and amino acids as glycogen, triglyceride, and protein, specifically when nutrient intake is adequate (Figure 10.11). These storage modes are accomplished by stimulation of the cellular uptake of glucose, amino acids, and lipids and by stimulation of lipogenesis, glycogenesis, and protein synthesis. Insulin is anti-lipolytic in mammals, but this action is not clear in avian species.

The actions of glucagon are opposite of insulin, at least in liver. These include:

- Mobilizing glucose (by increasing *glycogenolysis* and *gluconeogenesis*)
- Mobilizing fatty acids (by increasing lipolysis)
- Increasing amino acid catabolism.

The coordination of insulin and glucagon is key to the integration of metabolism. The pancreas of both mammals and birds produce insulin and glucagon, but insulin is the predominant pancreatic hormone in mammals, whereas glucagon is the predominant pancreatic hormone in birds. This difference probably explains some of the species variation observed in hormonal regulation of carbohydrate metabolism.

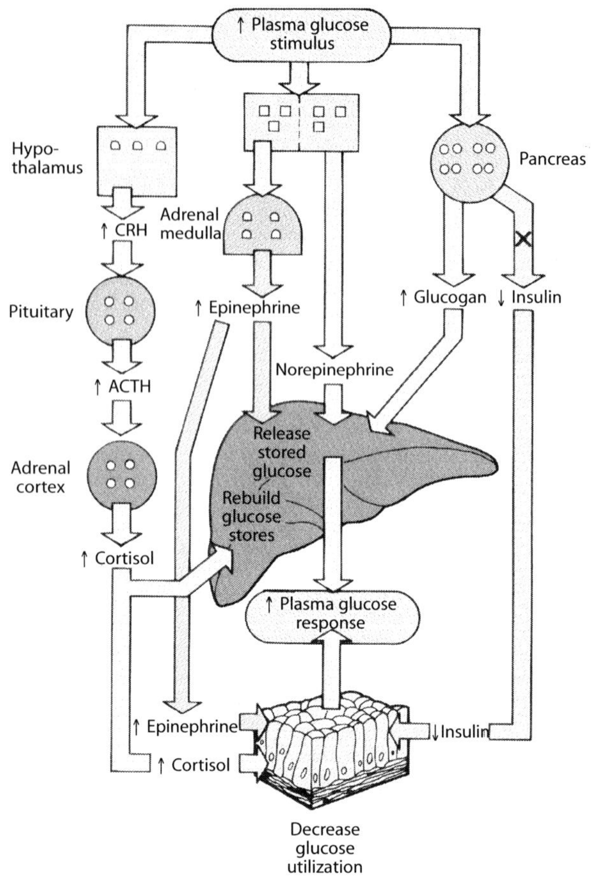

Figure 10.11 Hormonal and metabolic changes in response to hypoglycemia. CRH, Corticotropin-releasing hormone; ACTH, adrenocorticotropic hormone. From *Physiology, Second Edition* by Robert Berne & Matthew Levey, Copyright © 1988 by C.V. Mosby Company. Reprinted by permission.

Leptin

In recent years, leptin has been discovered to be a regulator of body energy stores. Leptin is a peptide

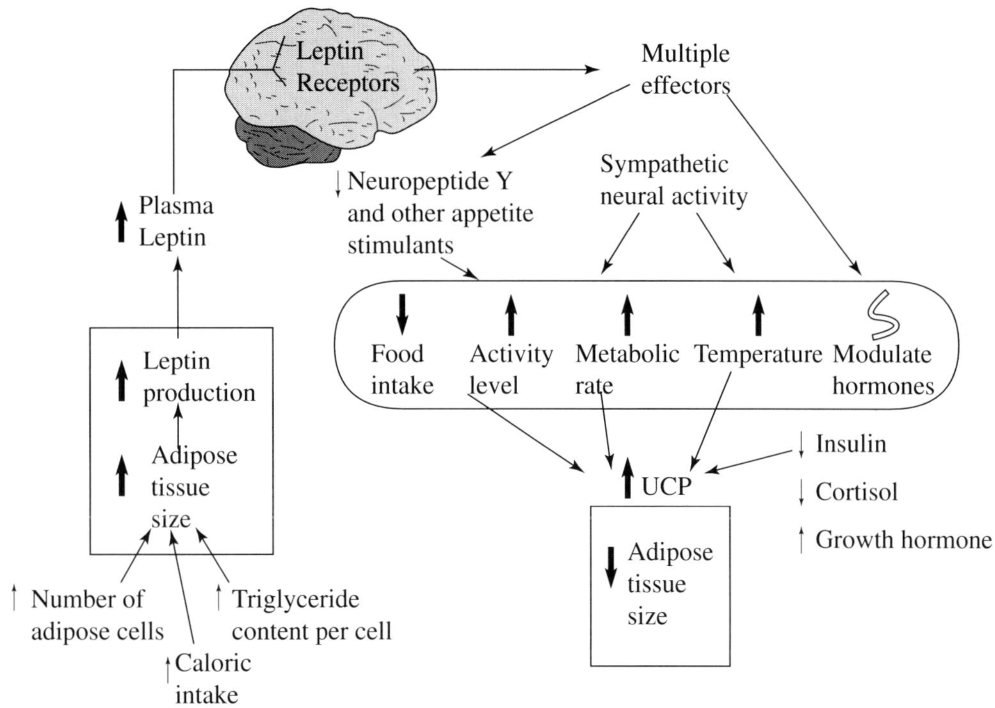

Figure 10.12 Regulation and action of leptin. UCP, uncoupling protein. From R. M. Berne and M. N. Levey, *Principles of Physiology, Third Edition,* 2000, Mosby, Inc.

hormone produced by adipose tissue (see Chapter 7). This finding has added support to the idea that adipose tissue is an endocrine gland, and not just a tissue for energy storage, and protection. Leptin appears to be important in providing signals to the hypothalamus that regulate multiple response mechanisms. An increase in adipose tissue mass through increased adiposity and triglyceride content results in concentrations of leptin that are sufficient to induce the hypothalamus to decrease food intake, increase energy expenditure, and modulate other hormones e.g., insulin, cortisol, and GH, that decrease lipogenesis and increase lipolysis (Figure 10.12). The increased energy expenditure is accomplished by increased levels of uncoupling proteins in adipose tissue mitochondria which increase heat production. Ultimately, these leptin-induced mechanisms result in decreased adipose tissue mass. The high correlation between plasma leptin concentration and adipose mass has led many to propose that leptin serves as a reporter of adipose mass to the hypothalamus. Understanding how leptin is regulated is an active area of research as its manipulation in some manner could provide possible strategies for altering growth and composition.

Glucocorticoids

The *zona fasciculata* (middle layer) of the adrenal cortex synthesizes *glucocorticoids* in response to ACTH from the pituitary. The major glucocorticoids (cortisol in mammals and corticosteroid in avian species) exert permissive effects by enhancing the action of other hormones on many tissues. They are important in metabolic regulation, especially during stress and injury. Although glucocorticoids enhance muscle cell growth and adipocyte differentiation in culture, they generally exhibit catabolic (tissue breakdown) effects *in vivo.* Glucocorticoids decrease muscle protein synthesis and increase muscle protein degradation in an attempt to make amino acids available for glucose production. They also increase lipolysis, in part, by enhancing GH- and catecholamine-stimulated lipolysis.

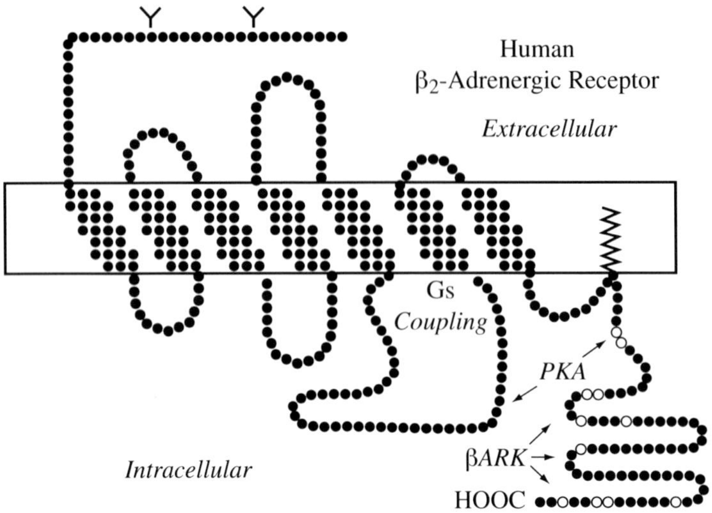

Figure 10.13 Illustration of the β_2-adrenergic receptor and its critical regions. *PKA,* protein kinase A; β*ARK,* beta adrenergic receptor kinase. From Hausdorff et al (1990).

Catecholamines

The catecholamines, *epinephrine* (or *adrenalin*) and norepinephrine, are stored in the adrenal medulla and are released when stimulated by nerve fibers. These hormones also play important roles in an animal's response to stress and induce significant metabolic effects on various tissues. These hormones (referred to as beta-adrenergic agonists) act through the *adrenergic receptors* (also referred to as *adrenoreceptors*). Several different receptor subtypes exist e.g., alpha$_1$, alpha$_2$, beta$_1$, beta$_2$, and beta$_3$, that allow the differential effects of norepinephrine and epinephrine and are important in tissue specificity. Epinephrine acts through the beta-adrenergic receptors, whereas norepinephrine acts through the alpha- and beta$_1$-adrenergic receptors. A diagram of a beta$_2$-adrenergic receptor is shown in Figure 10.13. Binding of the *agonist* to the receptor activates membrane-bound a*denylate cyclase* and the production of intracellular cAMP (Figure 10.14). The adenylate cylcase is stimulated to produce the cAMP through a GTP-binding protein (G_s) that is coupled to the receptor. The cAMP then activates protein kinases which phosphorylate various intracellular proteins. These phosphorylation events are key to the metabolic effects of the beta-adrenergic agonists.

Adjustments in metabolism (increased gluconeogenesis in liver, increased glycogenolysis in liver and muscle, and increased lipolysis in adipose tissue) that occur during stress are accomplished by the actions of epinephrine and norepinehrine. Epinephrine and norepinephrine maintain sufficient blood circulation by impacting the heart and blood vessels. Physiological effects of epinephrine include mobilization of glycogen for energy, various vascular changes (including increased heart rate and blood flow), increased body temperature, and increased respiration. Glucocorticoids participate in the stress response by exhibiting their permissive effects on metabolism (described earlier) and modulating some of the effects of epinephrine. Thyroid hormones also are important in regulating metabolic rate (discussed later).

Consequences of stress also are observed in skeletal muscle. Epinephrine induces muscle anaerobic glycolysis to meet the energy needs of the contracting skeletal muscle. Excessive muscle contraction generates heat, which if not dissipated by an adequate blood flow, leads to increased muscle temperature. If lactic acid accumulates, muscle pH may decrease. If the heart and liver are unable to metabolize the lactic acid, the accumulation of lactic acid in the blood will result in *metabolic acidosis,* which if severe enough, may be fatal. Stress that results in these metabolic changes in market animals also may

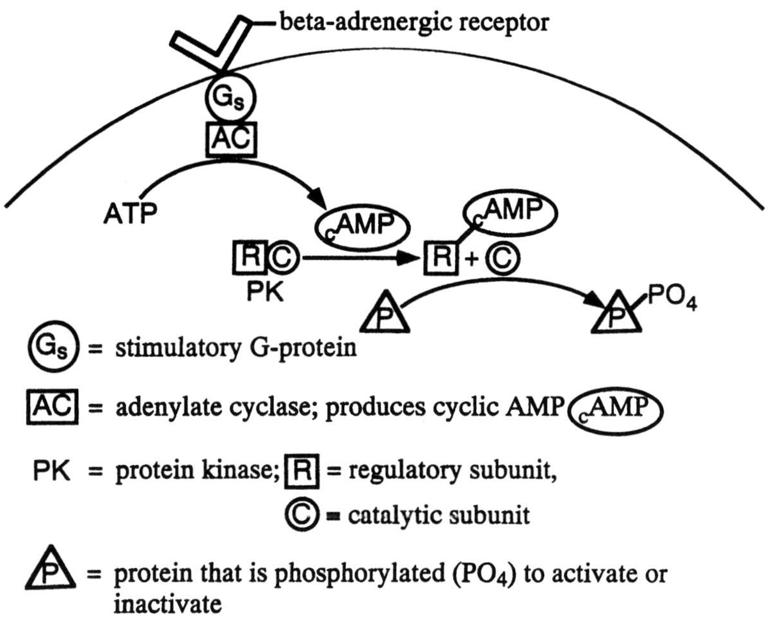

Figure 10.14 Intracellular signaling by the β-adrenergic receptors. From *The Biology of Fat in Meat Animals,* American Socity of Animal Science, Champaign, Illinois.

have significant effects on postmortem metabolism that occurs during processing. This altered metabolism will affect the muscle to a point that the structure and quality of the meat is detrimentally affected. Subjecting animals to long term stress (several days) will deplete glycogen stores prior to processing and generally leads to *dark cutting beef* or *dark, firm, dry* (DFD) pork. Acute stress just prior to processing often will accelerate the metabolism that occurs postmortem, leading to earlier onset of rigor, and often to *pale, soft, exudative* (PSE) pork. These stress-induced changes in metabolism may result from various environmental factors and genetic abnormalities (this is discussed in detail in Chapter 12).

Because of the effects of catecholamines on metabolism, synthetic compounds structurally similar to catecholamines have been investigated for their effects on growth and composition (Figure 10.15). These compounds, sometimes referred to as *phenethanolamines,* are classified as beta-adrenergic agonists because they act predominantly through beta-adrenergic receptors. Administration of synthetic beta-adrenergic agonists results in improved growth, feed conversion, efficiency and composition of both ruminants and pigs (Table 10.8). Synthetic beta-adrenergic agonists also are effective in poultry, but their effects are sometimes more limited.

Variable effects across species may be the result of different proportions and densities of beta-adrenergic receptor subtypes in target tissues, and possibly due to the type and dose of the compound used in the studies. The decreased fat deposition following feeding of beta-adrenergic agonists is the result of decreased lipogenesis and (or) increased lipolysis. Increased muscle mass is the result of increased protein synthesis rates and may involve changes in muscle protein degradation rates in some cases. Changes in protein degradation may result from alterations in the calpain proteolytic system. Muscle growth also may involve enhanced satellite cell recruitment for increasing muscle DNA content.

Ractopamine (Paylean® and Optiflex®, Eli Lilly and Co.) is the only beta-adrenergic agonist that has been approved by FDA for improving growth of meat animals in the US. It is approved for use in swine and cattle, respectively. Zilpaterol (Zilmax®, Hoechst Roussel Vet), another synthetic beta-adrenergic agonist, is approved for use in cattle in Mexico and South Africa and has significant effects on growth performance and carcass yield of feedlot steers (Tables 10.9 and 10.10). Beta-adrenergic agonists typically increase weight gain and feed conversion, but do not affect feed intake in feedlot cattle.

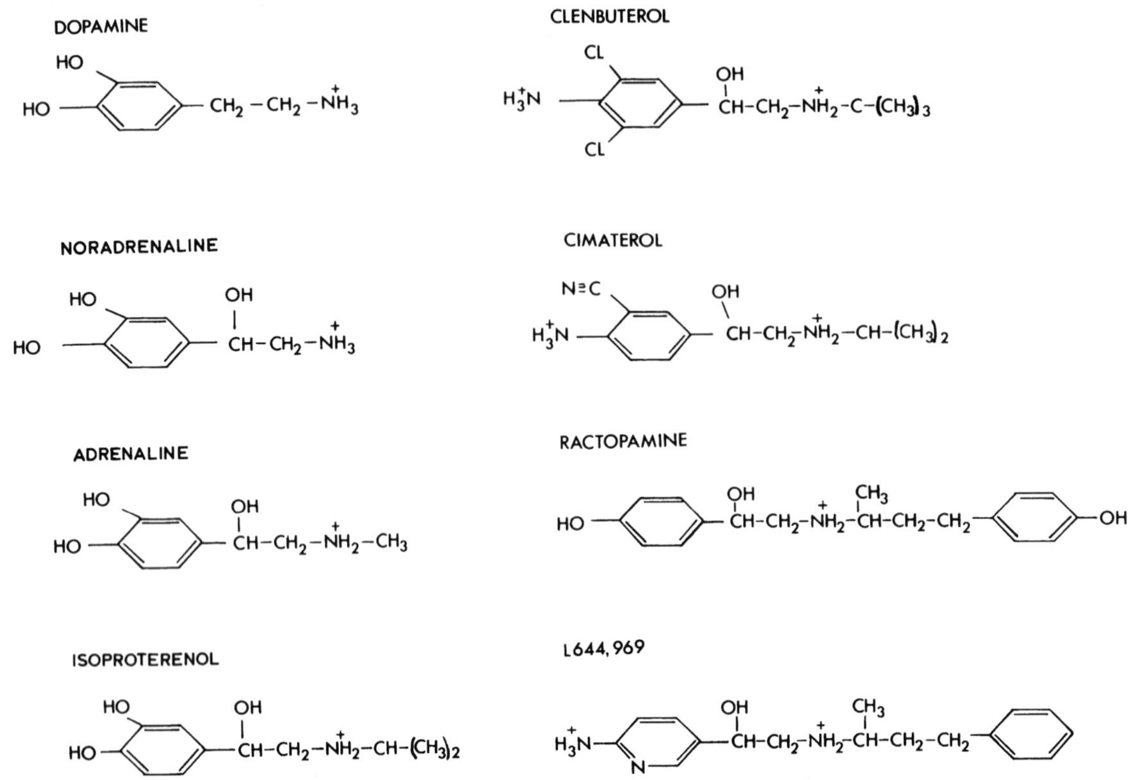

Figure 10.15 Structure of various adrenergic hormones and their agonists. Figure from *Muscle and Meat Biochemistry*, edited by A. M. Pearson & R. B. Young, copyright 1989, Elsevier Science (USA), reproduced with permission for the publisher.

TABLE 10.8 TYPICAL RESPONSES TO PHENETHENOLAMINE ADMINSTRATION IN FOOD PRODUCING ANIMALS

	Ruminants	Swine
Average daily gain	+10-30%	+0-10%
Feed conversion efficiency	+15-30%	+5-15%
Carcass fatness	–5-30%	–10-15%

From P. T. Anderson, Mechanisms by which metabolic modifiers alter growth rate and carcass composition of meat animals, Proceedings of the 53rd Annual Reciprocal Meat Conference, American Meat Science Association, Savoy, Illinois.

Thyroid Hormones

Two hormones synthesized by the follicular cells of the thyroid and stored in their lumens are *thyroxine* (T4) and *triiodothyronine* (T3). Thyroid hormones play important roles in metabolism, growth, and development. They potentiate insulin action on carbohydrate metabolism. They tend to increase protein synthesis and nitrogen balance possibly through increased GH release. Thyroid hormones stimulate lipid metabolism, both lipogenesis and lipolysis, but the effect on lipolysis is usually greater. Concentrations of thyroid hormones are very important in the way processes are influenced. Also present in the mammalian thyroid gland are *C cells (parafollicular cells)* which synthesize *calcitonin*. Calcitonin in birds is produced by ultimobranchial organs. Calcitonin is a hormone whose

TABLE 10.9 EFFECT OF ZILPATEROL ON THE GROWTH PERFORMANCE OF STEERS

Item	Control	Zilpaterol[a]
Days on test	42	42
Pen replicates	7	7
Live weight, kg[b]		
Initial	375	371
Final[c]	435	455
Weight gain, kg/d	1.42	1.94
Drymatter intake, kg/d	8.55	8.45
Drymatter intake/gain	6.08	4.37
Diet NE, Mcal/kg		
Maintenance	2.09	2.58
Gain	1.42	1.85
Observed/expected		
Maintenance	.99	1.23
Gain	.99	1.29

[a]6 mg zilpaterol/kg of air dry feed.
[b]Weights reduced 4% to account for digestive tract fill.
[c]Carcass adjusted final weight.
A. Plascencia, N. Torrentera, and R. A. Zinn, Influence of the β-agonist, zilpaterol, on growth performance and carcass characteristics of feedlot steers (1999). Proceedings, Western section, American Society of Animal Science, Savoy, Illinois.

TABLE 10.10 EFFECT OF ZILPATEROL ON BEEF CARCASS PERFORMANCE

Item	Control	Zilpaterol[a]
Carcass weight, kg	273	286
Dressing percentage	61.7	63.9
Rib eye area, cm^2	82.5	84.7
Fat thickness, cm	.51	.51
KPH, %	2.34	2.36
Marbling score	3.49	3.43
Retail yield	54.0	54.1
Liver abscess	17.1	7.1

[a]6 mg zilpaterol/kg of air dry feed.
[c]Coded: minimum slight = 3, minimum small = 4, etc.
A. Plascencia, N. Torrentera, and R. A. Zinn, Influence of the β-agonist, zilpaterol, on growth performance and carcass characteristics of feedlot steers (1999). Proceedings, Western section, American Society of Animal Science, Savoy, Illinois.

function is to decrease plasma calcium concentration (see Chapter 8).

Parathyroid Hormone

Predominant hormone-producing cells of the parathyroid glands are chief cells. *Chief cells* are responsible for synthesis of parathyroid hormone (PTH). PTH increases bone resorption to increase circulating calcium concentrations; thus, a major role of the parathyroid glands is to regulate calcium flux in the body (see Chapter 8).

Other Endocrine Glands

Other endocrine glands play important roles during animal growth and development. The thymus synthesizes and secretes substances important for normal development and maintenance of immunity. Many *biogenic amines* can be found in the *pineal gland* which is located above the thalamus. *Serotonin* and *melatonin* are found in high amounts in the pineal gland. The pineal may be involved in mediating the effects of photoperiod on animal behavior, reproduction, and performance. Two major hormones secreted by the placenta are estrogen and *placental lactogen.* Placental lactogen is involved in regulating the transport of nutrients from the maternal tissues to the fetus and also is involved in mammary growth. The gastrointestinal tracts synthesize several substances with hormone-like activities, including *gastrin, secretin, enterocrinin, pancreozymin, cholecystokinin,* and *enteroglucagon,* all of which are important in nutrient utilization.

Chapter 11

Nutritional Regulation of Growth

Introduction

Optimal growth and development of tissues is dependent on diets formulated with proper amounts and proportions of essential nutrients. In addition to primary nutrients (fatty acids, amino acids, carbohydrates, vitamins, minerals, etc.), the diet contains growth factors that are important in improving the utilization of the diet by the animal. When animals consume feed energy in excess of that required for body maintenance, growth and reproduction (including lactation and egg laying) may occur (Figure 11.1). Maximizing *feed efficiency* i.e., maximizing the proportion of nutrients available for weight gain, depends on maximizing the efficiencies of energy use for maintenance and production of tissue i.e., growth. The intent of this chapter is not to provide a detailed discussion of the principles and concepts of animal nutrition, but to discuss the concept of *nutrient partitioning* and how *nutritional status* (particularly protein and energy status) in prenatal and postnatal stages of growth may regulate body composition.

Nutrient Partitioning

A general term used to describe the nutritional status of animals is *plane of nutrition.* Plane of nutrition is broadly defined as the amount of nutrients consumed by an animal relative to the amount required for optimal growth and (or) performance. Therefore, one may understand why animals on high planes of nutrition exhibit more rapid rates of growth than animals on low planes of nutrition.

Dietary energy is a major determinant of plane of nutrition and its content is dependent on the amount of carbohydrate, fat, and protein present in the diet. *Energy density* of diets is important in optimizing energy intake when feed intake is limited, but proportions of energy and protein also are important in optimizing body growth and composition. Thus, to formulate diets with ingredients varying in nutrient composition and digestibility that satisfy requirements at different stages of growth and optimize body composition is a challenge for nutritionists. In doing so, it is important to understand the concept of *nutrient partitioning.*

Utilization of ingested nutrients is partitioned among various tissues and organs according to their metabolic rate and physiological importance. Various hormones and growth factors of the endocrine system coordinate distribution of nutrients among organs and tissues (Figure 11.2). Clearly, highest priority is placed on vital physiological systems. The order of priority, from highest to lowest, for various systems is:

- Nervous
- Circulatory
- Respiratory
- Digestive
- Reproductive.

The respective order for tissues is:

- Skeletal
- Muscle
- Adipose.

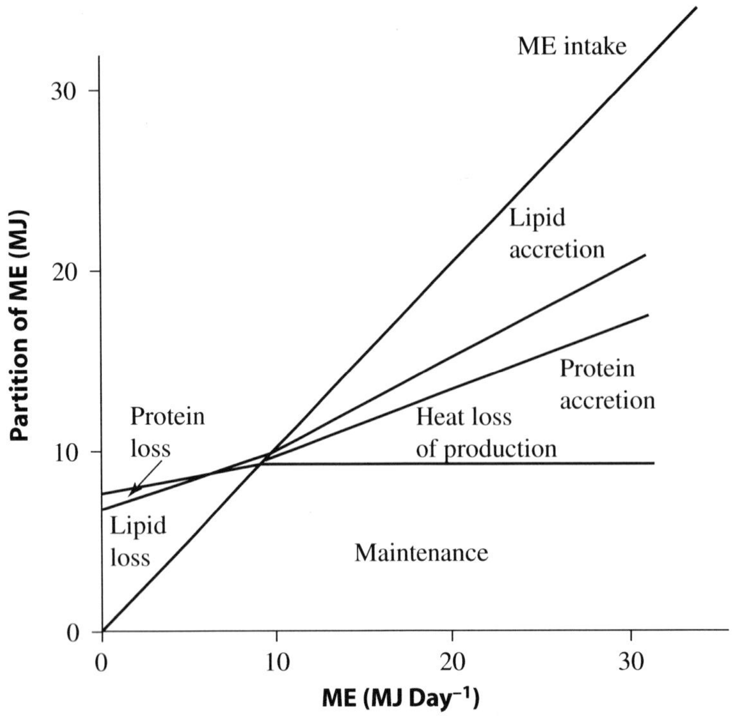

Figure 11.1 Energy partitioning in a pig (60 kg) fed different levels of metabolizable energy (ME). From T. L. J. Lawrence and V. R. Fowler, *Growth of Farm Animals,* University Press, Cambridge, London, 1997.

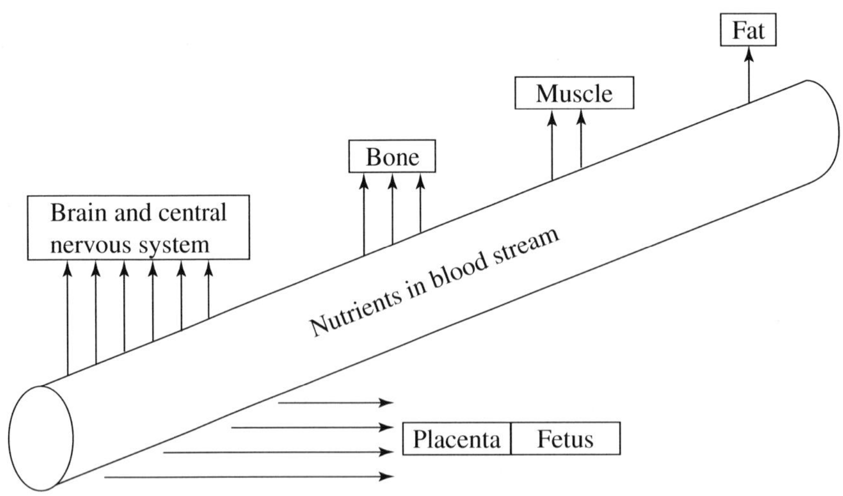

Figure 11.2 Priority and nutrient flow from the circulation to various tissues during growth. The number of arrows is proportional to the priority for nutrients. From T. L. J. Lawrence and V. R. Fowler, *Growth of Farm Animals,* University Press, Cambridge, London, 1997.

Adipose tissues may be ranked further into the various depots from highest to lowest priority according to their order of development:

- Mesenteric
- Perirenal
- Subcutaneous or Intermuscular
- Intramuscular.

The development of these adipose tissue depots has been discussed in Chapter 7.

In some circumstances, the regulation of nutrient partitioning among tissues is altered so that a new order of priority is established. From the rankings of tissues and organs for nutrient priority, one may see that vital tissues are maintained at the expense of others when food supply is limited. For example, during pregnancy, the fetus holds a priority similar to that of vital tissues and organs of the dam, and maternal nutrition and functional capacity of the placenta become key regulators of fetal growth. Another example is the partitioning of nutrients that occurs during transition from the dry (non-lactating) period to lactation period. The utilization of energy for gain and lactation occurs with different efficiencies (Figure 11.3).

Efficiency of tissue mobilization to support milk production is greater than the efficiency of tissue weight gain.

The time at which energy is stored and mobilized for milk production is also important in overall efficiency because energy is stored with greater efficiency during lactation than during the dry period.

Storage of energy during the dry period for subsequent mobilization during lactation is much less efficient than storage and utilization during lactation as well as the utilization of feed energy directly for milk production.

These differing efficiencies, which are affected by many factors, are involved in nutrient partitioning, which is largely due to the physiological state of the animal.

Regulation of tissue growth in a developmental manner involves differential expression of genes, many of which are developmentally regulated, tissue-specific, and important in proliferation, differentiation, and function of cells. The endocrine system plays a major role in these processes, and, as discussed in Chapter 10, hormones and growth factors are important in partitioning of nutrients among tissues and organs of the body. As stated earlier, long-term coordination of nutrient partitioning is referred to as homeorhesis. Perturbation of nutrient partitioning results in a change in body composition, growth efficiency, and feed:gain ratios (feed conversion or feed efficiency). If nutrients are repartitioned so that less fat and more muscle is produced, the composition of weight gain favors muscle. This shift in composition of gain is accompanied by improved feed:gain ratios because the amount of dietary energy required to deposit a unit of muscle is much less than that required to deposit a unit of adipose tissue. Consequently, *repartitioning agents* or strategies that alter the partitioning of nutrients among tissues may have important economical impacts on animal production.

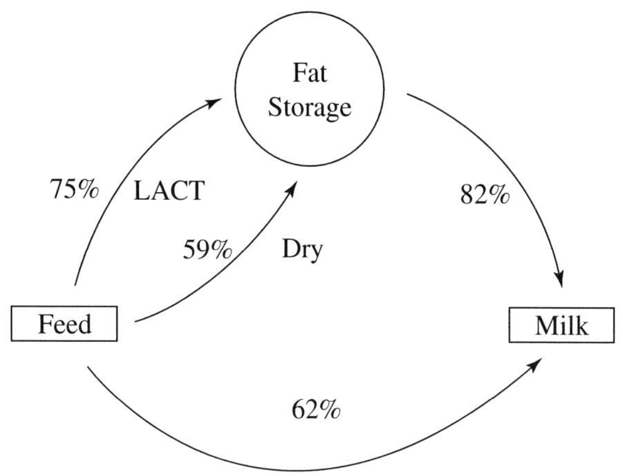

Figure 11.3 Efficiency of tissue storage and mobilization in lactating (LACT) and non-lactating (DRY) dairy cattle. From P. J. Van Soest, *Nutritional Ecology of the Ruminant,* O & B Books, Inc., Corvallis, Oregon, 1982.

When food is plentiful enough for body maintenance, growth, and fattening, an animal is considered to be in positive *energy balance.* A negative energy balance exists when food intake is not sufficient to meet these needs. A good example of a negative energy state is the energy status of dairy cows during peak lactation. At this peak period of lactation, during which daily milk production is at its highest, dairy cows cannot consume enough feed to maintain body tissue mass, even when energy-dense diets are formulated. Energy must be mobilized from body tissues to support production of milk during these peak production periods. It is not until later in lactation when milk production subsides that the cow enters a positive energy balance and tissue mass is replenished. Therefore, during periods of negative energy balance, animals lose weight, whereas in times of positive energy balances, they gain weight. Many examples of this phenomenon in other species exist. Shifts from positive to negative and back to positive energy balance during different stages of growth and physiological states is accompanied by repartitioning of nutrients within the body, indicating that energy balance is a key regulator of nutrient partitioning.

Undernutrition

Prenatal

The placenta plays an important role in the transfer of nutrients to the fetus. For some nutrients, transport requires energy because nutrients are transported against a concentration gradient. In other words, concentration of nutrients in the fetus is far greater than in maternal plasma. Metabolic activity of the placenta is, therefore very important in ensuring adequate nutrition for fetuses. Adequate nutrition during prenatal stages is important for fetal development, but prenatal undernutrition also may have significant effects on postnatal growth of animals. Low birth weights often are the result of prenatal undernutrition. Litter-bearing animals, on occasion, have one fetus that is much smaller than the other littermates. The cause of this is not completely known, but may be related to the position of the fetus in the uterus and (or) the size of accompanying placental tissues. In other words, inadequate blood or nutrient supply to the placenta and fetus may dramatically affect fetal growth.

As mentioned earlier, the runt pig is a good example of the consequences of prenatal undernutrition. Runt pigs exhibit birth weights that are two-thirds of the average birth weight of littermates. Vital organs are smaller and DNA content in various tissues is less in runt pigs than normal pigs (Table 11.1). Less DNA in skeletal muscle is a result of fewer muscle fibers (Chapter 5). Consequently, runt pigs possess less potential for postnatal muscle hypertrophy. If runt pigs are raised with normal pigs, requirements for lean growth are met sooner for runts than normal pigs. Thus, runt pigs grow slower and produce fatter carcasses because excess dietary energy is partitioned to adipose tissue (Table 11.2). If raised to similar carcass weights, carcasses from runts have lower percent muscle and higher

Table 11.1 Body Weight, Organ Weight, and Cellularity of Normal and Runt Pigs

	90 day fetus	At-birth runt	At-birth normal
Weight, gm	619	625	1606
Liver, gm	14.9	12.6	45.7
Quadriceps, gm	7.2	4.4	14.1
Heart DNA, mg	17.5	28.9	46.2
Brain DNA, mg	25.5	35.2	48.3
Quadriceps DNA, mg	25.4	16.4	33.3

From Widdowson et al. (1971).

Table 11.2 Carcass Composition of Runt and Normal Pigs at 96 kg Body Weight

	Normal	Runt 1	Runt 2
Fat depth, cm	2.39	2.49	3.02
Loin eye area, cm^2	36.6	35.6	29.8
Estimated % muscle	58.3	57.2	53.0
% Bone	13.8	14.0	14.3
Fat-free muscle, kg	17.8	17.6	16.1

From Powell and Aberle (1980).

percent fat. This change in composition also results in inferior feed:gain ratio in runts. Thus, prenatal undernutrition may result in restriction of postnatal growth potential and may significantly impact performance.

Postnatal

Many metabolic changes occur immediately before, during and after birth i.e., during *perinatal* stages of growth. This is probably driven by the fact that metabolic rates must increase to ensure survival of the animal. Therefore, nutritional requirements change drastically at birth. Glycogen and triglyceride is deposited in the fetus during late stages of gestation in preparation for this transition to the *neonatal* period or that period immediately after birth. Increased metabolic activity at birth is necessary for maintaining body temperature. Energy for this is derived initially from mobilization of large amounts of glycogen in liver and muscle. Some species may subsequently use adipose tissue as a source of energy when additional glycogen storage depots are empty. Large amounts of brown fat may be mobilized for heat production in some species, but livestock species contain little brown fat after birth. White adipose tissues may be used as energy sources, but in some species, especially pigs, there are only small amounts of fat present and available for use (Table 11.3). Consequently, colostrum plays an important role, not only in supplying immune factors, but also in supplying nutrients and factors necessary to improve survival of the newborn.

Weaning, or the time when offspring are removed from nursing their mothers, presents another major transition in which dramatic changes in quantity and composition of the diet occur. This transition often is accompanied by a transient decreased rate of growth, commonly referred to as *postweaning lag.* Fat tissue mass is usually affected during this restricted growth period, but lean tissue also may be lost. Formulating diets and management strategies to reduce or eliminate postweaning lag is a challenge for livestock producers. If postweaning lag is severe, it may potentially affect later body growth, and ultimate mature size.

In many animal production systems, animals are fed complete, balanced diets on an *ad libitum* or free choice basis. This ensures optimal lean tissue growth for meat-producing animals. However, excess energy beyond that required for lean tissue accretion results in excess fat deposition, which is detrimental to feed:gain ratio. One strategy to reduce fat deposition is to feed animals at amounts slightly less than ad libitum or on a *limit feed* basis. This has been shown to reduce fat deposition rates, but has little affect on lean tissue accretion (Table 11.4). The higher priority of muscle compared with adipose tissue ensures that muscle receives sufficient nutrients for maximum growth. This practice works well in individually housed animals where feed intake may be tightly con-

Table 11.3 Fat Content (g/100g Body Weight) of Various Newborns

Mouse	2.1
Rat	1.1
Cat	1.8
Pig	1.1
Lamb	3.3
Calf	2.8
Foal	2.6
Rabbit	5.8
Guinea pig	10.0
Human infant	15.0

[a]Taken from data by Widdowson (1950), Meyer and Ahlswede (1976) and Girard *et al.* (1979).
From E. M. Widdowson and D. Lister, Nutritional control of growth in Growth Regulation in Farm Animals, Advances in Meat Research, vol 7, A. M. Pearson and T. R. Dutson, editors, Elsevier Applied Science, New York, 1991.

Table 11.4 Effect of Feed Intake on Muscle and Fat Growth in Pigs

Feed intake (kg/day)	Muscle gain (g/day)	Fat gain (g/day)
3.0	600	850
2.5	590	750
2.0	375	450

[a]Adapted from Whittemore (1986).
From H. J. Mersmann, Regulation of adipose tissue metabolism and accretion in mammals raised for meat production in Growth Regulation in Farm Animals, Advances in Meat Research, vol 7, A. M. Pearson and T. R. Dutson, editors, Elsevier Applied Science, New York, 1991.

TABLE 11.5 COMPARISON OF LIVE WEIGHTS AND LIVE WEIGHT GAINS OF CATTLE FED A LOW PLANE OF NUTRITION FOLLOWED BY A HIGH PLANE OF NUTRITION VERSUS THOSE FED A HIGH PLANE OF NUTRITION THROUGHOUT THE STUDY

	Low Plane up to 350 kg High Plane after 350 kg	High Plane up to 350 kg High Plane after 350 kg
	Low	High
250–350 kg nominal weight range		
Initial weight, kg	259	258
Final weight, kg	357	356
Daily, initial to final weight	218	126
Gain, kg/day	0.45	0.78
350–400 kg nominal weight range		
Initial weight, kg	354	355
Final weight, kg	403	410
Daily, initial to final weight	36	56
Gain, kg/day	1.35	0.98
400–450 kg nominal weight range		
Initial weight, kg	410	407
Final weight, kg	447	458
Daily, initial to final weight	27	43
Gain, kg/day	1.38	1.20

From Wright and Russell, 1991, Animal Production, vol. 52, pg. 105–113.

trolled, but it is difficult to ensure consistent feed intake across animals that are fed in larger groups.

Compensatory Growth

Animals may exhibit *compensatory growth* after a period of nutrient restriction. Compensatory growth is growth exhibited by an animal after a period of nutritional stress. The rate of compensatory growth is often higher than that exhibited by a genetically-identical animal during normal growth (Table 11.5). It is also referred to as *catch-up growth* because the increased growth rate after the period of nutritional stress often results in body weights similar to animals whose growth was never compromised. Abnormally high growth rates are results of several physiological processes involving homeostasis and homeorhesis and may be explained as follows:

- A low basal metabolism is attained during the period of restricted growth because visceral weight is reduced.
- This low metabolic state is maintained for a time into the *realimentation* or re-feeding phase and allows more dietary protein and energy to be used for growth of major tissues rather than for basal metabolism.
- The increases in feed intake during the realimentation phase contribute to greater growth rates.

Precise mechanisms responsible for the biological basis of compensatory growth are unknown, but clearly the endocrine system is involved to regulate metabolism and nutrient partitioning among tissues.

Accumulative growth curves illustrating compensatory growth are shown in Figures 11.4, and 11.5. There are two types of compensatory growth:

- *Recovering Weight for Age*
- *Extended Growth Period.*

The former occurs when an animal "catches up" (weight) to the normal without requiring additional time. On the other hand, the latter situation occurs when feed restriction is so severe that the entire growth curve is extended to encompass greater time to reach the same weight endpoint. For example, if twin cattle are fed to 9 months on either a low or a high energy diet and then switched to a diet with a moderate energy level, the twin with significant reduced body weight at 9 months will exhibit sufficient compensatory growth to eventually catch up to its twin, and in some cases with surprisingly similar carcass composition (Table 11.6 and Figure 11.7). In yet more severe cases of nutrient stress, animals never 'catch up' even though additional time is provided. This type of nutrient restriction results in *permanent stunting* of growth. Although rare in production environment, severe nutrient deprivation in postnatal stages of growth commonly results in smaller mature body size (Figure 11.6 and Figure 11.8) and is often associated with excessive fat deposition.

Effects of compensatory growth on body composition depend on extent and severity of nutritional stress and the stage at which nutritional stress occurs. Stage of postnatal growth when nutritional deprivation is imposed affects the nature of response. For example, when undernutrition occurs early, long-lasting effects are greatest on the earliest maturing tissue (bone) and least on latest-maturing tissue (fat). Feeding a high plane of nutrition late in the lifetime of an animal when muscle and bone have matured will increase fat deposition. On the other hand, if pigs are started on a high plane of nutrition and switched to low level later, they produce carcasses with more muscle and less fat than those fed continuous high levels of nutrition. Such dietary regimes improve body composition and feed efficiency.

A classical experiment was performed by McMeekan and Hammond to demonstrate that changes in the shape of the growth curve by altering planes of nutrition result in changes in body composition (Figure 11.9). Pigs were grown on a low plane or a high plane of nutrition through 16 weeks of age to produce significant differences in body weights.

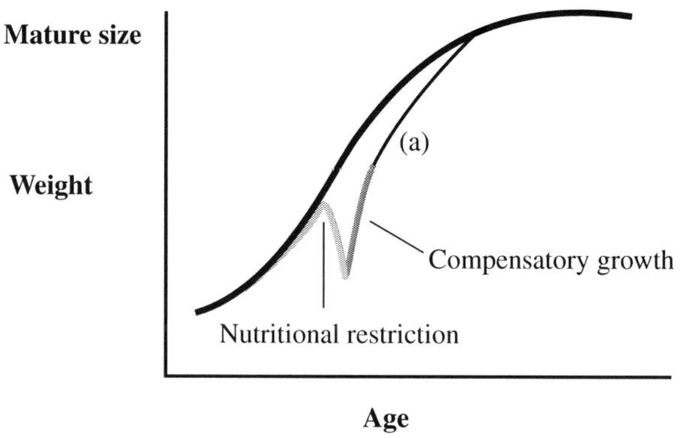

Figure 11.4 Compensatory growth curve: recovering weight-for-age. Figure from *Muscle and Meat Biochemistry,* edited by A. M. Pearson & R. B. Young, copyright 1989, Elsevier Science (USA), reproduced with permission from the publisher.

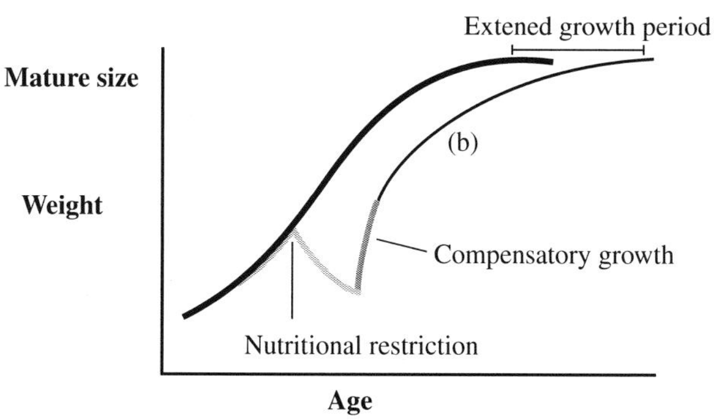

Figure 11.5 Compensatory growth curve: extended growth period. Figure from *Muscle and Meat Biochemistry,* edited by A. M. Pearson & R. B. Young, copyright 1989, Elsevier Science (USA), reproduced with permission from the publisher.

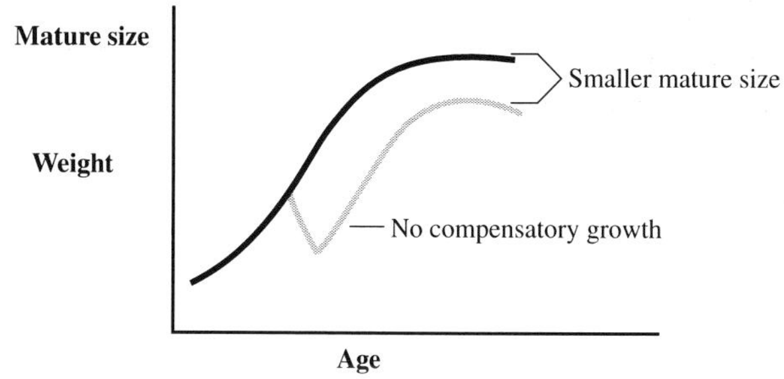

Figure 11.6 Permanent stunting of growth. Figure from *Muscle and Meat Biochemistry,* edited by A. M. Pearson & R. B. Young, copyright 1989, Elsevier Science (USA), reproduced with permission from the publisher.

TABLE 11.6 EFFECTS OF DIETARY ENERGY ON GROWTH AND COMPOSITION OF TWIN CATTLE

	Live Weight, kg			
	Energy Level			
Age	**High**	**Moderate**		**Low**
	323			92
1 yr.		383	154	
2 yr.		597	495	
4 yr.		760	782	
	Carcass Composition at 4 Years of Age			
Muscle, %		52.4	54.7	
Fat, %		31.9	28.8	
Bone, %		11.5	11.7	

From Bergström, P. L. Influences of extreme differences in feeding intensity until [i.e. until] 9 months of age on the growth rate and mature body composition in female identical twin calves. II, Body and carcass composition/P. L. Bergstrom and M. Dijkstra. *Zeist, [Netherlands]: Instituut voor Veeteeltkundig Onderzoek "Schoonoord", [1985]. 52 p.: ill.; 30 cm. Rapport/Instituut voor Veeteeltkundig Onderzoek "Schoonoord"; B–263. CAT89912980.*

Figure 11.7 Effects of restricted nutrient intake (top, left) and compensatory gain (bottom) in twin cattle. From Bergstrom, P. L. Influences of extreme differences in feeding intensity until 9 months of age on the growth rate and mature body composition in female identical twin calves. II, Body and carcass composition/P. L. Bergstrom and M. Dijkstra. *Zeist, [Netherlands]: Instituut voor Veeteeltkundig Onderzoek "Schoonoord", [1985]. 52 p.: ill.; 30 cm. Rapport/Instituut voor Veeteeltkundig Onderzoek "Schoonoord"; B–263. CAT89912980.*

Then, beginning at 16 weeks of age, pigs were grown on a high or low plane nutrition until they reached 90 kg body weight. Those on the high plane of nutrition for the duration of the experiment reached 90 kg by 180 days, whereas those on the low plane required 300 days. For pigs in which the plane of nutrition was interchanged at 16 weeks of age i.e., from low-to-high or from high-to-low, 240 days were required to reach 90 kg. Composition of carcasses from the pigs at 90 kg (Table 11.7), showed little difference in bone (as a percent of carcass weight) across treatments. This is not surprising because bone is an early-maturing tissue, so plane of nutrition has minimal effect on bone growth. Conversely, a high plane of nutrition late in development favors deposition of fat, a late-maturing tissue. Feeding a high plane of nutrition early followed by a low plane later, reduces fat deposition, yet allows muscle to develop at an acceptable rate, producing more desirable carcasses.

In some cases e.g., during times when quantity or quality of feed is limited, producers may be able to take advantage of compensatory growth because muscle is the predominant tissue deposited in the initial stages of compensatory growth. Frequently,

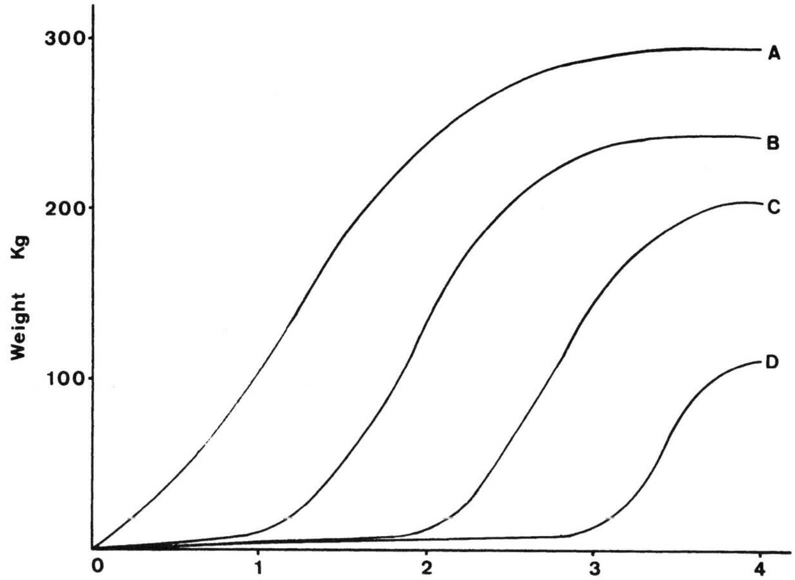

Figure 11.8 Cumulative growth curves of pigs fed levels of nutrition (A, maximal nourishment; B, undernourished for 12 months followed by maximal nourishment; C, undernourished for 24 months followed by maximal nourishment; D, undernourished for 36 months followed by maximal nourishment). Figure from *Muscle and Meat Biochemistry,* edited by A. M. Pearson & R. B. Young, copyright 1989, Elsevier Science (USA), reproduced with permission from the publisher.

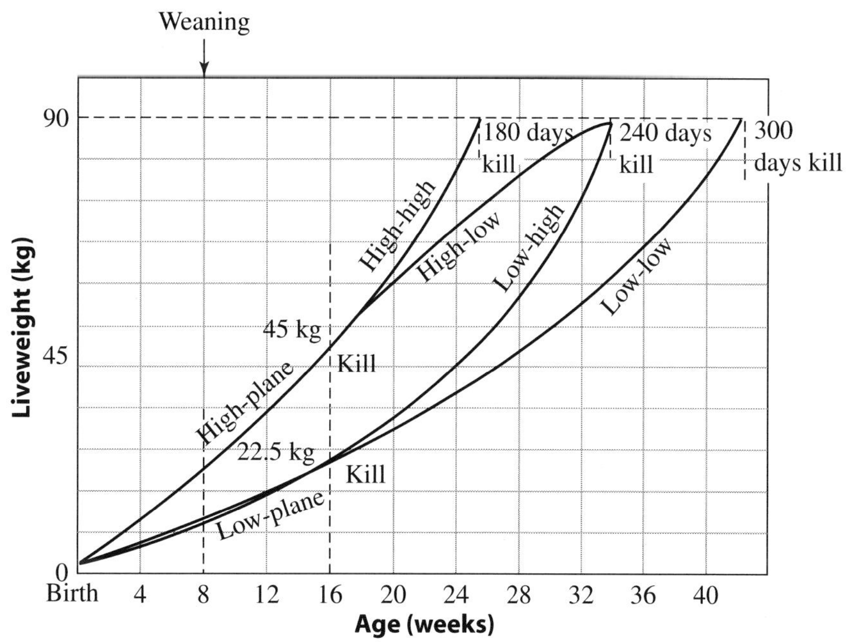

Figure 11.9 Effects of plane of nutrition on growth of pigs at different ages. From C. J. Hammond Jr, J. C. Bowman, and T. J. Robinson, Hammond's Farm Animals, Fifth Edition, Edward Arnold (Publishers) Ltd, London, 1983.

TABLE 11.7 EFFECTS OF PLANE OF NUTRITION (FROM FIGURE 11.9) ON CARCASS COMPOSITION OF PIGS AT 90 KG BODY WEIGHT

Percentages	High–High	High–Low	Low–High	Low–Low
1. Total weight basis				
Fat	36.9	28.7	40.1	26.6
Bone	11.1	11.6	10.4	12.6
Muscle	41.7	48.1	38.8	49.3
Skin, tendon, etc.	8.5	9.9	8.5	10.5
Dissection losses	1.8	1.7	2.2	1.0
Total	**100.0**	**100.0**	**100.0**	**100.0**
2. Fat free basis				
Bone	17.7	16.3	17.4	17.2
Muscle	66.1	67.6	64.9	67.2
Skin, tendon, etc.	13.5	13.9	14.2	14.3
Dissection losses	2.7	2.2	3.5	1.3
Total	**100.0**	**100.0**	**100.0**	**100.0**

From C. J. Hammond Jr, J. C. Bowman, and T. J. Robinson, Hammond's Farm Animals, Fifth Edition, Edward Arnold (Publishers) Ltd, London, 1983.

however, recovery of growth rate is greater for fat than for bone and muscle, and carcasses may become extremely fat. These changes in composition are likely due, in part, to changes in hormones that are involved in partitioning of nutrients and tissue accretion. Therefore, economical benefits of compensatory growth may require careful examination of composition of gain and ultimate carcass composition.

Compensatory growth is a phenomenon that occurs frequently throughout the growth of an animal. As pointed out in Chapter 9, accumulative growth curves are usually drawn as smooth sigmoidal curves, but in reality, growth curves are often more stair-step in appearance. Many environmental factors e.g., feed quality, disease organisms, or ambient temperature higher than the thermoneutral zone, may affect appetite or utilization and partitioning of nutrients. Consequently, these short periods of sub-optimal environmental conditions may result in slowing of weight gain (or even loss of tissue mass in severe conditions). Once the environment returns to a more optimal condition, appetite or nutrient utilization and partitioning returns to normal. This return to normalcy is accompanied by compensatory growth. Thus, the growth curve may be represented by a series of compensatory growth spurts, the frequency being dependent upon the environmental conditions.

Dietary Protein

Optimum dietary protein quantity and quality are necessary to maximize lean growth of animals. Both quantity and quality must be considered together because proteins are not utilized to the same extent. Protein quality is often defined by its *biological value,* and is calculated by determining the percentage of consumed protein that is actually retained by the body. Diets with sufficient energy, but insufficient protein for tissue protein deposition result in increased fat deposition in adipose tissue depots (Tables 11.8, 11.9, and 11.10). Furthermore, animals cannot synthesize tissue proteins beyond their genetic potential. Therefore, consumption of excess proteins results in surplus protein being metabolized for energy or excreted.

In ruminant animals, the quantity and quality of dietary protein is less critical than in monogastric animals because ruminants may utilize rumen microbial protein to satisfy some amino acid needs (Figure 11.10). The majority of the amino acids of ruminant feeds is extensively metabolized in the rumen to ammonia and carbon dioxide and volatile fatty acids (VFA). The ammonia and VFA are required for microbial growth. Sufficient nitrogen is necessary from the diet for microbial growth in the rumen, but some of this nitrogen may arise from *non-protein nitrogen* (NPN) sources. Optimal

TABLE 11.8 CARCASS COMPOSITION OF MALE BROILERS FED FINISHING DIETS VARYING IN PROTEIN AND AMINO ACID COMPOSITION

Treatments	Abdominal Fat (% Live Wt.)	Carcass Comp. Protein	(%DM) Fat
1 22% Protein	1.19	49.2	38.0
2 20% Protein	1.81	48.5	41.3
3 18% Protein	1.76	46.4	45.0
4 16% Protein	2.02	44.9	47.8
5 #4 + 0.1% Methionine	2.07	44.7	46.9
6 #4 = 0.1% Methionine + 0.1% Lysine	2.18	46.0	44.0

Summers and Leeson, 1985.

TABLE 11.9 EFFECT OF DIETARY PROTEIN (ENERGY HELD CONSTANT) ON BROILER COMPOSITION AT 49 DAYS OF AGE

Dietary Protein Level (%)	Carcass Composition		
	Fat (% D.M.)	Protein (% D.M.)	Moisture (%)
16	50.0	40.7	62.5
20	46.2	44.9	64.4
24	42.4	47.7	65.5
28	39.4	49.2	67.2
32	39.2	50.3	67.4
36	38.3	50.7	67.6

From Jackson et al., 1982.

growth of microbes is essential for maximizing efficiency of microbial protein synthesis. This microbial source of protein is not available in monogastrics, so diets for monogastrics must be formulated with protein of adequate quantity and quality.

Not all dietary protein in ruminant diets is subjected to ruminal degradation. The amount of protein that escapes the rumen without being digested and utilized is dependent on a number of factors including the degradability of the protein i.e., the rate and extent of protein degradation, the composition of the diet, feed intake level, and passage rate through the rumen. The protein that escapes the rumen is sometime referred to as *bypass protein,* although *rumen escape protein* is a more appropriate term. The

TABLE 11.10 EFFECT IN PIGS OF DIETARY PROTEIN

	*Protein Level of	
	17–15%	13–11%
Daily Gain, lb.	1.71	1.47
Feed/lb. Gain	3.16	3.57
Backfat, in.	1.28	1.39
LEA, sq. in.	3.75	3.31
% Lean Cuts	52.25	49.38

*Dietary protein in growing phase followed by dietary protein in finishing phase.

(Author unknown)

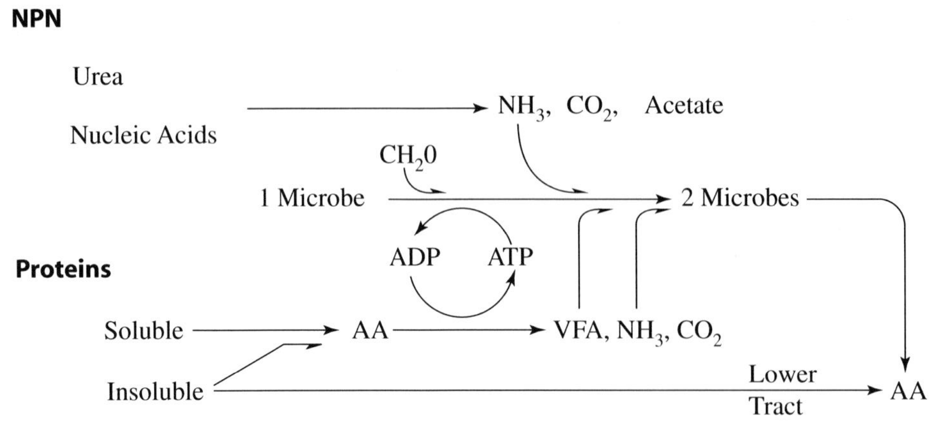

Figure 11.10 Nitrogen metabolism in ruminants. From Baldwin and Allison, 1993, Journal of Animal Sciences, Vol 57 (Suppl. 2), p, 470.

amount of protein that reaches the lower digestive tract of the ruminant is the combination of the rumen escape protein and microbial protein. Formulating ruminant diets that optimize both microbial protein synthesis and amount of rumen escape protein is a strategy for increasing the quantity and profile of amino acids available for absorption in the small intestine and may result in increased animal growth and performance. Characterizing protein degradability, i.e., proportion of protein that is *rumen degradable* versus *rumen undegradable,* of various feedstuffs has received considerable attention in attempts to improve formulation of diets for improved growth and performance of ruminants, especially those that have been genetically-selected for increased growth and performance.

Animals that exhibit increased muscle protein accretion rates require more protein. This is true of animals that have been genetically-selected for increased muscling or have been administered metabolism modifiers for the intent of increasing muscling. When comparing pigs with and without porcine somatotropin treatment, treated pigs have significantly greater protein requirements (Figure 11.11) because of the increased protein accretion in somatotropin-treated pigs.

Requirements for limiting amino acids in diets may be met by supplementing diets with synthetic amino acids. Formulating corn-soybean meal-based diets for swine based on total protein requirement may result in lysine deficiency, so supplementing swine diets with synthetic lysine is an economical way to meet lysine requirements. Amino acid supplementation prevents amino acid imbalance that may lead to restricted muscle growth and increased fat deposition (Table 11.11). An amino acid imbalance may result when there is a deficiency or excess of any one of the essential amino acids in the formulated diet. Amino acid imbalance also may result from addition of an amino acid other than the limiting one. This may cause decreased feed intake and growth. An appropriate balance of amino acids is important for efficient and optimal protein nutrition, and is optimized when amino acids are presented in proportions represented by proteins to be formed. Thus, diets are formulated in an attempt to accomplish ideal amino acid balance at various stages of growth. When an animal is unable to use all of the amino acids for protein synthesis because of an imbalance, the unused amino acids are *deaminated* or broken down and urea excretion increases. This results in a state of *amino acid deficiency* and is accompanied by decreased feed intake. Furthermore, catabolism of unused amino acids requires energy, which reduces energy available for tissue growth.

Dietary Energy

The energy required for maintenance increases with body size. That needed for growth varies with the rate and composition of body weight gain. Optimum dietary energy level may be assessed by increasing

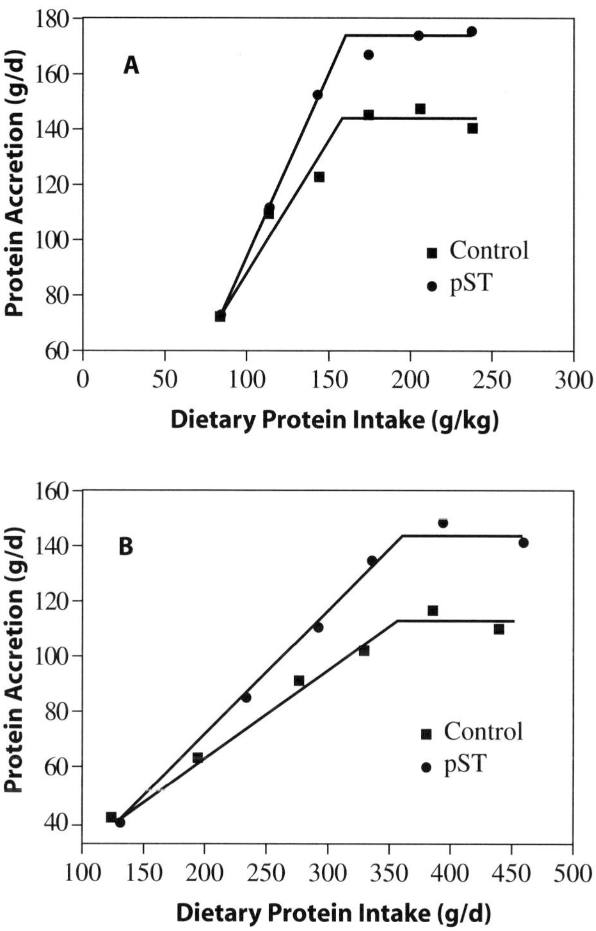

Figure 11.11 Effects of exogenous porcine somatotropin (pST) on protein requirements of growing (25 to 55 kg) pigs. From D. H. Beermann, Implications of biotechnologies for control of meat animal growth, in Proceedings of the 43rd Annual Reciprocal Meat Conference, National Live Stock and Meat Board, Chicago, Illinois, 1990.

Table 11.11 Amino Acid Supplementation of Low Protein Diets for Pigs

	Supplementation	
	Without	**With**
Daily Gain, lb.	1.36	1.44
Feed/lb. Gain	3.76	3.36
Backfat, in.	1.41	1.26
LEA	3.24	3.74
% Lean Cuts	49.20	52.70
Marbling Score	high	low

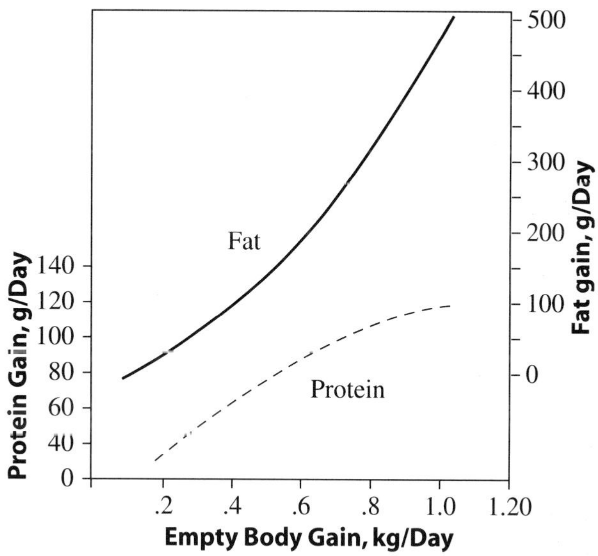

Figure 11.12 Theoretical protein and fat deposition rates in an animal with different body weight gains.

dietary energy intake (which increases body weight gain) to a point at which rate of lean gain is maximized (Figure 11.12). Although fat deposition will occur to some extent, it will be relatively lower until energy intake is increased beyond the point of maximum lean gain, at which point fat gain will increase significantly because the excess energy is partitioned to fat depots. Because fat contains much more energy than protein, energy requirement per unit body gain increases with this increased fat content. Consequently, composition of weight gain has a significant impact on total dietary energy requirements and feed conversion rates. This phenomenon is important to understand during the selection of animals. Animals that exhibit the highest growth rates are often genetically-selected in anticipation that such animals, or their offspring, will reach market weight sooner. However, high body growth rates may be a consequence of greater feed intakes, greater energy intakes, and may be a reflection of greater rates of fat deposition. Increases in body weights and rates of gain are very useful measures of growth, but they do not show the nature of the tissues being

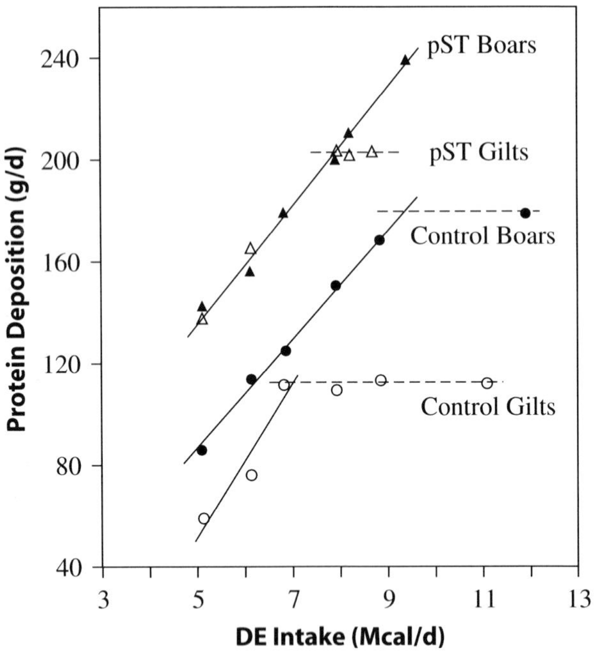

Figure 11.13 Effects of digestible energy (DE) intake on protein deposition in control and porcine somatotropin (pST)-treated gilts and boars. From R. G. Campbell, R. J. Johnson, M. R. Taverner and R. H. King, 1991, Interrelationships between exogenous porcine somatotropin (PST) administration and dietary protein and energy intake on protein deposition capacity and energy metabolism in pigs, *Journal of Animal Science*, vol. 69, pg. 1522.

formed. Monitoring body composition is important when animals are selected based on weight gain.

Maximum tissue protein accretion also requires an optimum amount of dietary energy. For example, when comparing gilts and boars, boars have greater capacity for depositing protein, and therefore, have a higher dietary energy requirement (Figure 11.13). Similarly, increased tissue protein deposition that results when treating pigs with somatotropin also increases the dietary energy requirement. Formulating diets for animals with such extreme protein deposition often requires the use of *supplemental fat* sources, such as *tallow.* Dietary fat is used by the animal for energy and is often used to supplement diets so that energy density may be increased in diets.

When energy content is high, but dietary fat is limited in diets, *de novo* synthesis of fatty acid will occur as described in Chapter 7. Carbohydrates will induce expression of the genes encoding the enzymes responsible for fatty acid synthesis e.g., fatty acid synthase and acetyl CoA carboxylase. However, if dietary fatty acids are available from the diet, expression of enzymes responsible for *de novo* fatty acid synthesis will decrease. Monogastrics can *assimilate,* or build, and deposit these dietary fats in a relatively unchanged form. This means that carcasses from monogastrics will have fat deposits of similar chemical composition to dietary fat. For example, pigs and chickens fed a diet high in *unsaturated fat* i.e., fat with a high ratio of unsaturated: saturated fatty acids, will have soft, oily carcass fat. The fatty acids are not altered and are used in the deposition of triglycerides in the fat depots of the animal. However, rapid turnover of depot fats permits replacement of deposited fat by dietary changes. Incorporation of fats with an increased proportion of saturated fatty acids in diets will tend to change the carcass fat from soft to more firm, thus the name *fat hardening diets.*

It is difficult to modify the fatty acid composition of fat depots of ruminants via diet because of rumen microbial action. Much of the dietary fat does not escape the rumen without first being subjected to microbial hydrogenation. Although the differences among nonruminants in composition of deposited fat are due to differences in dietary fat composition, differences between ruminants and non-ruminants in composition of deposited fat are explained largely by the difference in the gastrointestinal tract.

Antibacterial Feed Additives

Ionophores are commonly used as feed additives for cattle because of increased growth efficiency that results. The commercialized ionophores used for this purpose are *carboxylic polyether ionophores.* Although their mode of action is not completely clear, most appear to alter cell membrane properties, and consequently, play a major role in altering microbial populations in the digestive system. In fact, some ionophores have been used as effective *coccidiostats* (anticoccidial drugs) in poultry. In ruminants, ionophores change the populations and metabolism of rumen microbes (*bacteria, protozoa,* and *fungi*). Consequently, rumen fermentation is altered and digestion and absorption of nutrients is changed to such an extent that the animal grows more efficiently (improved feed conversion) and with some ionophores, the animal grows faster; however, body and carcass composition are not changed significantly.

Antibiotics also are used as poultry, swine, and cattle feed additives at therapeutic levels to treat sick animals, at prophylactic levels to prevent disease, and at subtherapeutic levels in healthy animals to improve feed efficiency and growth rate. Treatment and prevention of disease reduces the amount of feed that must be used to provide energy to the immune system and allow more feed to be partitioned towards tissue growth. The major mode of action by which subtherapeutic levels improve growth involves changing populations or growth of gastrointestinal microorganisms, which alters nutrient utilization and metabolism. Use of antibacterial feed supplements (including both antibiotics and ionophores) results in significant reductions in animal production costs. Little change in composition occurs by feeding antibiotics.

Regulation of Protein Accretion

The effect of nutritional status on overall body protein accretion is accomplished through alteration of rates of protein synthesis and degradation. Although synthesis and degradation rates may be coordinated, they appear to be regulated independently. This regulation depends on the type of nutritional stress. For example, when comparing pigs fed high carbohydrate and low carbohydrate diets, increased whole body protein accretion rate in pigs fed the high carbohydrate diet is accomplished through increased whole body protein synthesis with little change in whole body protein degradation (Table 11.12). Alternatively, altering total pro-

TABLE 11.12 TOTAL BODY PROTEIN TURNOVER IN PIGS FED LOW AND HIGH CARBOHYDRATE DIETS

Dietary carbohydrate[b]	Protein synthesis (g/day)	Protein degradation (g/day)	Protein accretion (g/day)
Low	112.2	99.5	12.7
High	131.1	96.7	34.4

[a]Data from Reeds *et al.* (1989).
[b]To provide 215 and 320 kcal/kg$^{0.75}$/day.
From W. G. Bergen and R. A. Merkel, Protein accretion, in Growth Regulation in Farm Animals, Advances in Meat Research, vol 7, A. M. Pearson and T. R. Dutson, editors, Elsevier Applied Science, New York, 1991.

TABLE 11.13 DIFFERENTIAL EFFECTS OF PROTEIN QUALITY AND QUANTITY ON PROTEIN BALANCE AND TURNOVER IN GROWING PIGS

Diet	Protein balance (g protein/day)	Protein synthesis (g protein/day)	Protein degradation (g protein/day)
Low protein	49	655	606
Low protein + lysine	104	637	533
High protein	128	706	578
High protein + lysine	117	757	640

Calculated from Fuller *et al.* (1987); protein = N × 6.25.
From G. R. Hollis, Growth of the Pig, CAB International Wallingford, UK, 1993.

tein concentration versus protein quality of the diet may have differential effects on synthesis and degradation (Table 11.13).

Rates of synthesis and degradation occur at different rates in different tissues of the body. Effects of low nutritional status on muscle protein accretion tend to be the result of decreased muscle protein synthesis rates. Muscle protein degradation rate may remain constant unless the nutritional insult persists to a point where degradation rate must decrease to reserve tissue protein. Such regulation is not common across all tissues. Skeletal muscle protein synthesis is very sensitive to nutritional stress compared with other tissues, especially in growing skeletal muscle. Amino acid supply and insulin concentrations are important in mediating this nutritional regulation of muscle protein metabolism. Conversely, protein accretion rates in liver appear to be regulated by nutritional status through changes in protein degradation rates rather than through changes in synthesis rates. Thus, coordination of protein synthesis and degradation rates among tissues is a key component for nutrient partitioning and adaptation to nutritional stresses.

The differential regulation of protein synthesis and degradation has significant impacts on the energetic efficiency of protein deposition. Although theoretical biochemical energetic efficiency of protein deposition is near 80%, actual efficiency is much lower (near 40%). The difference between the theoretical and actual efficiencies is due to the continual turnover of protein that occurs in the body. If degradation rates are increased, more protein synthesis is required to deposit the same amount of protein. For example, the higher protein turnover in cattle compared to the lower turnover in pigs results in significant lower energetic efficiency of protein deposition in cattle. This results in dramatic differences in feed conversion rates i.e., feed efficiency or feed:gain. The development of strategies to minimize protein turnover is important for optimizing ratio feed efficiency and energetic efficiency of protein deposition.

Chapter 12

Relationships among Growth, Composition, and Meat Quality

INTRODUCTION

Optimizing or enhancing growth of animals affects overall production efficiency. In the case of the meat-animal industry, the ultimate goal of such production systems is to produce an abundant, high quality source of protein as meat for human consumption. Two ways to improve animal productivity from an end product standpoint are by increasing the quantity of meat produced or by improving the quality. These are related, and in some cases, improving one attribute will result in a corresponding reduction in the other. Balancing the quantity and quality of products produced is challenging and often is determined by consumer demands and economic conditions. Therefore, it is necessary to develop an appreciation for characteristics and parameters used to assess and predict product quality and quantity.

The quantity of meat protein produced may be determined by quantifying the amount of muscle in animal at the time of marketing. Carcass dissections are extremely time consuming and require much effort. More acceptable alternatives are preferred for evaluating animal composition. Assessing product quality, is also difficult as characteristics that predict ultimate quality are difficult to measure. Furthermore, quality may mean different things to different consumers. Understanding the methods by which quality and quantity of animal products are assessed is beyond the scope of this chapter because there are other textbooks devoted solely to this field. However, fundamentals and principles underlying these factors must be discussed and understood, as any means of altering growth and productivity of meat animals will likely alter one, if not both of these parameters.

The Grading System

In an effort to standardize the identification of meat quality and quantity across the industry and to segregate and identify carcasses and fresh meat products according to palatability and compositional yields, the federal government developed and established a grading system. The *grading* system consists of a set of standards that must be met so that a particular level of quality or yield may be assigned. In most cases, the grading system requires human judgement for assessing various characteristics and parameters; however, automated, objective technologies are quickly gaining access in the industry. Once these characteristics or parameters have been evaluated for accuracy, they are subjected to a set of standards or equations to derive ultimate grades.

Started as an act of Congress in 1925, the Federal Meat Grading Service was primarily responsible for developing and providing meat grading standards and other associated services. The grading system relies heavily on the evaluation of various characteristics that are indicative of ultimate quality and quantity. After assessment of these indicators, final quality and yield grades are applied, respectively (Figure 12.1). Over the years these indicators have been modified and adjusted, partially to

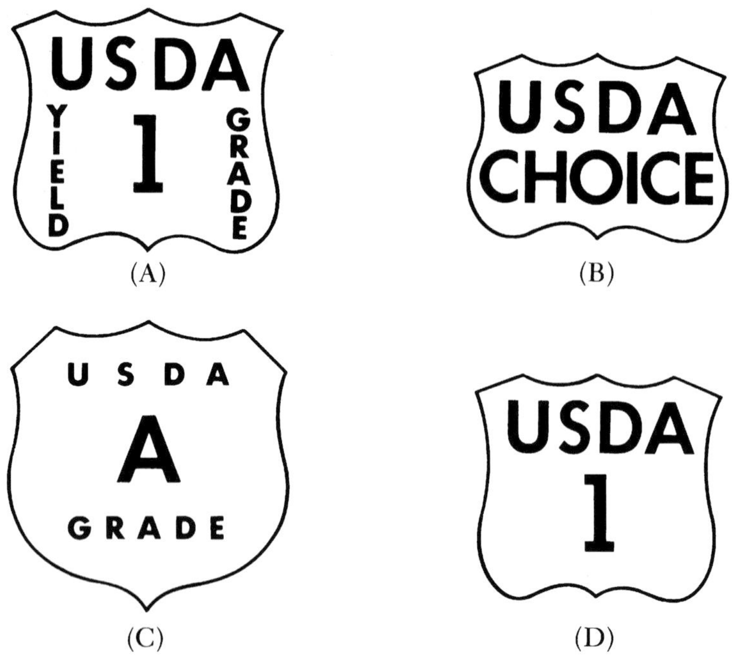

Figure 12.1 Identifying marks used for the U.S. Department of Agriculture grades for meat; (A) beef, lamb, and mutton yield grades; (B) beef, lamb, and mutton quality grades; (C) poultry grades; and (D) pork carcass grades.

improve the accuracy of the grading system, but also to reflect changes in consumer attitudes. Currently, the meat grading service is administered by the United States Department of Agriculture, Agricultural Marketing Service, Livestock and Seed Program, Meat Grading and Certification Branch.

Quality Grades

Meat quality, per se, is difficult to define and assess because of the inherent differences among different consumers. Essentially, this term is used to reflect consumer acceptability of the final product. Although meat quality is broad enough to entail processed meats, for the purpose of this text, our discussion will be confined to *fresh meat quality*, or that muscle tissue derived from meat animals that has not undergone further processing of any means, including freezing, grinding, chopping, etc.

Factors Affecting Quality Grades

Kind and Class First and foremost, grading meat quality requires an assignment of *kind*, which simply refers to the species of origin. Next, kind is subdivided into *class*, which is related to sex and maturity. Understandably, kind and class have a profound effect on the overall eating quality. Differences in palatability due to species and sex condition are sometimes striking. Therefore, kind and class are extremely important factors affecting the quality grade (Table 12.1).

Maturity Meat from younger animals is more tender than meat from older animals. Although the exact basis for this age-related decrease in tenderness is not known, changes in meat quality due to maturity are related to connective tissue properties. Therefore, maturity plays an important role in predicting overall palatability of meat.

Maturity, as discussed in an earlier chapter, may be defined in different ways. In reference to quality grades, maturity really implies physiological age. Recall from Chapter 9 that all animals do not mature at the same rate, and therefore, chronological age or absolute age is not as important as physiological age in predicting quality. Furthermore, even though chronological age may be of some value in predicting meat quality, this information is rarely available to meat grading personnel.

TABLE 12.1 UNITED STATED DEPARTMENT OF AGRICULTURE GRADES FOR MEAT

Kind	Class	Grade names
Beef quality grades	Steer, heifer, cow	Prime, Choice, Select, Standard, Commercial Utility, Cutter, Canner
	Bullock	Prime, Choice, Select, Standard, Utility
Beef yield grades	All classes	1, 2, 3, 4, 5
Calf quality grades		Prime, Choice, Good, Standard, Utility, Cull
Veal quality grades		Same as Calf
Lamb and mutton quality grades	Lamb, yearling mutton	Prime, Choice, Good, Utility
	Mutton	Choice, Good, Utility, Cull
Lamb yield grades	All classes	1, 2, 3, 4, 5
Pork carcasses	Barrow, gilt	U.S. No. 1, U.S. No. 2, U.S. No. 3, U.S. No. 4, U.S. Utility
	sows	U.S. No. 1, U.S. No. 2, U.S. No. 3, U.S. No. 4, U.S. Utility, Cull
Chicken	Rock Cornish game hens	A, B, C
	Broiler or fryer	A, B, C
	Roaster	A, B, C
	Capon	A, B, C
	Stag	A, B, C
	Cock or rooster	A, B, C
	Hen, stewing chicken, fowl	A, B, C
Turkey	Fryer or roaster	A, B, C
	Young hen	A, B, C
	Young tom	A, B, C
	Yearling tom	A, B, C
	Mature or old turkey	A, B, C
Duck	Broiler or fryer	A, B, C
	Mature or old duck	A, B, C
Geese	Young goose	A, B, C
	Mature or old goose	A, B, C
Pigeon	Squab	A, B, C
	Pigeon	A, B, C

Therefore, maturity or the physiological age of an animal is ascertained by evaluating various tissue characteristics of carcasses.

Carcass maturity is determined by viewing physical changes in bone and cartilage (Figure 12.2). As outlined in Chapter 8, when cartilage matures, it becomes ossified. In addition, bone morphology changes with age. These changes are clearly visible in cartilage located throughout the carcass and are especially apparent in some key locations. The most distal aspects of the dorsal processes of the vertebrae contain cartilaginous segments, the extensions of which are processes called *buttons* that are prominent on *split carcasses* i.e., carcasses separated into right and left halves along the median plane. In addition to the dorsal processes of individual vertebrae, cartilage is evident at the growth or epiphyseal plates of young animals and also between the *sacral verte-*

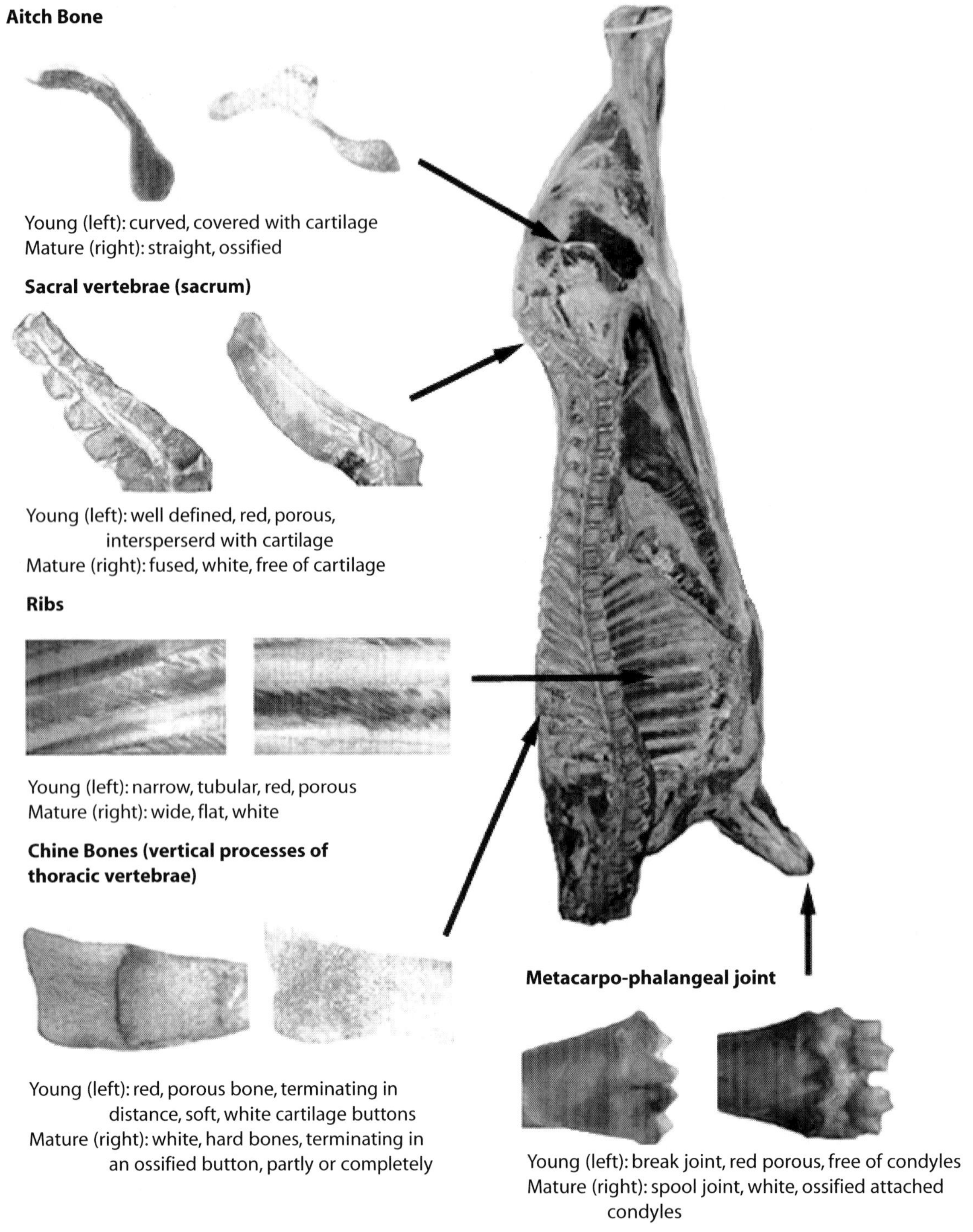

Figure 12.2 Indices of maturity in meat animal carcasses. A comparison of immature and mature characteristics. Courtesy of Amanda Weaver, Purdue University, West Lafayette, Indiana.

brae (i.e., vertebrae located in the region of the pelvis). In young animals, cartilage appears very white with a bluish tinge; as animals age, the cartilage begins to adopt a reddish tinge, mainly from increased infiltration by blood vessels. After additional time, ossification begins and bone forms in the center of the buttons. For reasons not understood, this process of ossification occurs in a rostral (tail) to caudal (head) direction along the vertebral column, meaning that the buttons in the sacral vertebrae ossify prior to those in the lumbar and thoracic regions of the vertebral column.

The most distal epiphyseal plates of the metacarpals located in the front legs of sheep also are important in determining class (Chapter 8). Again, during slaughter, pressure can be applied to the distal metacarpal of a sheep carcass. If the epiphyseal plate contains cartilage, the force will result in separation, or breaking, of the diaphysis from the epiphysis, commonly known in lay terms as a *break joint* (Figure 12.2). If the animal has matured to a point that sufficient ossification of this cartilage has occurred, the long bone will not break at the epiphyseal plate but rather will break more distally at the metacarpophalangeal joint. The resulting smooth, rounded anatomical structure exposed is the distal epicondyle of the metacarpal or a *spool joint*. Sheep carcasses yielding a break joint are considered *lamb*, whereas those containing at least one spool joint are graded as *mutton or yearling mutton* (Figure 12.3).

The optical properties of lean tissue change with maturity as well. Lean tissue color becomes darker or more intensely red with increased maturity as a consequence of increased concentrations of myoglobin. Because darkening may occur from other non age-related causes such as stress, changes in lean color are used only to fine-tune maturity estimates.

Muscle texture also is taken into account when assessing maturity. Muscle texture is related to the distribution of muscle bundles throughout the muscle. As animals age, the bundles become more prominent, larger and coarser in appearance than those from younger animals. The finer the texture, the more tender the meat, which is related to connective tissue amount and maturation. Therefore, texture of the lean is yet another indicator of increased maturity. Individual and genetic variations have pronounced affects on lean texture and may be misconstrued for differences in maturity. However, when used in conjunction with other indicators, lean texture may provide insight into maturity.

Marbling As pointed out in Chapter 7, fat deposits located within muscles are referred to as marbling. Individual *marbling flecks* are located between muscle bundles beneath perimysial connective tissue. An increased presence of marbling

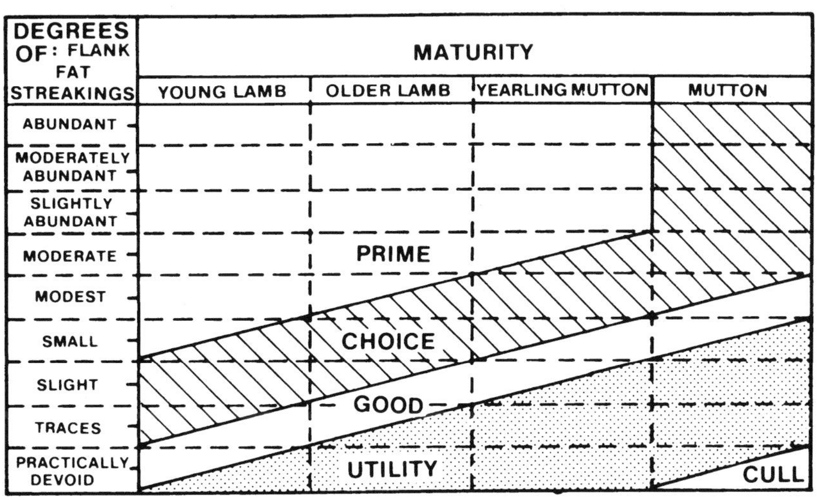

Figure 12.3 Relationships among flank fat streaking, maturity, and quality as used to establish lamb, older lamb, and yearling mutton carcass quality grades. From *Principles of Meat Science, Fourth Edition* by Aberle et al. Copyright © 2001 by Kendall/Hunt Publishing Company. Reprinted by permission.

within muscle is associated with increase palatability. Although most consumers believe that marbling improves meat tenderness, marbling simply improves the juiciness and flavor of the cooked product. Furthermore, marbling may improve mastication or lubrication during chewing, thereby improving perceived tenderness.

Marbling within the muscle provides crucial information about the history of an animal. In particular, marbling provides the grading system information regarding how the animal was fed during the later stages of its life. Animals provided maximal nutrient intakes for extended periods prior to slaughter generally have more marbling, and consequently, are considered of higher quality. Marbling scores are subjectively evaluated through visual appraisal of the longissimus dorsi muscle during grading (Figure 12.4). In the case of beef and pork carcasses, there are pictorial standards available for assigning quality scores. In most commercial settings, pork and lamb carcasses are not *ribbed,* a process whereby the fore and hind portions of the carcasses are separated exposing the cross-sectional area of the longissimus muscle. Nonetheless, the need for indicators that ensures that animals have deposited sufficient fat for palatability dictates the necessity for using another means for evaluating fatness in these species. As an alternative means of evaluating the degree of *finish* on lamb carcasses, the amount of *flank streaking* and *feathering* is ascertained (Figure 12.5). Flank streaking, as the name implies, is the appearance of fat in the rectus abdominus muscle or flank of the animal carcass. After processing, these muscles may be exposed for visual assessment. Because the rectus abdominus muscle contains alternating regions of muscle and connective tissue, greater and greater amounts of fat deposited in this region begin to obscure this architecture. Loss of this unique pattern in the muscle means more fat has been deposited on the carcass. Feathering, on the other hand, is fat that accumulates in the muscles between the ribs or *intercostal* muscles. The degree to which fat accumulates in this area is similar to marbling. In other words, fat accumulates in this location only when animals have been fed adequate nutrition over an extended period of time. Accumulation of feathering begins in the intercostal muscles closest to the ribs and, as fat deposition accumulates and fills the space between the ribs, the leading edge of fat resembles feathers.

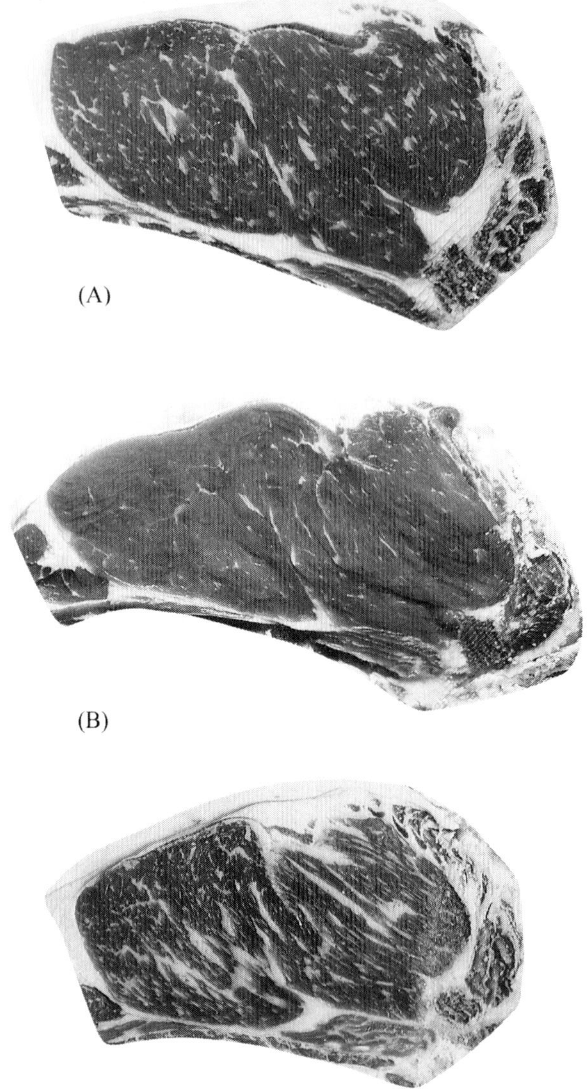

(A)

(B)

(C)

Figure 12.4 Marbling in beef longissimus muscle. (A) Ideal marbling; a modest amount uniformly distributed and finely dispersed in the muscle. (B) A muscle deficient in marbling. (C) A muscle containing excessive marbling. From *Principles of Meat Science, Fourth Edition* by Aberle et al. Copyright © 2001 by Kendall/Hunt Publishing Company. Reprinted by permission.

Firmness A characteristic that provides additional clues to overall quality of product is *firmness.* This characteristic is quite subjective in nature but is understandably very important to overall quality. Firmness of lean tissue is related to several attributes but essentially is the ability of the tissue to resist pressure. Obviously, factors such as temperature and fat content may affect firmness without directly

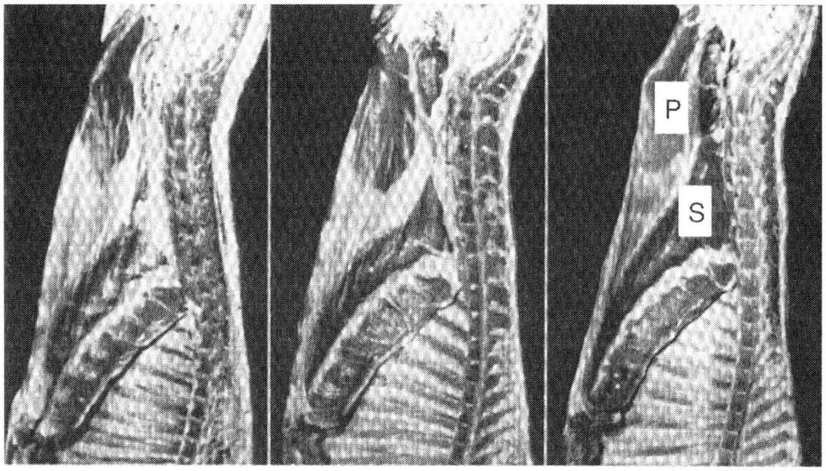

Figure 12.5 Illustrations showing high, intermediate and low degrees of flank fat streakings in the primary (P) and secondary (S) flank. From *Livestock & Carcasses* by Boggs et al. Copyright © 1998 by Kendall/Hunt Publishing Company. Reprinted by permission.

affecting quality. However, those carcasses possessing extremely soft lean are considered to be of inferior quality to consumers both from palatability and cooking characteristics. Firmness, or lack thereof, is related to the ability of the lean to hold or retain moisture. Therefore, lean that possesses a firm texture results in cooked product with more moisture and resultant desirable mouth-feel and juiciness. In contrast, those products lacking firmness lose much moisture and become dry and unsavory.

As stated earlier, firmness may be related to fat content as well as the composition of fat contained within the meat. Soft product has recently become a concern for pork processors because bellies have become soft. This softness is due largely to modern-day genotypes that do not deposit large amounts of fat anywhere in the carcass. As a result, the belly, which traditionally contains significant amounts of fat, is extremely lean. Once removed from a pork carcass, the main characteristic providing firmness to a belly is the fact that the fat "sets-up" or hardens. Therefore, in the case of lean pigs, bellies are soft and pliable at refrigerated temperatures. During bacon processing, bellies are pressed and formed to facilitate uniform and successful bacon slicing operations. For those bellies lacking firmness, the pressing operation fails to accomplish its task and slicing is difficult. To circumvent this problem, many pork processors have established minimums on belly thickness, wherein those animals failing to have minimum amounts of fat are discounted.

Firmness is not reserved only for lean tissue, as fat firmness is subject to evaluation as well. Recall from Chapter 11 that dietary factors are extremely important when considering carcass firmness. Unsaturated fat is less firm and more oily in appearance. This condition is also discounted severely.

Color Color of lean varies across species as well as within species. Color is important to quality, a characteristic that may be affected by processes in the *transformation of muscle to meat* or those biochemical reactions occurring immediately after the animal is slaughtered. In much the same way as firmness, consumers expect and prefer meat with a desirable color. Deviations from the norm result in decreased demand and subsequent losses to retailers.

Conformation, Fleshing and Finish As is the case for evaluating live animals, conformation is the relative proportionality of the product. Consider carcasses or a portion of a carcass that is extremely long and lacking in a proportional amount of width and thickness (Figure 12.6). These carcasses lack conformation or proportionality when compared to a carcass that may, in fact, be shorter but much thicker and wider in the other dimensions. Remember from

Chapter 9 that superior live animal conformation is related to increases in muscle to bone ratio and fatness of the carcass. Therefore, those carcasses or meat cuts with greater conformation generally have greater muscling, but may be fatter, especially considering intermuscular fat depots.

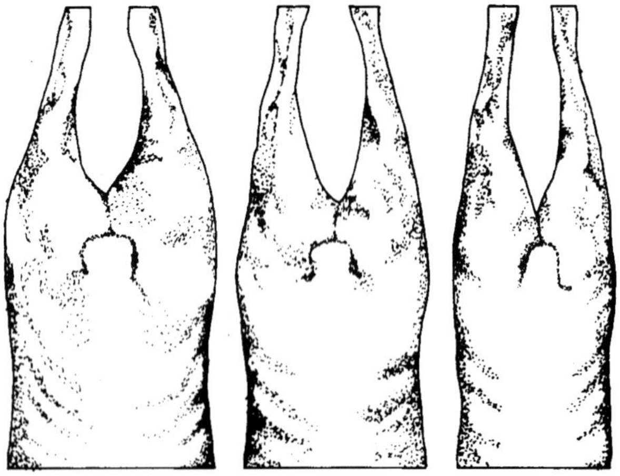

Figure 12.6 Relationship between conformation and leg scores in lamb carcasses. From *Livestock & Carcasses* by Boggs et al. Copyright © 1998 by Kendall/ Hunt Publishing Company. Reprinted by permission.

Poultry grades make extensive use of conformation. In particular, grading personnel are required to evaluate the conformation or *fleshing* of the entire bird (Figure 12.7). In this case, the muscling in breasts, legs, wings and thighs leads to more desirable conformation scores.

Defects Discounting quality due to carcass defects occurs primarily in the poultry industry. Poultry carcasses are not capable of attaining the premier distinction of "Grade A" if there are major pieces of flesh (skin or muscle) missing from the carcass. Whole birds with parts missing such as a leg or wing are unappealing even though they are wholesome and may be high quality. Any parameter that may affect the sensory properties likely will be used to evaluate quality.

Assigning a Final Quality Grade

Assigning a final quality grade to a carcass or product is based on standards that include the aforementioned characteristics. Once various parameters or indicators are collected, establishing a final quality grade is simply a matter of identifying the grade that best reflects the combination of parameters assessed. These parameters are quite different depending on the species (Figures 12.3, 12.8 and Table 12.2). Quality grades, per se, are a collection of descriptive terms that captures the essence of the product attributes and provides some insight to consumers as to the overall acceptability of the product. Terminology used to reflect those grades is not the

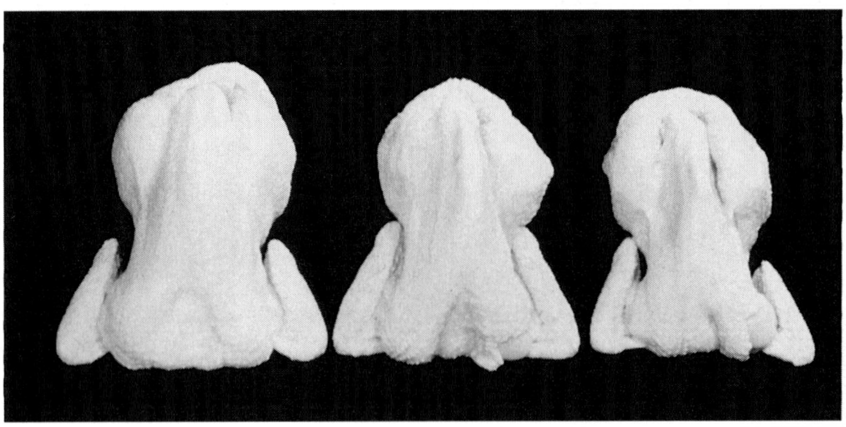

Figure 12.7 Front view of stewing chickens illustrating from left to right A, B, and C quality grades. From *Principles of Meat Science, Fourth Edition* by Aberle et al. Copyright © 2001 by Kendall/ Hunt Publishing Company. Reprinted by permission.

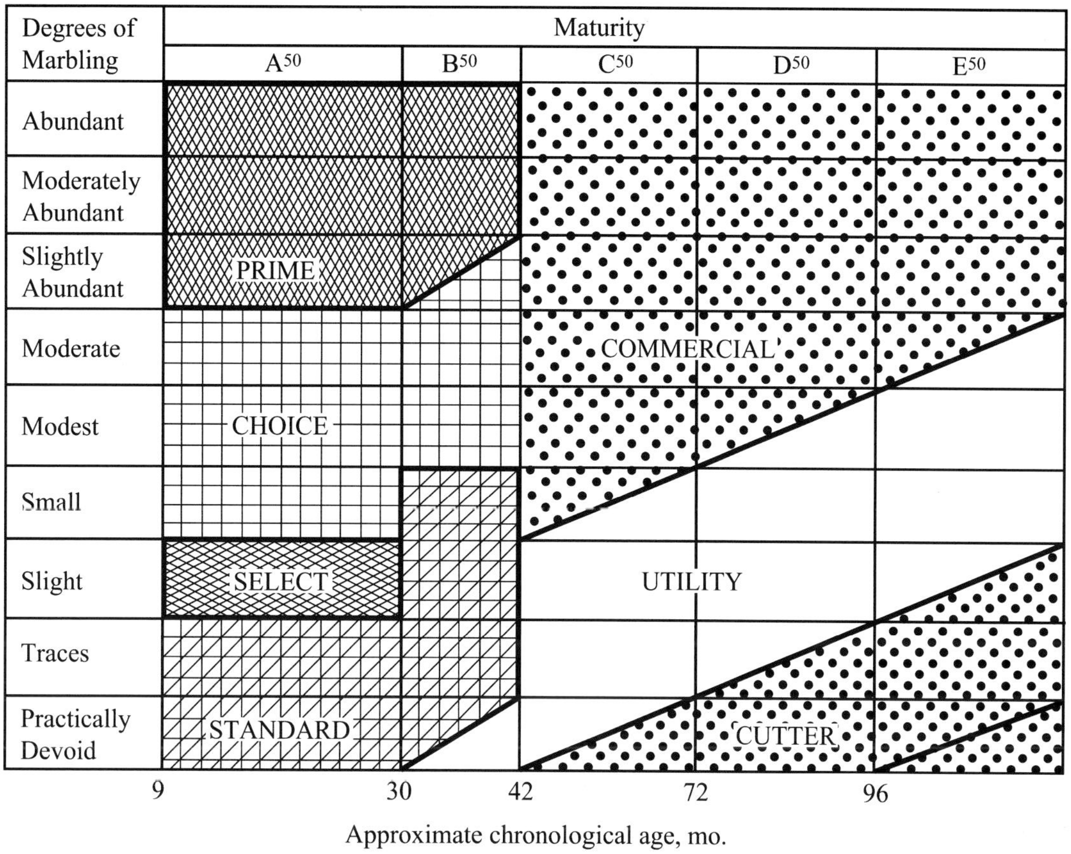

Figure 12.8 Relationships among marbling, maturity, and quality as used to establish beef carcass quality grades. From *Principles of Meat Science, Fourth Edition* by Aberle et al. Copyright © 2001 by Kendall/Hunt Publishing Company. Reprinted by permission.

same for all species. For example, beef quality grades consist of terminology that attempts to capture in some manner the overall acceptability of the meat. In contrast, quality grades for poultry products (A, B and C) are less descriptive (Table 12.2) and quality grades for pork are even more generic, wherein pork is either considered acceptable or unacceptable, depending on minimums and maximums for various quality characteristics. However, given the concern over diminished pork quality over past years, more precise standards have been developed, not by USDA, but by the industry through its commodity board (Figure 12.9) as shown on the inside front cover of this text.

Yield Grades

The other major attribute that the grading system attempts to convey to consumers is how much of the product is edible. This information is best captured in the *yield grade* and reflects the *cutability* or composition (leanness) of a carcass or product. Yield grades, much like quality grades, are assigned after various parameters are evaluated. These grades are used to segregate carcasses based on their *retail yield.* Therefore, retail yields designated for specific yield grades are expressed as a percentage (Table 12.3). In essence, yield grades provide insight into the amount of closely trimmed, retail cuts that may be expected from a given carcass. Carcass composition is essentially related to two very important factors: the amount of fat and the amount of lean in the product.

Amount of Muscle

Predicting the contribution of lean to the overall composition of the carcass is accomplished by estimating the muscularity of the carcass using a muscle measurement as a predictor. Recall from Chapter

TABLE 12.2 SUMMARY OF SPECIFICATIONS OF QUALITY FOR INDIVIDUAL CARCASSES OF READY-TO-EAT POULTRY

Quality		**A**			**B**			**C**
CONFORMATION		Normal			Moderate deformities			Abnormal
Breastbone		Slight curve or dent			Moderately dented, curved or crooked			Seriously curved or crooked
Back		Normal (except slight curve)			Moderately crooked			Seriously crooked
Legs and Wings		Normal			Moderately misshapen			Misshapen
FLESHING		Well fleshed, moderately long, deep and rounded breast			Moderately fleshed, considered kind, class and part			Poorly fleshed
FAT COVERING		Well covered—especially between heavy feather tracts on breast and considering kind, class and part			Sufficient fat on breast and legs to prevent distince appearance of flesh through the skin			Lacking in fat covering over all parts of carcass
PINFEATHERS								
Nonprotruding pins and hair		Free			Few scattered			Scattering
Protruding pins		Free			Free			Free
EXPOSED FLESH[1]	*Carcass Weight*	*Breast and Legs*	*Elsewhere*	*Part*	*Breast and Legs*[2]	*Elsewhere*[2]	Part	
Minimum	Maximum							No Limit
None	1 ½ lbs.	None	¼"	Slight trim on edge	¼"	1 ½"	Moderate amount of the flesh normally covered	
Over 1 1/2 lbs.	6 lbs.	None	1 ½"		1 ½"	3"		
Over 6 lbs.	16 lbs.	None	2"		2"	4"		
Over 16 lbs.	None	None	3"		3"	5"		
DISCOLORATIONS[3]								
None	1 ½ lbs.	½"	1"	¼"	1"	2"	½"	No Limit[4]
Over 1 1/2 lbs.	6 lbs.	1"	2"	¼"	2"	3"	1"	
Over 6 lbs.	16 lbs.	1 ½"	2 ½"	½"	2 ½"	4"	1 ½"	
Over 16 lbs.	None	2"	3"	½"	3"	5"	1 ½"	
Disjointed bones		1			2 disjointed and no broken or			No limit
Broken bones		None			1 disjointed and 1 non-protruding broken			No limit
Missing parts		Wing tips and tail[5]			Wing tips, 2nd wing joint and tail Back area not wider than base of tail and extending half way between base of tail and hip joints			Wing tips, wings and tail Back area not wider than base of tail extending to area between hip joints.
FREEZING DEFECTS (When consumer packaged)		Slight darkening over the back and drumsticks. Few small ⅛" pockmarksfor poultry weighing 6 lbs. or less and ¼" pockmarks for poultry weighing more than 6 lbs. Occasional small areas showing layer of clear or pinkish ice.			Moderate dried areas not in excess of ½" in diameter. May lack brightness. Moderate areas showing layer of clear, pinkish or reddish colored ice.			Numerous pockmarks and large dried areas.

[1]Total aggregate area of flesh exposed by all cuts and tears and missing skin, not exceeding the area of a circle of the diameters shown.
[2]A carcass meeting the requirement of A quality for fleshing may be trimmed to remove skin and flesh defects, provided that no more than one-third of the flesh is exposed on any part and the meat yield is not appreciably affected.
[3]Flesh bruises and discolorations such as blue back are not permitted on breast and legs of A quality birds. Not more than one-half of total aggregate area of discolorations may be due to flesh bruises or blue back (when permitted), and skin bruises in any combination.
[4]No limit on size and number of areas of discoloration and flesh bruises if such areas do no render any part of the carcass unfit for food.
[5]In ducks and geese, the parts of the wing beyond the second joint may be removed, if removed at the joint and both wings are so treated.

9 that, for the most part, muscle distribution does not change greatly with any management practice. Therefore, by estimating the size of one muscle, overall muscularity of the entire carcass may be evaluated. The cross-sectional area of the *longissimus* muscle has been chosen as a predictor of muscle mass. This muscle, at least in some species, is easily accessible by grading personnel. Furthermore, much of the carcass value comes from cuts made from the *longissimus* muscle, and therefore, changes in the size of this muscle are economically important.

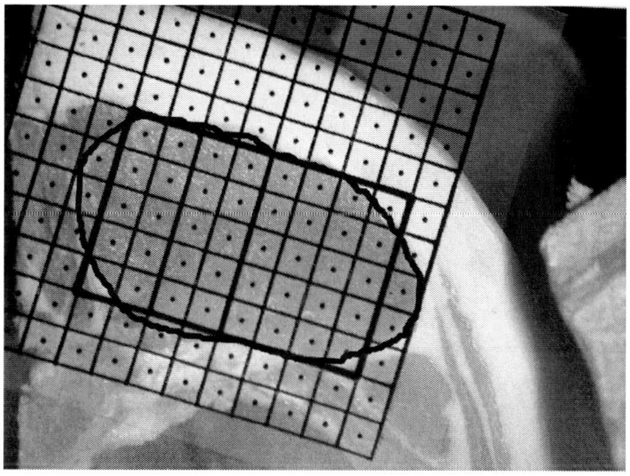

Figure 12.10 Photograph illustrating the method for assessing loin muscle area in a pork carcass. From *Pork Composition & Quality Assessment Procedures* by National Pork Board. Copyright © 2000 by National Pork Producers Council. Reprinted by permission.

The location for measurement of *longissimus muscle area, ribeye area* or *loin-eye area,* differs with species. In cattle, this measurement is made after carcasses are separated between the 12th and 13th ribs. Lamb carcasses are evaluated at the same location. Evaluating pork carcass muscling, on the other hand, is achieved using the same muscle but is performed between the 10th and 11th ribs (Figure 12.10). Because some carcasses are not ribbed commercially (pork and lamb), there must be another means of evaluating overall muscularity in the carcass. Although not as accurate as longissimus muscle area, subjective muscle thickness scores are used. The role of the evaluator in this case is to place the pork carcass in one of three muscling designations: *thick, thin* or *average* (Figure 12.11) and to assign a descriptive conformation grade to lamb carcasses.

Amount of Fat

The amount of fat on a carcass, obviously, has a profound effect on the yield or cutability of a carcass. In fact, if only one measurement could be taken from a carcass to predict composition, a measurement of some fat-related characteristic would be the most accurate choice. The number and type of measurements needed depend on the species being evaluated. Recall from Chapter 7 that the amount of fat

TABLE 12.3 EXPECTED RETAIL YIELD FOR BEEF AND LAMB YIELD GRADES AND EXPECTED PERCENT LEAN CUTS FOR PORK CARCASS GRADES

Beef or lamb yield grade	Retail yield		Pork grades	Lean cuts[c]
	Beef	Lamb[b]		
1.0	54.6	51.0	U.S. No. 1	60.4 and greater
2.0	52.3	49.7	U.S. No. 2	57.4-60.3
3.0	50.0	48.4	U.S. No. 3	54.4-57.3
4.0	47.7	47.1	U.S. No. 4	Less than 54.4
5.0	45.4	45.8	U.S. utility	Unacceptable quality

[a]Percent boneless, closely trimmed, retail cuts from the round, loin, rib, and chuck.
[b]Percent boneless shoulder, semiboneless leg (femur bone remaining), and bone-in rib and loin trimmed to 0.25 cm of external fat.
[c]Percent trimmed ham, loin, picnic shoulder, and Boston shoulder based on chilled carcass weight.
From *Principles of Meat Science, Fourth Edition* by Aberle et al. Copyright © 2001 by Kendall/ Hunt Publishing Company. Reprinted by permission.

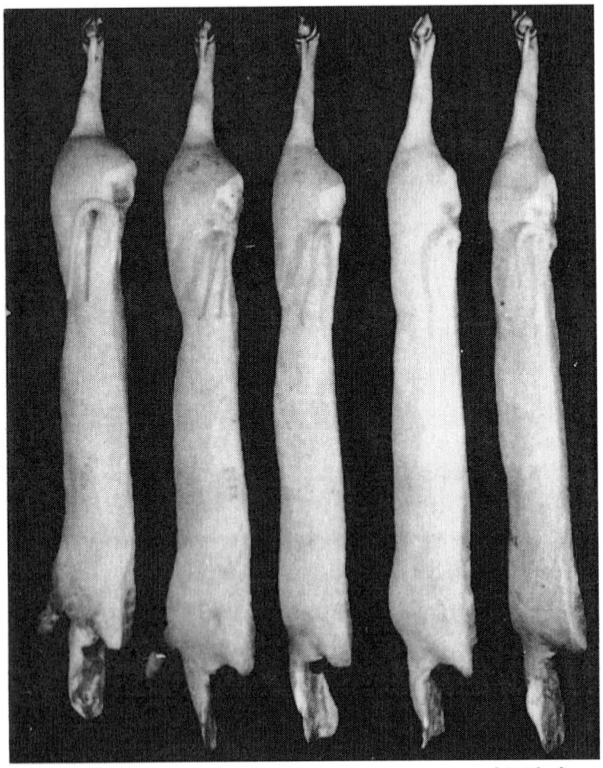

Figure 12.11 Degrees of muscle development for pork carcasses. From *Principles of Meat Science, Fourth Edition* by Aberle et al. Copyright © 2001 by Kendall/ Hunt Publishing Company. Reprinted by permission.

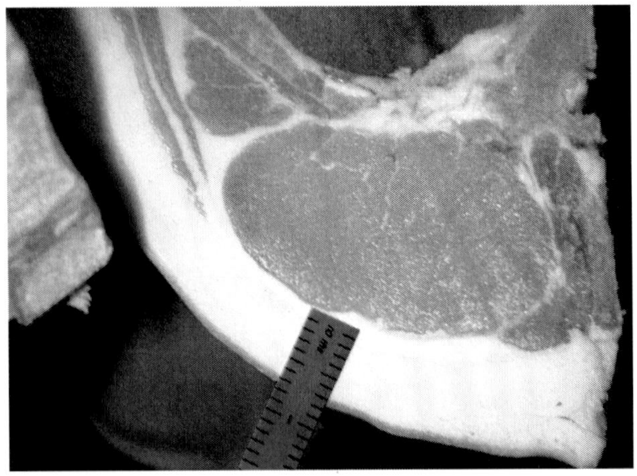

Figure 12.12 Photograph illustrating the method for assessing fat thickness at the 10th rib in a pork carcass. From *Pork Composition & Quality Assessment Procedures* by National Pork Board. Copyright © 2000 by National Pork Producers Council. Reprinted by permission.

in each depot differs with species. In particular, the percentage of fat located in the subcutaneous depot of pigs is extremely high compared to ruminant species. Therefore, any single measurement that accurately estimates the contribution of fat in this depot would likely provide great insight into the amount of fat in the carcass. In contrast, ruminants have considerable amounts of fat in the peri-renal (around the kidneys) depot, in addition to that located in the subcutaneous depot. Therefore, it is logical that any prediction of carcass composition would necessitate a measurement from each depot. To that end, the amount of fat located in the subcutaneous depot adjacent to ribeye area at the 12th and 13th ribs is quite useful in determining cutability in cattle and sheep. In pork carcasses, a subcutaneous *fat depth* or *fat thickness* measurement is made at the location where the loineye muscle area is evaluated, between the 10th and 11th ribs (Figure 12.12). In addition, another useful predictor of the amount of fat in the subcutaneous depot is known as the *backfat* measurement, which is determined at the midline or split surface of the carcass across from the last rib. The other means of evaluating fat content in ruminants is by evaluating the amount of kidney, pelvic and heart fat in beef carcasses, and kidney and pelvic fat in lamb carcasses. The amount of this fat is generally determined as a percentage of the carcass weight; therefore, graders must be well trained to make such determinations (Figure 12.13).

Carcass Size

As animals (or carcasses) become larger, the relative proportion of each component (muscle, fat and bone) changes. This concept was discussed in some length in Chapter 9. In particular, recall that as animals grow, the relative contribution of fat to the overall gain increases with time, and body composition changes. In the same manner, as carcass weight changes, so does composition. Therefore, it is imperative that any means of evaluating carcass composition include an estimate of carcass size. Carcass weight can be easily measured.

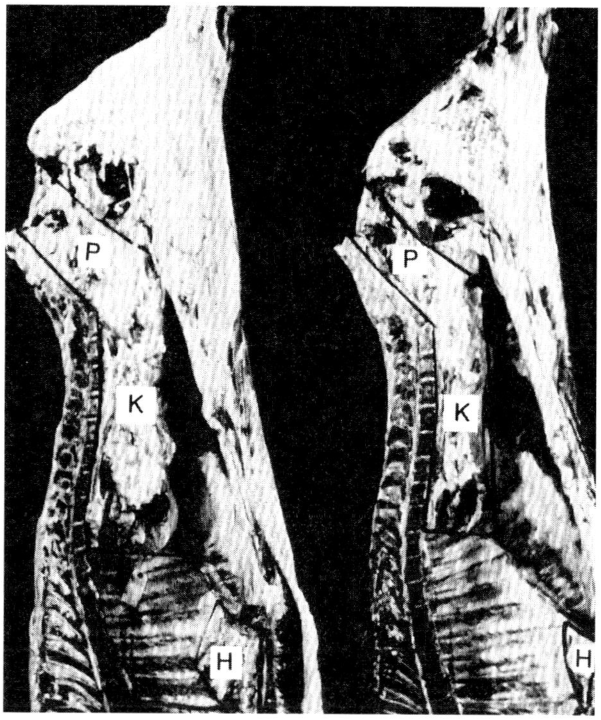

Figure 12.13 Beef carcasses showing the outline (in black) of the kidney, pelvic, and heart fat. The kidney fat is labeled with a "K", the pelvic fat is labeled "P" and heart fat is labeled "H". The carcass on the left has 5% kidney, pelvic, and heart fat. The carcass on the right as 2.5% kidney, pelvic and heart fat. From Livestock & Carcasses by Boggs et al. Copyright © 1998 by Kendall/Hunt Publishing Company. Reprinted by permissions.

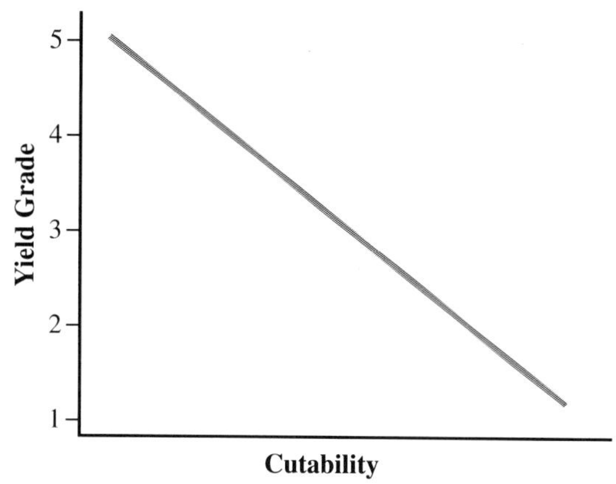

Figure 12.14 Relationship between beef yield grades and cutability.

Assigning Yield Grade

In some respects, assigning yield grade is perhaps less subjective than assigning quality grade. Yield grades are calculated using various measurements collected from the carcasses during grading.

Beef yield grades are calculated using the equation:

2.50 + (2.50 × fat thickness, inches)
– (0.32 × ribeye area, square inches)
+ (0.20 × percent kidney, pelvic and heart fat)
+ (0.0038 × hot carcass weight, pounds).

Lamb yield grades are calculated using:

0.4 + (10 × fat thickness, inches).

Depending on the values placed in the equation, calculated yield grades may vary widely and may be estimated to several decimal points. However, the industry only recognizes the numerical yield grades of 1, 2, 3, 4 and 5. Any yield grade falling between two whole numbers is always rounded down to the lower (numerical) yield grade. For example, a lamb or beef carcass possessing a 2.8 yield grade will always be considered a yield grade 2 carcass.

The larger the numerical yield grade the lower the cutability of the carcass (Figure 12.14). A greater fat thickness results in a larger numerical yield grade, but lower cutability and reduced percentage of lean yield. Fat occupying the areas around the kidneys, in the pelvis and in the thoracic cavity, must be removed during the fabrication of the carcass into retail cuts and, therefore, has a negative impact on yield grade, but a positive impact on cutability. Finally, carcass weight negatively impacts cutability by adding to the numerical yield grade. Referring to an earlier discussion in this and Chapter 9, the growth impetus of fat is greater than muscle during the later stages in the lifetime of an animal and even though increasing carcass weight is not truly an age related phenomenon; these are closely related.

Assigning pork carcass grades is quite different than that used to calculation of grades for beef and lamb carcasses, mainly because pork carcasses are not fabricated in the same manner. Intact or unribbed pork carasses are assigned a grade consistent with their last rib backfat thickness, and then it is adjusted for the relative thickness of muscling in the carcass. Pork carcass yield grade is calculated as:

(4 × last rib backfat thickness, inches) – (muscle score).

Muscle scores for pigs range between thick and thin, which corresponds to 3 and 1, respectively (Figure 12.11). If necessary, muscle scores may be further expanded by adding intermediate muscles scores 2.5 and 1.5. In reality, this is a rather imprecise means of reporting composition for pork carcasses. Another more accurate system is by calculating the absolute quantity of *fat free lean.* This type of prediction equation is similar to assigning yield grades for beef and lamb carcasses in that it contains some of the same measurements. The end result of this equation, however, is different in that it does not provide a grade (1–5) but rather an absolute amount of lean contained in the carcass. The amount (in pounds) of fat free lean is calculated:

8.588 + (0.465 × warm carcass weight, pounds)
+ (3.005 × loin eye area, square inches)
– (21.896 × fat depth at the 10th rib, inches).

The amount of muscle in a pork carcass may be further expressed as a percentage of carcass weight.

Changing Muscle into Meat

Conversion of living, functioning tissue into a highly desirable food source is a complex event that is extremely sensitive to a number of environmental factors. Of particular concern is the preslaughter biological status of the animal, which is particularly vulnerable to changes in animal growth performance. The intent of this section is to provide an appreciation for mechanisms associated with the transformation of muscle to meat and to identify some aberrations where augmentation of animal growth may change this process.

As previously mentioned (Chapter 10), the ability of an animal to maintain *homeostasis* or a relative state of equilibrium is mainly under control of the central nervous system and essentially adjusts various physiological parameters to compensate for deviations from the norm (Figure 12.15). This process of maintaining homeostasis or resisting effects of stressors is instrumental in regulating the conversion of muscle to meat.

Any and all activities that challenge the homeostatic status of an animal during *antemortem* handling are considered stressors. When challenged by a stress, animals mobilize energy and increase their heart rate to compensate for unsettling situations. Release of hormones that aid in the mobilization of energy occurs early in the process. Therefore, the homeostasis of an animal prior to slaughter is critical because energy reserves are extremely sensitive to antemortem handling.

In the slaughter facility, animals are rendered unconscious by *stunning.* Immediately after stunning, they are *exsanguinated* (i.e., blood is released from the circulatory system). In response to a rapid loss of blood pressure, the animal responds by constricting blood vessels in an attempt to preserve blood volume. Soon after, however, the circulatory system is lost and body temperature rises.

At the tissue level, muscle attempts to maintain homeostasis. As a result, huge amounts of ATP are utilized and the tissue begins to replenish dwindling reserves. Because tissue lacks the ability to garner oxygen, normal *aerobic metabolism* is not possible (Figure 12.16). Therefore, the tissue reverts to *anaerobic metabolism.* This biochemical pathway is known as *anaerobic glycolysis* and *glycogenolysis.* (Figure 12.17). Glycogenolysis is the process of cleaving *glucose residues* or *molecules* from *glycogen,* the primary energy store in skeletal muscle. This process is mediated by the enzyme *glycogen phosphorylase* and is stimulated by increased cytosolic calcium and ADP levels, the latter being a breakdown product of ATP. Once glucose residues, in the form of *glucose-1-phosphate,* are cleaved from glycogen, they are further metabolized to *pyruvate* via glycolysis. Because the *TCA* or *Kreb's cycle* is incapable of functioning without oxygen, pyruvate is metabolized to lactate. As hydrogen ions continue to accumulate from various biological reactions, lactate, in turn, becomes lactic acid. As a result, there is a progressive reduction in muscle pH postmortem until metabolism ceases, at which point pH remains relatively constant (*ultimate pH*).

Deviations from Normal Postmortem Metabolism

The biochemical processes outlined previously are quite normal and necessary for muscle to assume the normal characteristics of meat. Depending on the antemortem status of the animal and postmortem conditions, the rate and magnitude of postmortem metabolism may be altered, which may severely impact ultimate quality of meat. The direct cause of this change in meat quality is directly related to changes in muscle pH and temperature during the postmortem period (Figure 12.18).

Dark Cutting Meat

If animals are exercised rigorously or administered epinephrine (see Chapter 10) prior to slaughter,

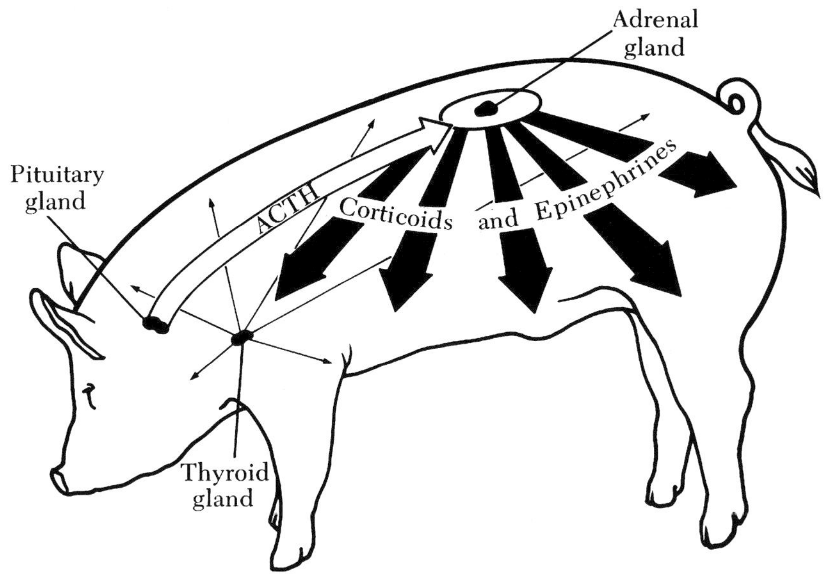

Influences of the "Stress" Hormones

ACTH—stimulation of corticoid release
Corticoids—synthesis of carbohydrates; maintenance of cell response and ion balance
Epinephrines—maintenance of circulation; breakdown of glycogen and lipids
Thyroid hormones—stimulation of oxidative metabolism

Figure 12.15 A simplified diagram of the sites of production, and the general actions of the major hormones related to stress reactions. From *Principles of Meat Science, Fourth Edition* by Aberle et al. Copyright © 2001 by Kendall/ Hunt Publishing Company. Reprinted by permission.

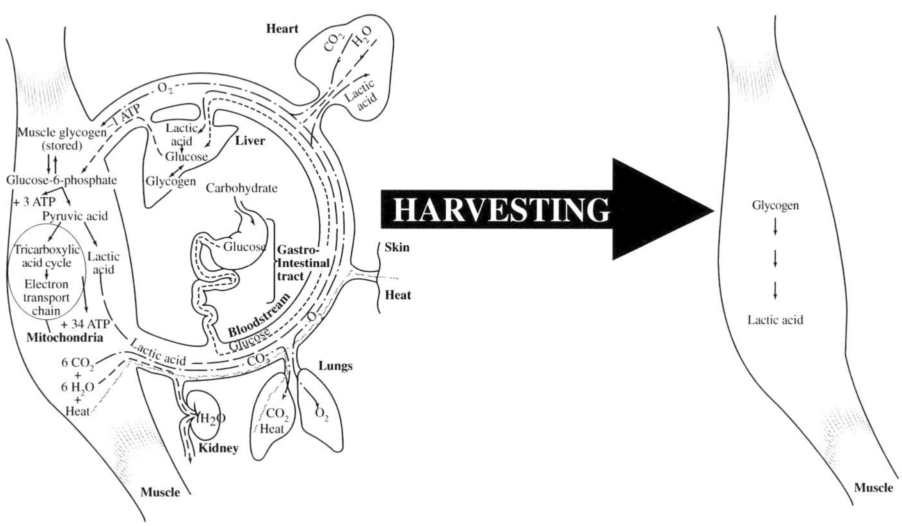

Figure 12.16 Schematic illustrating the metabolism and movements of metabolites in living tissues (left) and muscle (right) immediately postmortem. From *Principles of Meat Science, Fourth Edition* by Aberle et al. Copyright © 2001 by Kendall/ Hunt Publishing Company. Reprinted by permission.

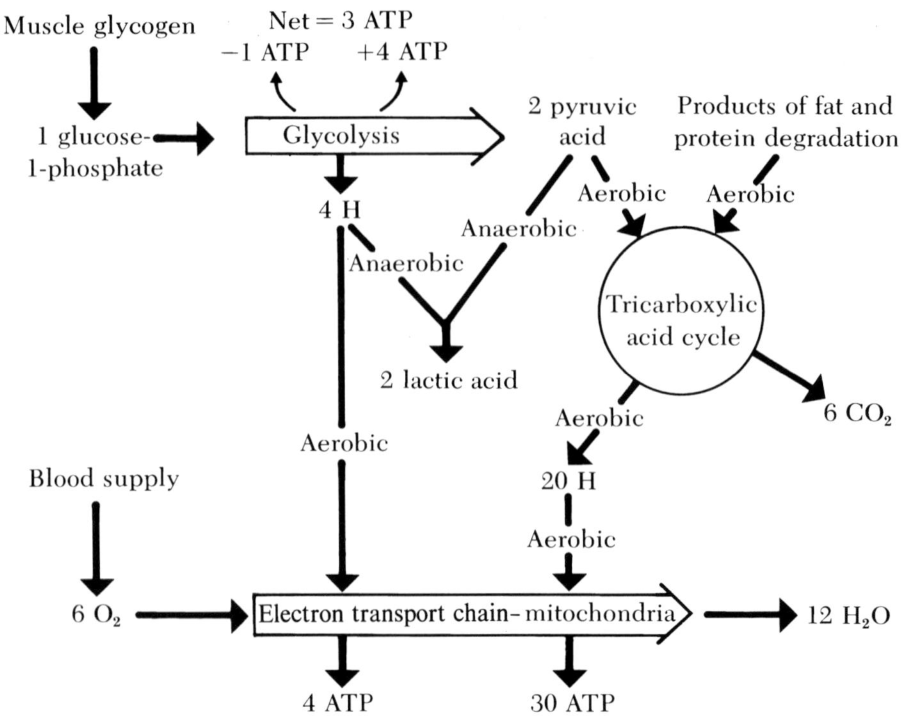

Figure 12.17 A diagrammatic illustration of the pathways that supply energy for muscle function. One molecule of glucose 1-phosphate split from glycogen is degraded to CO_2 and H_2O in the glycolytic pathway, tricarboxylic acid cycle, and electron transport chain. The energy yield (in terms of molecules of ATP produced) is indicated at each step. When oxygen is limited, energy may be supplied by glycolysis and conversion of pyruvic acid to lactic acid.

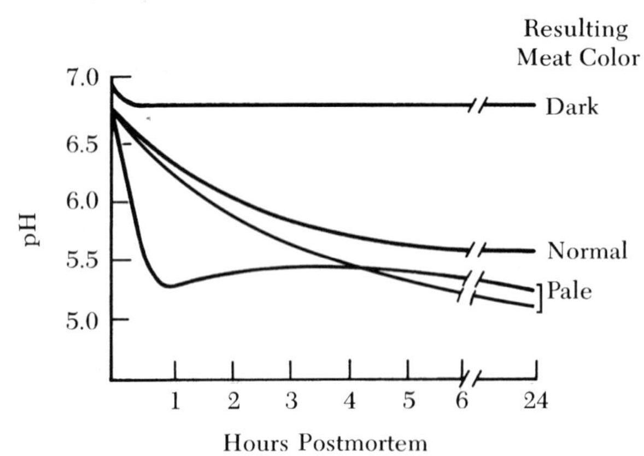

Figure 12.18 Postmortem pH decline curves. Modified from Briskey. E. J., "Etiological Status and Associated Studies of Pale, Soft, Exudative Porcine Musculature," *Adv. Food Research 13,* 89 (1964) Academic Press, Inc.

muscle glycogen content may be reduced to a point that little lactic acid is generated in the muscle postmortem. In this case, the ultimate pH of the muscle reaches about 6.8, only a few tenths below normal physiological muscle pH (7.2). Animals generating such a small reduction in muscle pH are considered *dark cutters,* mainly because the color of the resulting lean is quite dark red. The ramifications of this dark color are reduced consumer acceptability and shelf-life (Figure 12.17). Therefore, meat quality may be altered simply by altering the amount of substrate (energy), in this case glycogen, in muscle prior to slaughter.

Pale, Soft, Exudative Meat

In contrast to development of dark cutters, animals with adequate levels of glycogen in the muscle prior to slaughter that experience rapid rates of glycolysis will exhibit pale, soft, exudative (PSE) meat (Figure 12.18), a characteristic discussed in earlier chapters. This type of meat results not solely from rapid rates of energy metabolism postmortem, but rather a pH/temperature interaction. If the pH of the muscle decreases too rapidly before carcass temperature is sufficiently reduced by normal chilling practices, muscle proteins will denature and lose their native structure. This loss of structure translates into a loss of the proteins' ability to bind and hold water. Loss of *water holding capacity* translates into greater *purge* from the cut surface of meat cuts. Purge or exudate is rich in proteins and represents a loss to the industry. Furthermore, cooked products made from meat lacking adequate water holding capacity loose much moisture when cooked and are relatively dry when eaten. Pale, soft, exudative meat is most prevalent in the pork and poultry industry; however, any practice that leads to a rapid pH decline in muscle early postmortem may negatively impact meat quality.

Acid Meat

Another meat quality defect related to postmortem metabolism is the development of *acid meat*, (Figure 12.17). During normal transformation of muscle to meat, an ultimate pH of about 5.6 to 5.8 normally occurs. Even pale, soft, exudative meat achieves a normal ultimate pH. In acid meat, however, the ultimate pH of muscle reaches levels of 5.5 or lower. Characteristics of acid meat are essentially identical to pale, soft, exudative meat. The only difference between these quality-altering conditions is the way in which they manifest themselves during the postmortem period. In contrast to an increased rate of postmortem metabolism, muscle undergoing the development of acid meat does so through a protracted or extended period of postmortem metabolism. The mechanism underlying this phenomenon is not understood, but it is known that the muscle contains greater amounts of glycogen than normal. This condition also seems to be more prevalent in pork carcasses, mainly due to a genetic mutation that will be covered later.

Meat Tenderness

The most difficult quality parameter to predict or assess is meat tenderness. Nonetheless, tenderness is very important to consumers. Tenderness is the perceived ease of *mastication* (chewing). Muscle contains two major entities that contribute to tenderness. The filamentous or fibrous nature of muscle contributes to chewing resistance (see Chapter 2). Likewise, muscle contains connective tissue, which adds strength to the tissue and may alter tenderness. Both of these components of muscle are extremely important to final eating quality of meat and can be affected by changes in animal growth. Little can be gained by improvement of animal growth if the resulting quality of the endproduct is undesirable to consumers.

Connective Tissue

Recall from Chapter 2 that muscle tissue contains an elaborate network of connective tissue (Figure 12.19). The type and amount of connective tissue in muscle affects meat tenderness. Much of the tenderness difference observed between different muscles may be attributed to connective tissue abundance. Connective tissue properties change with animal age, too, and contribute to the meat from older animals being less tender than that from younger animals. This toughness of meat from older animals is due to the overall age or maturity of the connective tissue. As animals mature physiologically, the number of heat-stable crosslinkages increases and become more stable, and the strength of the collagen increases (Figure 12.20). As a result, the solubility of connective tissue decreases with age and the amount of collagen that may be broken down during cooking decreases.

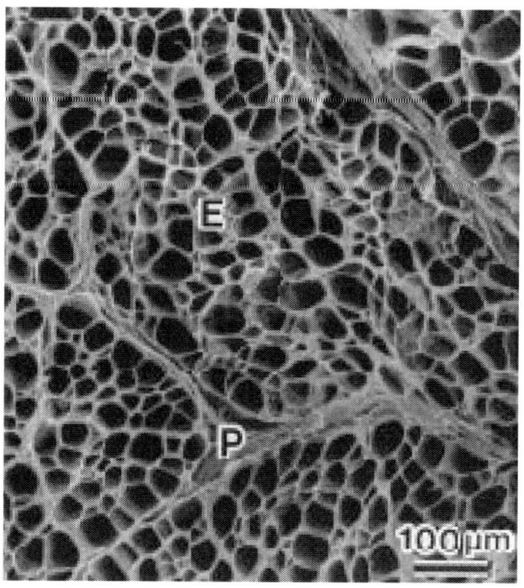

Figure 12.19 Micrograph illustrating the importance of the connective tissue framework in beef muscle (E, epimysium; P, perimysium). From T. Nishimura, A. Hattori, and K. Takahashe (1999) Structural Changes in Intramuscular Connective Tissue During the Fattening of Japanese Black Cattle: Effect of Marbling on Tenderization, JAS 77:93.

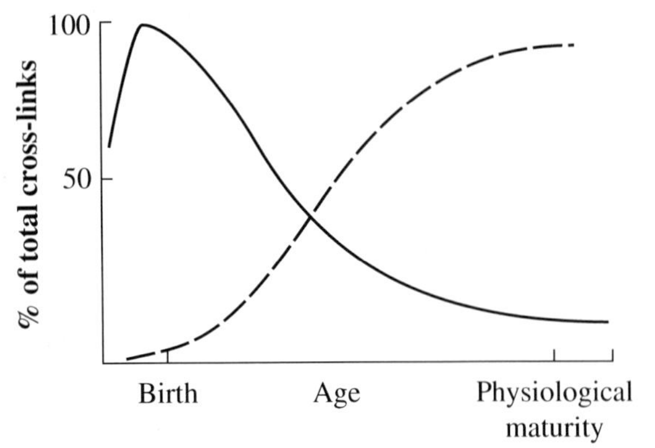

Figure 12.20 Effect of ageing on reducible (- - - -) and mature (——) cross-link formation in skeletal muscle. From A. J. Bailey and N. D. Light, *Connective Tissue in Meat and Meat Products,* Elsevier Applied Science, London (1989).

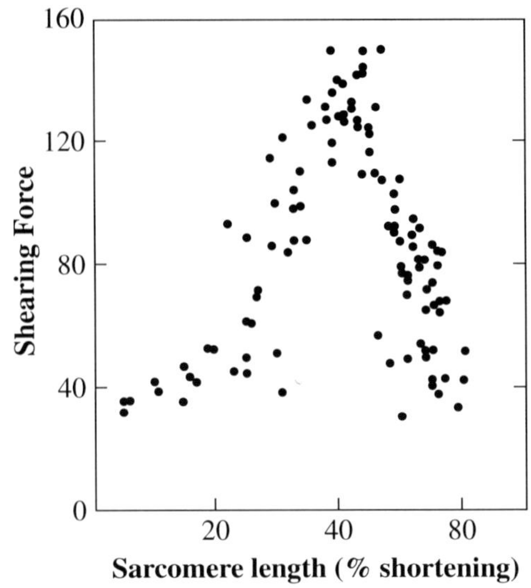

Figure 12.21 Relationship between meat tenderness (shear force) and sarcomere length. From B. B. Marsh and N. G. Leet. *Studies in Meat Tenderness. III. The Effects of Cold Shortening on Tenderness* (1966). J. Food Tech. 31:450.

Myofibrillar Protein Interactions

The innate structure of muscle is not conducive to physical disruption. Recall from Chapter 2 that muscle fibers are filamentous and provide a means of decreasing tenderness. Further, a multitude of myofibrils are aligned in register across the muscle fiber, which provides added strength to the structure. Finally, and most importantly, there are overlapping structures within the thick and thin filaments of the sarcomeres. The degree of overlapping between myosin and actin filaments is related to tenderness and is expressed as *sarcomere length* (Figure 12.21).

Protein Degradation and Tenderization Meat cuts from carcasses at 24 h postmortem compared with cuts taken several days later from the same carcasses, will be less tender, even though sarcomere length and connective tissue content is equal (Figure 12.22). The reason for this age-related change in tenderness is related to the action of various enzyme systems that are active postmortem and act to dismantle or degrade the contractile proteins (Figure 5.21, Chapter 5). As a result of this destruction, meat becomes more tender. Of particular interest to meat tenderness is the calpain enzyme system, which orchestrates protein turnover in living muscle tissue as that outlined in Chapter 5. In postmortem aging of muscle, calpains

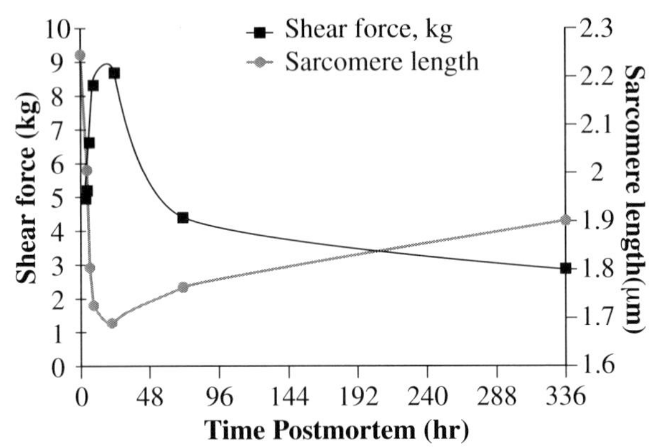

Figure 12.22 Effect of postmortem ageing on beef tenderness (boxes) and sacromere length (circles). From M. Koohmaraie, M. E. Doumit and T. L. Wheeler *Meat Toughening Does Not Occur when Rigor Shortening is Prevented* (1996) JAS 74:2935.

are responsible for the destruction of myofibrils by attacking proteins that hold myofibrils together and those located at the Z-lines. During cooking, the resulting protein matrix is disrupted and meat becomes tender.

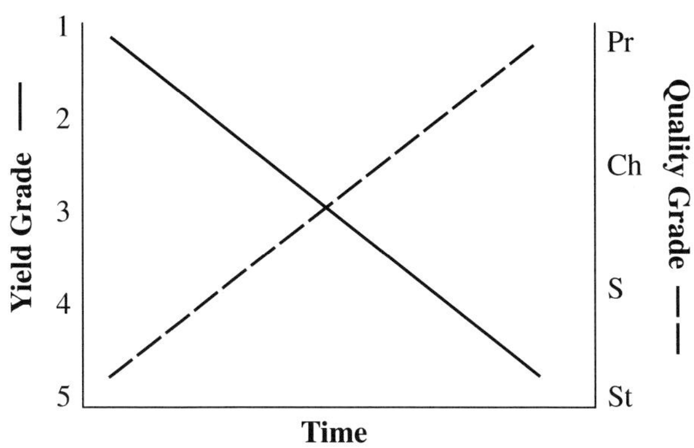

Figure 12.23 Relationship between beef quality and yield grades.

Augmenting Growth and Altering Quality

Quality versus Quantity

As mentioned earlier, in some cases the ability to provide quality is in direct conflict with providing quantity of meat. In particular, the major parameter used to establish beef quality grade, marbling, represents one of the last fat depots to develop. As a result, meat quality tends to improve as an animal becomes fatter, but carcass cutability decreases (Figure 12.23). Moreover, quality as indicated by color, firmness and structure may also suffer if yield is too excessive. In turkeys selected for high breast weights, color and texture is impacted. Therefore, any process that leads to an aggressive increase in yield of muscle may ultimately impact meat quality.

Sex

One of the easiest ways to improve efficiency, in terms of feed efficiency and growth rate is to eliminate castration. Recall from Chapters 9 and 10 that intact males are generally the latest maturing gender of most species. As a result, intact males are generally leaner and heavier at any point during their lifetime. Feeding intact males is not, however, without major drawbacks. As much as the practice improves carcass cutability, it reduces the quality of meat produced. In part this is due to insufficient amounts of fat and therefore, reduced quality grades.

Lean from carcasses of intact males generally has less desirable color. Protein content, especially myoglobin, is greater in meat of intact males compared to their castrated counterparts. Myoglobin is responsible for giving meat a characteristic red color and greater amounts of this protein result in a deeper red colored lean. Intact animals also tend to be more susceptible to development of the dark cutting condition. Recall that anabolic steroids such as testosterone increase protein deposition but also are responsible for increased aggression that is expressed immediately prior to slaughter (Chapter 10). This increased activity decreases the amount of energy in the muscle and consequently reduces the amount of glycogen available for metabolism postmortem. Thus, dark-cutting meat often results from intact males.

Carcasses from intact males tend to have fat that is less firm than that of castrates. For example, fat on carcasses of rams tends to be soft and oily. Fat depots are not as developed in intact males as those in castrated animals because of the actions of testosterone. The fat depots of rams tend to have an increased proportion of unsaturated fatty acids which results in relatively soft fat.

Meat from intact males sometimes has a strong odor and (or) flavor. This perceived flavor is due partially to the aforementioned change in fatty acid profiles. However, in some cases, especially in pork carcasses, the undesirable aroma, or *boar taint,* emanates from phermonal steroid compounds that result from steroid metabolism. These compounds, one of which is *5α-androst-16ene-3-one,* are synthesized by the gonads, stored in salivary glands, released with saliva, ingested, and ultimately deposited in fat depots. *Skatole,* another fat-soluble compound, is derived from tryptophan metabolism in the hind-gut. The concentration of this compound in pork is highly correlated to the incidence of boar odor and its source may explain the presence of boar odor in pork from male pigs. Although some consumers are more sensitive than others, this onion-like, pungent odor is mostly detected in freshly cooked, warm meat items from which the compounds may be volatilized. The incidence of boar taint is higher after pigs reach 200 days of age. Producers, especially those in Europe and Australia, have attempted to circumvent such off-flavor by marketing intact boars at an early age before they reach sexual maturity.

Meat tenderness also is negatively impacted in intact males but the differences seem greatest in beef. Regardless of species, meat from intact males is less tender or more variable in tenderness. Muscle of bulls tends to have more connective tissue that is more mature i.e., has more heat-stable crosslinks. This variability seems to manifest itself in bulls that are greater than 24 months of age.

Absence of Fat

A tenderness factor that stems from an interaction between postmortem temperature and overall leaness of the animal is *cold-shortening*. Cold-shortening is muscle shortening that occurs during its transformation to meat when the rate of carcass chilling is rapid. When this happens, increased levels of calcium are allowed into the sarcoplasm of muscle fibers resulting in contraction. If *rigor mortis* (the irreversible binding of myosin to actin in the absence of ATP) has not yet occurred, or is in the process of occurring, the muscle will shorten. As a result, sarcomere lengths are shorter and the meat is less tender (Figure 12.24). This type of toughening may occur in any carcass with reduced subcutaneous fat if subjected to rapid chilling rates. One way to circumvent this is to decrease chilling rates of carcasses with minimal fat thickness.

Age

Age may improve apparent meat quality of well-fed animals because time allows more marbling (intramuscular fat) to accumulate. However, age-related changes negatively impact tenderness via changes in connective tissue characteristics as outlined above. Therefore, in most animals, advanced age reduces meat quality.

Rendement Napole Gene

Rendement napole (RN) is French for "yield technology", which is a quality control test used in meat processing plants in France. Some years ago, a group of pigs were identified that produced hams with inferior processing yields. In addition, these pigs produced carcasses with meat having a lower than normal ultimate pH (< 5.5), or acid meat. Animals with the RN condition are characterized as having increased *glycolytic potential values.* Glycolytic potential is an index, which reflects the ability of the muscle to metabolize energy. This value reflects the collective amount of glycogen, glucose, glucose-6-phosphate and lactate in muscle tissue. Rendement Napole pigs have glycolytic potential values ≥ 200 mmol per gram of muscle tissue. The incidence of RN gene seems to be the greatest in Hampshire pigs and, therefore, is sometimes termed the *Hampshire effect.*

The gene responsible for this condition encodes for adenosine monophosphate-activated protein kinase (AMPK). This enzyme is particularly important in development of meat quality because it essentially acts to monitor the energy balance of the cell. As the name implies, this enzyme is activated in the presence of increased cellular AMP concentrations. Recall that ATP is first hydrolyzed to ADP and then further to AMP. Only after extreme bouts of exercise would AMP increase appreciably in muscle fibers. At this point, AMPK becomes activated and essentially increases those cellular functions that increase energy output to re-phosphorylate AMP to ADP and finally ATP. Not only does 'activated' AMPK increase ATP by increasing biochemical pathways, it increases the flow of glucose into the cell. This latter function, most likely is responsible for the exaggerated glycogen accumulation observed in RN pigs that possess AMPK that is always working. As a result and once the animal is slaughtered, there is more glycogen available and greater amounts of lactate are produced in the muscle resulting in pale, soft and exudative muscle. A complete understanding of how increased levels of glycogen can severely impact meat quality is poorly understood. However, this serves as a clear indication that any modification of the animal's energy balance in muscle tissue prior to slaughter could impact ultimate meat quality.

Halothane Gene

Several years ago a genetic condition became evident in pigs that were extremely lean and muscular. These pigs are especially attractive for improving productivity because of the improved carcass leanness and meat content of offspring. Unfortunately, these pigs are incapable of coping with most types of stress. This malady was widely spread throughout the world and was referred to as the *porcine stress syndrome.* Pigs having this syndrome responded to virtually any stress by becoming cyanotic (blue) and lethargic (slow moving). Respiration would become shallow and rapid, and muscle tremors were exhibited. Core body temperature would ultimately increase, a condition known as *malignant hyperthermia.*

Malignant hyperthermia exists in humans and is characterized by severe reaction to the anesthetic, *halothane.* Malignant hyperthermia is not lethal as long as actions are taken quickly on patients during its onset. Thus, many producers began using halothane as a means to test pigs for the presence of the "stress gene". This led to the term, *halothane positive* pigs.

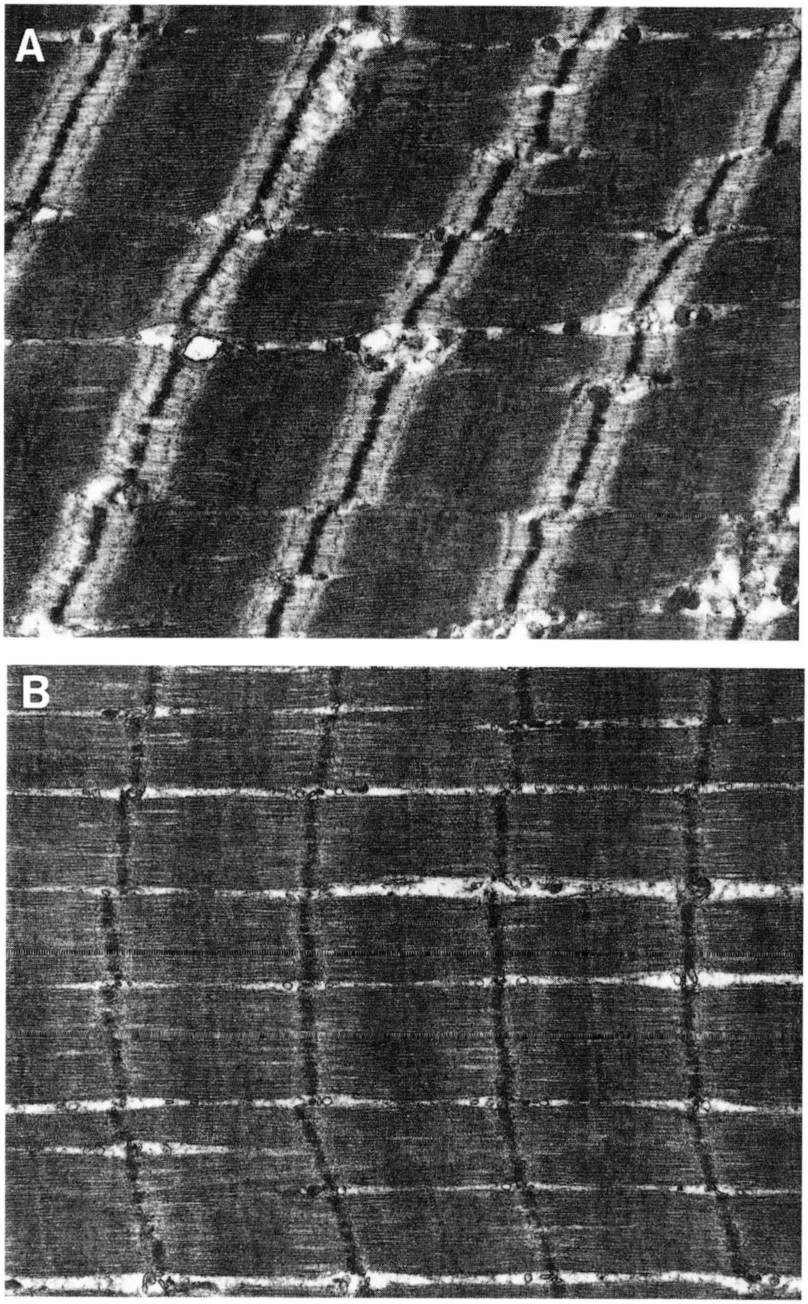

Figure 12.24 Electron micrographs of (A) relaxed and (B) cold-shortened beef muscle. From Pearson and Dutson, *Adv. Meat Research 1;* 185 (1985). Courtesy of Van Nostrand Reinhold. From *Meat Research* by Pearson & Dutson. Copyright © 1985 by Pearson & Dutson. Reprinted by permission.

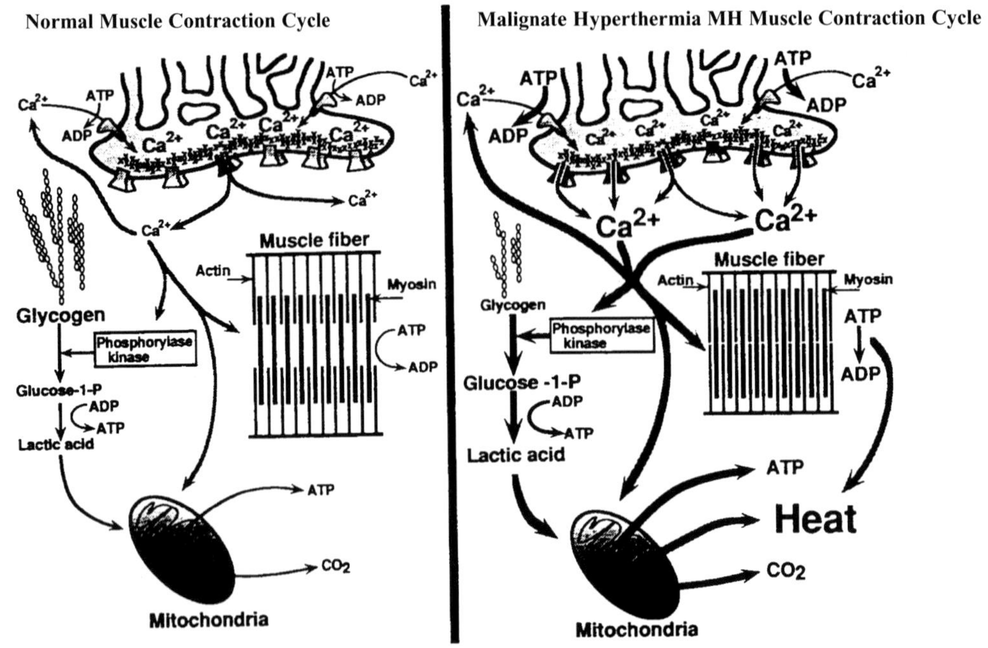

Figure 12.25 Role of enhanced sarcoplasmic calcium release on various cellular functions in muscle cells. Reprinted with permission from *Science*, Vol. 256, May 8, 1992 by D. H. MacLennan & M. S. Philips. Copyright 1992 American Association for the Advancement of Science.

The genetic mutation responsible for this condition has been identified and is due to impaired calcium regulation. Specifically, halothane positive pigs possess a mutated calcium release channel that is located in the sarcoplasmic reticulum. This large sarcoplasmic protein is known as the *ryanodine receptor* because it binds the plant alkaloid ryanodine.

Postmortem metabolism occurring in these pigs is a hallmark for development of pale, soft, exudative pork as outlined earlier in this chapter. The increased rate of metabolism occurs because cytoplasmic calcium is high early postmortem. Increased ATP consumption occurs because of increased activity by the sarcoplasmic calcium pump and actomyosin ATPases (Figure 12.25). In addition, calcium stimulates activity of glycogen phosphorylase, which speeds up glycogen breakdown. This increased rate of metabolism increases the rate of pH decline when carcasses temperatures are elevated, further compromising the integrity of muscle specific proteins and decreasing water-holding capacity. The history of this genetic condition demonstrates that changing muscularity of animals without regard to ultimate quality may have serious ramifications to the industry world-wide.

Double-Muscling

As outlined in earlier chapters, doubling muscling improves the overall muscularity in cattle. As a result of this genetic condition, meat yield is greatly improved. Yield grades calculated using classical yield grading formulas for carcasses of double muscled cattle have yield grades less than 1. Unfortunately, these carcasses have little fat and flavor. Tenderness of meat from double muscled cattle is inconsistent across many laboratories located in many countries, likely a result of the methods used to prepare the meat. In the US, traditional dry-heat cookery methods are less forgiving for meat cuts lacking sufficient fat. However, in other countries, in which cuts of meat are much thinner and moist cookery methods rely less on protective effects of fat, improved tenderness is often reported in cuts derived from double muscled cattle. Nonetheless, this genetic condition raises the question of whether ten-

TABLE 12.4 RELATIONSHIP AMONG DIFFERENT PORCINE MUSCLE FIBER TYPES AND RESULTING MEAT QUALITY ATTRIBUTES

Relationship		Correlation	n
Fiber type 1 (%)	Shear force	0.61	16
Fiber type 2B (%)	Shear force	–0.59	16
Glycogen content	$pH_{45\ minute}$	0.38	38
	$pH_{24\ hour}$	0.66	20
	Drip loss	–0.54	20
	Light reflectance	–0.47	20
Glycogen-depleted 2B Fibers	$pH_{45\ minute}$	–0.41	34
	$pH_{24\ hour}$	–0.85	20
	Drip loss	0.68	20
	Light reflectance	0.40	19

[1]A. Karlsson, Ph. D. Dissertation, Uppsala, Sweden, 1993. Porcine Muscle Fibers: Biochemical and Histochemical Properties in Relation to Meat Quality

TABLE 12.5 CORRELATION BETWEEN MUSCLE FIBER TYPE AND MEAT QUALITY CHARACTERISTICS

	Muscle Fiber Type		
Meat Quality Attribute	Type 1	Type 2A	Type 2B
pH 30 min	0.39	0.93	–0.46
pH 24 hr	—	—	–0.62
Color*	–0.29	—	0.55

Minolta L value (higher value equals lighter color)
From C. Larzul, L. Lefaucheur, P. Ecolan, J. Gogué, A. Talmaut, P. Sellier, P. LeRoy, and G. Monin. (1997). JAS 75:3126.

derness is compromised in those animals selected for increased tissue. In double muscling, the type of connective is not different from that of normal meat, but the amount of connective tissue is less because of its dilution by increased amount of myofibrillar proteins. Although this would favor tenderness, carcasses from double muscled cattle are leaner and more subject to cold-shortening than normal beef.

CHANGING MUSCLE FIBER TYPE

There is a strong relationship between muscle fiber type and meat quality (Tables 12.4 and 12.5). As pointed out in earlier discussions, muscles consisting of abundant red muscle fibers have desirable red color. Muscle consisting of high proportions of white or glycolytic muscle fibers possesses a rapid and more extensive change in muscle pH shortly after slaughter, characteristic of development of PSE muscle in pigs. This type of adverse meat quality development is mainly observed in pigs and poultry, primarily because they have a more glycolytic muscle than do sheep or cattle.

NUTRITION

As pointed out earlier, one of the most crucial ingredients for ensuring optimal meat tenderness is to produce meat from young animals. Therefore, animals fed quantities of feed that delay attainment of marketable weight could be less tender. However, reducing tenderness by increasing animal age is not the only means by which nutrition may impact quality. Cattle fed high energy for extended periods prior to slaughter yield more tender meat than those fed low energy over the same time (Table 12.6). Connective tissue concentration remains constant, suggesting the increased growth rate does not result in dilution of the connective tissue with myofibrillar

TABLE 12.6 MEAT PALATABILITY AND MUSCLE PROPERTIES OF CATTLE GROWN AT DIFFERENT RATES

	Growth rate, kg/d		
Item	**1.42**	**.77**	**.34**
Sensor panel scores[a]			
Tenderness—LD	6.19	5.64	5.39
Tenderness—SM	4.59	3.78	3.29
Juiciness—LD	5.11	4.85	5.04
Juiciness—SM	4.60	4.44	4.51
Warner-Bratzler shear, kg			
LD	3.83	4.45	4.74
SM	4.14	4.48	4.70
Semimembranosus collagen			
Total, mg/g	5.29	5.56	6.71
Soluble, %	10.39	10.00	7.51
Thermal shrinkage temperature, C			
LD	59.98	62.26	61.67
SM	62.78	63.18	63.17
Sarcomere length, μm			
LD	1.94	1.87	1.76

[a]Descriptive scale: 1 = extremely tough – dry, 8 = extremely tender – juicy.
From V. Fishell, E. D. Eberle, M. D. Judge and T. W. Perry. (1985) *Palatability and Muscle Properties of Beef as Influenced by Preslaughter Growth Rate.* JAS 61:151.

proteins. Therefore, meat from animals fed diets designed to maximize growth rates may have more soluble collagen that lacks extensive cross-linking. Furthermore, muscle experiencing more rapid growth likely has different amounts of proteases than muscle from animals growing at a slower rate, which may enhance the postmortem ageing process.

Nutrition also may affect the transformation of muscle to meat and subsequent quality by altering the amount of glycogen in muscle. Producers generally remove feed from animals targeted for slaughter. In the case of pigs, *feed withdrawal* may have beneficial effects on development of ultimate muscle pH, primarily by lowering the amount of glycogen stored in the muscle prior to slaughter. However, if feed withdrawal time is too long, the majority of glycogen may be depleted from the muscle tissue, and in cattle this can result in dark cutting beef.

REPARTITIONING AGENTS

As discussed in the previous chapter, various compounds have been utilized to improve animal performance. Metabolism modifiers have been quite impressive in their ability to decrease fat deposition and facilitate protein accretion, but have varied effects on meat quality. Consistent with the idea that gains made in performance often are at the cost of product quality, these compounds have been shown to affect a variety of meat quality characteristics in several species (Table 12.7). However, these affects may not be serious enough to offset the gains in growth rate and feed efficiency.

Hormone Implants

Hormone-based implants have been widely used in the beef industry for many decades. Depending on the combination of hormones delivered, hormone implants may improve animal performance by increasing fat or protein deposition. As pointed out in Chapter 10, hormone implants improve live animal performance. However, improved performance does not necessarily improve the final product and implanted cattle have significantly lower ultimate quality grades (Table 12.8). In addition to lowering the amount of intramuscular fat (marbling) in the lean, meat from implanted cattle is less tender than

TABLE 12.7 MEAT QUALITY PARAMETERS OF ANIMALS TREATED WITH SOMATOTROPIN AND BETA-ADRENERGIC AGONISTS

Waterholding capacity		
Somatotropin	Pigs	No differences
Beta-agonists	Pigs	No differences
	Cattle	Increased drip loss
Tenderness		
Somatotropin	Pigs	No difference to a decrease in tenderness
Beta-agonists	Pigs	No difference to 35% increase in shear force
	Cattle	19% increase in shear force
	Sheep	No difference to 60% increase in shear force
Juiciness		
Somatotropin	Pigs	No difference to slight decrease
Beta-agonists	Pigs	No difference
Color		
Somatotropin	Pigs	Slightly paler
Beta-agonists	Pigs	No difference
	Cattle	No difference
	Sheep	No difference visually
Marbling		
Somatotropin	Pigs	24–62% decrease
Beta-agonists	Pigs	No difference to slight decrease
	Cattle	No difference to slight decrease
	Sheep	No difference to slight decrease

From Merkel, Proc. Nutr. Inst., NFIA (1989)

TABLE 12.8 EFFECT OF IMPLANTS ON QUALITY GRADE

	Quality Grade (No.)				
Implant Strategy	**Prime**	**Choice**	**Select**	**Standard**	**% Prime/Choice**
No Implant/No Implant	2	49	2	1	94.44
Encore & Component T-S/No Implant	0	41	3	0	93.18
Ralgro/Synovex Plus	0	41	12	0	77.36
Ralgro/Revalor-S	0	39	9	0	81.25
Revalor-S/Revalor-S	0	32	16	0	66.67
Revalor-S/No Implant	1	39	12	0	76.92
No Implant/Synovex Plus	1	40	11	0	78.85
Synovex Plus/No Implant	0	39	12	1	75.00

From Roeber et al. (1999).

TABLE 12.9 EFFECT OF TYPE OF IMPLANT ON MEAT TENDERNESS (WARNER-BRATZLER SHEAR FORCE)

First Implant	Second Implant	Third Implant	WBS[b], lb.
Non-implanted			8.00
ME[d]	—	—	+.10
ME	ME	ME	+.93
A	—	—	+1.30
ME/A	ME/A	—	+1.57
SE	—	—	+.94
SE	SE	—	+.97
SE/A	—	—	+1.08
SE/A	SE/A	—	+1.40
MC	—	—	+.25
MC	MC	—	+1.70
SC	—	—	+1.70
SC	SC	—	+1.30

[a]Source: OSU Implant Data Base.
[b]WBS: Warner-Bratzler shear force value, lb.
[c]Change in WBS compared to nonimplanted controls.
[d]Implant classification: ME, SE, A, MC and SC are mild estrogen, strong estrogen, androgen, mild combination and strong combination, respectively.
From Morgan (1997).

that of non-implanted cattle and is related to the implant strategy (Table 12.9). This scenario is hallmark of the delicate balance between growth enhancement of live animals and potential drawbacks on meat quality. Even so, use of hormone implants has dramatically improved productivity in the animal industry and will likely continue in the foreseeable future because even though there is a reduction in perceived tenderness, overall acceptance of the product is not impacted significantly (Table 12.10).

TABLE 12.10 CONSUMER SENSORY PERCEPTION OF MEAT FROM IMPLANTED AND NON-IMPLANTED CATTLE

	Implant Strategy							
Consumer Sensory Response	No Implant/ No Implant	Encore & Component T-S/No Implant	Ralgro/ Synovex Plus	Ralgro/ Revalor-S	Revalor-S/ Revalor-S	Revalor-S/No Implant	No Implant/ Synovex Plus	Synovex Plus/ No Implant
Overall L/D	4.25	4.47	4.61	4.71	4.61	4.42	4.69	4.73
Flavor L/D[b]	4.14	4.34	4.47	4.52	4.34	4.33	4.71	4.47
Flavor Intensity[c]	5.23	4.97	4.95	4.83	5.10	4.20	4.80	5.14
Tenderness L/D[d]	3.94	4.25	4.48	4.42	4.54	4.27	4.48	4.46
Level of Tenderness[e]	3.90	4.25	4.50	4.49	4.53	4.45	4.48	4.54
Juiciness Level[i]	4.87	5.06	5.01	5.27	5.26	4.94	5.28	5.21

[a]Overall like (L) to dislike (D) rating by consumers: 1 = like extremely, 0 = dislike extremely.
[b]Like to dislike rating of flavor by consumers: 1 = like extremely, 9 = dislike extremely.
[c]Intensity of flavor rated by consumers: 1 = none or extremely bland, 9 = extremely intense.
[d]Like to dislike rating of tenderness by consumers: 1 = like extremely, 9 = dislike extremely.
[e]Level of tenderness rating by consumers: 1 = extremely tender, 9 = extremely tough.
[i]Level of juiciness rating by consumers: 1 = extremely juicy, 9 = extremely dry.
From Roeber et al. (1999).

GLOSSARY

A

alpha (α)-red muscle fiber – fast contracting muscle fiber with intermediate (oxidative/glycolytic) metabolism.

alpha (α)-white muscle fiber – fast contracting muscle fiber with predominantly glycolytic metabolism.

5α androst-16ene-3one – a major metabolite of testosterone metabolism and partially responsible for boar taint in pork products.

A-band – dark appearing band of the myofibril consisting mainly of thick filaments and associated actin.

abattoir – location or facility where animals are slaughtered.

absolute growth rate – rate of growth expressed in units of mass per unit time.

absolute muscle fiber number – number of muscle fibers in a given muscle.

acetate – volatile short chain fatty acid (also produced by rumen microbes).

acid meat – meat (generally pork) with a abnormally low pH.

actin – thin filamentous protein of the myofibril.

ad libitum – unrestricted access to feed.

adherins – cell surface proteins that aid in anchoring cells to a given location.

adipoblasts – cell precursors of adipocytes.

adipocytes – mature fat cells.

adipogenesis – process of adipocyte development.

adipose tissue – specialized connective tissue responsible for storing energy in the form of triglyceride.

adiposity – fatness and (or) number of adipocytes.

adult myoblasts – muscle precursor cells found in adult skeletal muscle; also referred to as muscle satellite cells.

aerobic metabolism – biochemical reactions requiring oxygen.

affinity – attraction between two molecules e.g., between a hormone and its receptor.

aitch bone – portion of the pubis visible on a split carcass.

allantois – membrane of the placenta that represents the vascular bed of the hind gut of the developing organism.

alpha-actinin – major protein of the Z-line.

ameboid cells – cell type capable of movement.

amino acid – basic building block of protein.

amino acid deficiency – insufficient dietary intake or utilization of an essential amino acid.

amino acid imbalance – proportions of dietary amino acids that result in sub-optimal performance.

amino acid supplementation – addition of a synthetic amino acid to a diet.

amnion – membrane of the placenta forming a closed sac in which the developing embryo is immersed.

anabolic – related to increasing tissue production.

anaerobic metabolism – biochemical reactions in the absence of oxygen.

animal pole – region of a fertilized egg in which the developing animal forms.

antemortem – prior to death.

antibiotics – substance that can kill or inhibit growth of microorganisms.

anuclear – without a nucleus.

apical – portion of the cell located towards the lumen, opposite the basement membrane.

apical ectodermal ridge – portion of the outermost surface of an embryo that forms a limb.

aponeurosis – thick connective tissue, other than tendon, that facilitates muscle attachment.

apparent muscle fiber number – estimation of muscle fiber number using mean packing density.

appositional growth – increased mass by addition of material to the outer surface.

area pellucida – transparent region of a developing embryo.

articular cartilage – cartilage associated with a joint.

articulation – moveable joint between two bones.

astrocytes – star-shaped glial cells.

ATPase – enzyme capable of hydrolyzing ATP.

atrophy – decrease in size.

autocrine – self-regulating; action of a hormone or growth factor on the same cell that produced the factor or on an adjacent cell of similar type.

autolysis – self cleaving or inactivating.

axon – protrusion of a nerve cell responsible for delivering impulses away from the cell body.

B

backfat – fat located on the dorsal aspects of an animal just underneath the skin.

basement membrane (basal lamina) – layer of filaments or fibers that attach epithelium to underlying connective tissue.

beta (β)-red muscle fiber – slow contracting muscle fiber with predominantly oxidative metabolism.

biological value – the ability of a given source of dietary protein to support body functions.

blastocoele – hollow, fluid-filled cavity of the blastocyst.

blastocyst – early embryo consisting of multiple cells arranged in a hollow, ball-like structure.

blastomeres – cells of a developing blastocyst.

blastulation – process of developing a blastocyst.

blood and lymph – specialized connective tissue responsible for transporting nutrients, hormones, waste, etc. throughout the body.

bloodstream conveyance – transport of hormones or growth factors via the blood circulation.

boar taint – off-odor or off-flavor normally associated with pork from gonadally-intact male pigs.

bone lining cells – relatively inactive cells associated with the surface of bones.

bone marrow – tissue that fills the internal cavities of bone.

bone remodeling – process of bone resorption and formation resulting in a change in bone shape and function.

break joint – epiphyseal cartilage of the distal end of the metacarpal bones of lamb carcasses.

buttons – cartilage located at the dorsal most aspects of the thoracic vertebrae.

butyrate – volatile short-chain fatty acid produced by rumen microbes.

bypass protein – protein that passes through the rumen without being degraded.

C

C protein – protein of the thick filament that binds the myosin molecules together.

Callipyge – genetic condition of sheep associated with muscle hypertrophy.

calpain system – a major proteolytic system for protein breakdown.

calpastatin – member of the calpain system that serves as an inhibitor of proteolysis.

calsequestrin – protein located in the sarcoplasmic reticulum and responsible for binding calcium.

canaliculi – canals that allow cytoplasmic extensions of osteocytes for diffusion of nutrients and wastes in bone.

cancellous bone – spongy bone.

cap Z – protein located at the end of filamentous actin to inhibit elongation.

capillary density – number of capillaries per unit area.

cardiac muscle – contractile tissue of the heart.

cardiocytes – heart muscle cells.

carnucles – out growth of tissue associated with placenta attachment.

castrate – animal with ovaries or testes removed.

catabolic – related to tissue breakdown.

catch-up growth – compensatory growth; accelerated growth occurring after a period of sub-optimal growth.

cathepsins – proteinases associated with the lysosomes of a cell.

caudal – posterior aspects of an animal.

caul fat – fat located around the gastrointestinal tract.

cell cycle – progression of a cell through the phases required for cell division.

cell division – process of replication.

cell lineage – cells that arise from a common progenitor.

cell-surface receptors – structures on cell membranes responsible for binding molecules and mediating a biological response.

cell-adhesion molecules – extracellular structures responsible for attaching cells to one another.

cell replication – division.

central nervous system – the brain and spinal cord.

chemical maturity – stage of development defined on the basis of chemical composition.

chemotaxis – process of directing cellular movement via a chemical trail.

chondroblasts – chondrocyte precursor cells.

chondrocytes – mature cartilage cells.

chondroitin sulfate – proteoglycan located in cartilage.

chorion – the outer membrane of mammalian placenta and is associated with the allantois and forms the placenta.

chronic – constant presence.

chronological age – age assigned using a measurement of time.

ciliation – containing cilia.

cleavage – process of increasing cell number.

cold-shortening – shortening of muscle that occurs when pre-rigor muscle is chilled too rapidly.

collagen – major protein of connective tissue.

committed – state in which the fate of a cell has been decided; determined.

compact – dense.

compaction – process during blastocyst formation in which the inner cells become closely associated.

compensatory growth – accelerated growth occurring after a period of sub-optimal growth.

compositional maturity – stage of development based on body or carcass composition.

conception – fertilization.

conformation – the proportionate development of animal or carcass parts as reflected in the shape of the animal or carcass.

connective tissue – a primary tissue type involved in providing structural support for tissues and organs.

connective tissue proper – a subtype of connective tissue.

costamere – structure responsible for anchoring myofibrils to the sarcolemma.

crosslinking – covalent binding that occurs within connective tissue fibrils.

crown-rump length – measurement used to determine fetal age.

Culard – French word meaning double muscled.

cutability – the amount of closely trimmed, retail cuts that may be obtained from carcasses.

cytokinesis – separation of the cytoplasm following nuclear division.

cytoskeleton – framework and system supporting the internal structure of a cell.

D

daily gain – increased mass expressed on a daily basis.

dairy type – animals genetically-selected to produce large amounts of milk.

dark cutter – carcass with dark purple lean.

daughter cells – genetically identical cells resulting from cell division.

deaminated – removal of ammonia from an amino acid.

delamination – one of the types of cell movements during gastrulation.

denaturation – alteration of natural protein structure by non-proteolytic changes.

dendrites – sensory processes of a neuron.

dense connective tissue – connective tissue proper in which the extracellular fibers are densely packed.

dense zones (dense bodies) – structures in smooth muscle that are involved in linking actin filaments to each other and to the cell membrane.

dermamyotome – compartment of the somite that eventually forms the dermatome and myotome.

dermatome – compartment of the somite that forms dermis.

desmosine – unique amino acid of elastin.

determined – state in which the fate of a cell has been decided; committed.

diabetic – state of high blood glucose or insulin-insensitivity.

diaphysis – the shaft portion of a long bone.

differentiate – becoming more specialized during development.

diphasic growth – allometric growth with a change in growth coefficient.

division – replication

DNA unit – amount of cytoplasm that may be supported with a given amount of DNA.

Doppellender – German word meaning double muscled.

Double muscled – genetic condition in cattle reulating increased muscle mass.

E

early maturing – attainment of maturity at a relatively young age.

ectoderm – one of the three primary germ layers.

elastic cartilage – cartilage composed of large amounts of elastin fibers.

elastin – a connective tissue protein that provides elasticity to connective tissue fibers.

embryonic – stage of development prior to fetal stage.

endochondral ossification – bone growth that results from the replacement of cartilage with bone.

endocrine – action of hormones that are conveyed by the bloodstream.

endoderm – one of the three primary germ layers.

endomysium – loose connective tissue surrounding individual muscle fibers.

endoneurium – loose connective tissue surrounding nerve fibers.

endosteum – epithelial lining of the medullary cavity of bones.

endothelium – epithelium that lines inner surfaces of vessels.

endpoint – point of reference or a time at which a specific measurement or observation is made.

energy balance – measure of gain or loss of energy by the body.

energy metabolism – biochemical reactions that involve production and utilization of energy.

epiboly – one of the types of cell movements resulting in multiple layer dormation during gastrulation.

epimysium – loose connective tissue surrounding a muscle

epineurium – connective tissue that surrounds a peripheral nerve bundle.

epiphyseal closure – complete ossification of the epiphyseal cartilage.

epiphyseal plate – cartilage located between the diaphysis and epiphyses of long bones.

epiphysis – enlarged head of a long bone.

epithelium – primary tissue type that serves as linings for organs and vessels.

exercise-induced hypertrophy – increased muscle cell size as a result of exercise.

extended growth period – prolonging the time required for an animal to reach mature size.

external fat – fat associated with the subcutaneous adipose tissue depot.

extracellular – outside of a cell.

extraembryonic – outside of the embryo.

F

F actin – filamentous (fibrous) actin formed by polymerization of G actin; principal protein of the thin filament.

fascia – connective tissue located between muscles and other organs.

fat cell – adipocyte.

fat depth – thickness of subcutaneous adipose tissue at a specific site.

fat endpoint – time at which a specific amount of fat deposition has occurred.

fat free lean – method of expressing lean content of carcasses, cuts or boneless trim on a lipid-extracted basis.

fat hardening diets – diets containing lipid with a high proportion of saturated fatty acid relative to unsaturated fatty acid that may harden adipose tissue depots in some animals.

fat thickness – measure of subcutaneous adipose tissue thickness at a specific site.

fatback – subcutaneous fat.

feathering – fat streaks in the intercostal muscles.

feed withdrawal – withholding feed from animals.

fetal – stage of prenatal development following the embryonic stage.

fetal myoblasts – lineage of muscle cells that participate in secondary muscle fiber formation.

fibroblast – predominant cell of connective tissue proper.

fibrocartilage – cartilage containing an abundance of collagen fibers.

fill – gut contents.

finish – subcutaneous fat.

flank streaking – streaks of fat on the primary and secondary flank muscles.

fleshing – thickness associated with muscling.

fractional rate – rate expressed as the fraction of the total per unit time.

frame scores – numerical system used to classify the size of the skeleton of animals.

frame size – relative size of the skeleton.

free choice – feed access on an ad libitum basis.

G

G actin – globular monomer of the protein actin.

gap junctions – connections between cells that allow for communication.

gastrulation – formation of the three germ layers of the embryo.

genetic potential – growth or performance that is possible genetically.

genetics – the study of heredity.

genotype – the genetic complement of an individual.

girth – circumference.

globular head – secondary and tertiary protein folding of myosin molecules.

glycolytic metabolism – biochemical reactions involving the catabolism of glucose or glycogen.

glycolytic potential values – summation of muscle glucose equivalents.

gonads – organs that produce gametes.

grading – identification of carcasses or cuts of meat based on expected yield or quality.

ground substance – a subtype of connective tissue proper.

growth factors – small proteins or peptides that affect the growth or metabolism of cells.

growth plate – epiphyseal plate.

H

H zone – center region of the A band in which the thin filaments do not overlap the thick filaments.

halothane – an anesthetic.

halothane positive – pig that is stress-susceptible when exposed to halothane gas.

Hampshire effect – adverse pork quality induced by the presence of the Rendement Napole gene.

Haversian system – basic functional unit of compact bone.

heavy meromyosin – fragment of the myosin molecule.

histamines – compounds involved in conveying an inflammation response.

homeorhesis – coordination of nutrient partitioning.

homeostasis – maintenance of a constant internal environment.

hyaline cartilage – type of cartilage composed of large amounts of collagen fibers.

hyaluronic acid – proteoglycan in the matrix of connective tissues.

hydroxyapatite – family of calcium phosphate salts.

hydroxyproline – amino acid present in collagen.

hypoblast – ventral most layer of the embryonic disc.

hyperphagia – increased appetite.

hyperplasia – increase in cell number.

hypertrophy – increase in cell size.

hypothalamus – region of the brain involved in coordination of the endocrine system.

I band – light band on the myofibril bounded by the ends of the thick filaments and bisected by the Z line.

implantation – process by which developing embryos attach to maternal tissues.

in ovo – in the egg.

in utero – in the uterus.

inclusions – amorphous materials in the cell existing in local areas of high concentration.

indeterminate – not fixed.

ingression – one of the types of cell movements during gastrulation.

inner cell mass – tightly condensed group of cells during early embryonic development

insertion – point of attachment of the movable end of a muscle to a bone.

intercalated discs – connections between cardiac muscle cells.

interdigitated – interlocked.

intermuscular fat – adipose tissue depot located between muscles; associated with epimysium.

interstitial fluid – fluid between the cells of a tissue.

interstitial growth – growth occurring from within.

intracrine – action of growth factor in which secretion of the growth factor from a cell is not required.

intrafasicularly terminating fibers – muscle fibers that terminate in the body of a muscle bundle.

intramuscular fat – adipose tissue depot located between bundles of muscle fibers; associated with perimysium.

invagination – one of the types of cell movements during gastrulation.

involution – one of the types of cell movements during gastrulation.

ionophores – organic substances that alter the transport of ions through membranes of certain cell types.

isodesmosine – unique amino acid of elastin.

K

keratan sulfate – type of proteoglycan.

kidney fat – adipose tissue located around the kidney.

Koller's Sickle – site of rapid cell proliferation on the blastocyst.

L

lacunae – small cavities in which osteocytes reside.

lamellae – concentric layers of bone within an osteon.

large-framed – animal with a relatively large skeleton.

late maturing – animal that reaches maturity at a relatively advanced age.

leaf fat – adipose tissue between the lining of the thoracic cavity and the ribs of mammalian animals; associated with the abdominal fat pad and intestines in poultry.

leptin – hormone produced by adipocytes that regulates appetite and energy metabolism.

light meromyosin – fragment of the myosin molecule.

limit feed – restricted intake of feed.

lipase – an enzyme that breaks down triglycerides.

lipogenesis – formation of triglyceride.

lipolysis – breakdown of triglyceride.

local conveyance – action of hormones and growth factors near their site of secretion.

long bones – bones with a diaphysis and epiphyses.

loose connective tissue – connective tissue proper in which the extracellular fibers are loosely woven.

lysl oxidase – enzyme partially responsible for collagen crosslinking.

lysosomal system – proteolytic system that uses enzymes of the lysosomes for protein degradation.

lysosomes – small vesicles in the sarcoplasm that contain proteolytic enzymes known as cathepsins.

M

M line – dark band bisecting the A band representing a structure that holds the thick filaments in proper array and/or guides the ends of the thin filaments during sliding motion.

M protein – protein of the M line.

malignant hyperthermia – condition in animals when body temperature increases uncontrollably.

marbling – intramuscular fat visible in cuts of meat associated with the endomysium.

mast cells – cell of connective tissue.

mastication – chewing.

maturation – becoming fully developed.

maturity – stage at which an animal becomes fully developed.

micro (μ)-calpain (calpain 1) – calcium-dependent enzyme involved in protein degradation.

milli (m)-calpain (or calpain 2) – a calcium-dependent enzyme involved in protein degradation.

mean packing density – number of cells within a given area.

meat type – animals genetically-selected to produce high amounts of meat.

medullary cavity – space within a bone that contains marrow.

meiosis – cell division that produces gametes.

membrane potential – electrical potential difference that results from uneven distribution of ions on opposite sides of a cell membrane.

mesoderm – one of the three primary germ layers.

microfilaments – protein filaments of the cytoskeleton.

microphages – neutrophils and eosinophils.

microtubules – components of the cytoskeleton.

microvilli – extensions of the epithelial cell membrane.

mitosis – division of a cell that produces two identical cells; M-pahse of the cell cycle.

monophasic growth – allometric growth with no change in growth coefficient.

morphology – form and structure.

morula – early embryo consisting of multiple cells produced through mitotic divisions of a zygote.

motor units – muscle cells controlled by a single neuron.

MRF-4 – one of the myogenic regulatory factors involved in muscle cell development.

mucous – presence of mucus.

multipotent – retaining the ability of developing multiple fates.

muscle cell lineage – muscle cells that arise from a common progenitor and exhibit specific characteristics.

muscle fascicle – a group (bundle) of muscle fibers surrounded by the perimysium, a layer of loose connective tissue.

muscle fiber – muscle cell.

muscle fiber type – muscle cell characterized by a specific type of metabolism and speed of contraction.

muscle fiber typing – process of characterizing muscle fiber type.

muscle regulatory factor genes (MRFs) – genes that encode proteins involved in regulating muscle cell development.

muscle tissue – primary tissue type characterized by its ability to contract.

myelin sheath – layers of glial cell membrane around an axon.

myf-5 – one of the myogenic regulatory factors involved in muscle cell development.

myoblast – muscle precursor cell.

myocyte – muscle cell.

Myo-D – one of the myogenic regulatory factors involved in muscle cell development.

myofiber – mature muscle cell.

myofibril – basic functional unit of the muscle fiber contractile system.

myofibrillar – relating to the myofibril.

myofibrillogenesis – formation of the myofibril.

myofilament – basic sub-unit of the myofibril that exists as two types, the thin (actin) and thick (myosin) filaments.

myogenesis – formation of muscle tissue.

myogenin – one of the myogenic regulatory factors involved in muscle cell development.

myoglobin – oxygen-binding pigment in muscle.

myometrium – layer of smooth muscle in the uterus.

myosin – principal protein of the thick filament.

myostatin – member of the transforming growth factor-beta family that regulates muscle development.

myotendinal junction – connection of tendon to the ends of muscle fibers of a muscle.

myotome – compartment of the somite.

myotube – immature muscle fiber.

N

negative feedback – regulation of hormone synthesis and (or) secretion by a downstream hormone or growth factor.

neonatal – newborn.

nervous tissue – major tissue of the central and peripheral nervous system.

net lipid accretion – difference between lipid deposition and breakdown.

net protein accretion – difference between protein deposition and breakdown.

net protein deposition – difference between protein deposition and breakdown.

neural crest – apical or uppermost aspects of primitive groove.

neural ectoderm – outermost layer of cells of the primitive groove.

neural tube – precursor of the spinal cord located just dorsal to the notochord of a developing embryo.

neuralation – formation of spinal cord precursors.

neurocrine – action of a hormone that is synthesized in a neuron, stored in an axon, and secreted into the bloodstream to act on distant cells.

neuroendocrine system – hormonal action that coordinated the endocrine and nervous systems.

neurohormone – hormone produced by a neuron.

neuromuscular junction – the point of contact of a nerve axon and the sarcolemma of a muscle fiber; structurally called the motor end plate.

neuron – a nerve cell.

neuroglia – non-neural cells of the nervous system.

neutral fat – neutral lipid (triglyceride).

Nodes of Ranvier – area on an axon lacking a myelin sheath.

non-protein nitrogen (NPN) – nitrogen-containing compounds that are not proteins.

notochord – along with the neural tube, forms the spinal cord of a developing embryo.

O

obese – state of excessive fat deposition.

oocyte – cell that gives rise to an ovum.

organelle – structured element of the cell.

origin – one of the points of attachment of a muscle to a bone; usually the least movable end of the muscle.

osteoblast – bone cell precursor.

osteoclast – bone cell involved in bone resorption.

osteocyte – mature bone cell.

osteocytic osteolysis – mobilization of minerals out of bone.

osteon – basic functional unit of compact bone.

ovum (pl. ova) – egg; gamete produced by the female reproductive system.

ovaries – gonads of the female reproductive system.

oxidative metabolism – biochemical reactions requiring oxygen.

P

papain – proteolytic enzyme isolated from papaya that is used to tenderize meat.

paracrine – action of hormones and growth factors near their site of secretion, but on cells of different type.

peptidases – enzymes that cleave peptides and liberate amino acids.

perichondrium – outer epithelial layer of cartilage.

perimysium – loose connective tissue surrounding individual muscle fasciculi.

perinatal – occurring around the time of birth.

perinerium – connective tissue surrounding bundles of nerve fibers.

periosteum – outer epithelial layer of bone.

peripheral nervous system – all neural tissue outside the CNS.

perirenal fat – adipose tissue depot located adjacent to kidneys.

permanent stunting – reduced body size and weight resulting from animal inability to exhibit compensatory growth following a severe nutritional stress.

phasic – distinguished, repeated pattern.

phenotype – outward expression of genotype.

phospholipids – membrane lipid.

physiological age – stage of development of animals based on the physical state of tissues.

physiological maturity – stage of development of animals based on the physical state of tissues.

pituitary gland – central endocrine gland.

plasma – fluid portion of blood.

white blood cells – immune cells of the blood.

platelets – component of blood participating in the clotting process.

polypeptide – chain of several amino acids.

porcine stress syndrome – condition in pigs characterized by the inability to maintain homeostasis when presented with a stressor.

postmitotic – no longer able to undergo cell division.

postmortem metabolism – biochemical reactions that occur after death.

postweaning lag – reduced body growth rate that occurs for a period of time following weaning.

primary bundle – collection of muscle fibers by perimysial connective tissues.

primary muscle fibers – embryonic muscle fibers located in close proximity to capillaries and of a permanent aerobic metabolic type.

primary myotubes – first myotubes formed in the developing embryo.

primary ossification center – intitial site of bone formation in the diaphysis of a long bone.

primitive groove – depression in the primitive streak of the developing embryo.

primitive streak – thickened structure that forms parallel to the long axis of the early embryo.

primordial – earliest formed.

progenitor – precursor.

proliferation – replication.

proliferative zone – region of the growth plate of long bones where chondrocytes proliferate.

propionate – volatile short chain fatty acid produced by rumen microbes.

protein breakdown – proteolysis.

protein degradation – proteolysis.

protein synthesis – formation of proteins.

protein turnover – process of continual synthesis and degradation of protein.

proteinaceous – composed of protein.

proteoglycan – connective tissue component composed of a polysaccharide complex attached to a small protein.

proteolytic – action of breaking down proteins to peptides or amino acids.

pseudo H zone – center region of the H zone of a myofibril probably represented by the region on the thick filaments having no myosin heads.

purge – juices that exude from fresh, cooked and cured meats and accumulate inside the package during shipping and storage.

quiescent – dormant.

R

realimentation – re-feed.

recovering weight for age – compensatory growth that results in an animal catching up to its normal weight and size without requiring increased time.

red blood cell – erythrocyte.

red muscle fibers – aerobic muscle fibers.

relative growth rate – rate of growth in relation to total mass (rate per unit weight).

releasing and inhibiting hormones – hormones produced by an endocrine gland that regulate synthesis and (or) secretion of other hormones in other endocrine glands.

Rendement Napole – genetic condition that results in exaggerated glycogen stores in muscle.

replication – duplication.

replication competent – able to duplicate.

reserve zone, resting zone, or stem cell zone – region of the growth plate of long bones that serve as a source of chondrocytes for the proliferative zone.

resorption – bone breakdown.

resting potential – electrical potential across the membrane of a resting muscle fiber.

retail yield – amount of product available after fabrication.

reticular fibers – extracellular fibers of connective tissue proper composed of reticulin.

ribbing – fabrication of beef carcasses into fore and hind quarters by cutting between the 12th and 13th ribs.

ribeye area – cross-sectional area of the *longissimus dorsi* muscle exposed on a ribbed carcass.

ribosome – site of protein synthesis in cells.

rigor mortis – muscle stiffening after death that results from depletion of ATP.

ruffled border – region of contact between an osteoclast and bone.

rumen degradable protein – protein that is capable of being degraded in the rumen.

rumen escape protein – protein that passes through the rumen without being degraded.

rumen undegradable – protein that is unable to be degraded by the rumen.

runt – low birth weight animal.

Ryanodine receptor – sarcoplasmic calcium channel found in muscle cells.

S

sarcolemma – outer membrane of the muscle fiber.

sarcomere – longitudinal repeating contractile unit of the myofibril that lies between two adjacent Z lines in striated muscle.

sarcomere addition – formation of new sarcomeres.

sarcomere length – distance between Z-lines of a sarcomere.

sarcoplasm – cytoplasm of muscle fibers.

sarcoplasmic reticulum – membranous system of tubules and cisternae that forms a network around each myofibril.

satellite cell – mononucleated muscle cell present in skeletal muscle after birth.

Schwann cell – glial cell that surrounds an axon in the peripheral nervous system.

sclerotome – compartment of the somite that develops bone.

seam fat – intermuscular adipose tissue.

secondary center of ossification – site of initial ossification in the epiphyses of long bones.

secondary muscle fibers – fibers arising from secondary myotubes.

secondary myotube – second wave of myotubes that are formed in a fetus following the formation of primary myotubes.

senescence – degeneration of cells that are not completely replaced or repaired.

sex odor – boar taint.

sexual maturity – stage of development when an animal is capable of reproducing.

signal transduction – cascade of intracellular events that occur beginning with receptor binding.

simple epithelium – epithelium consisting of a single layer of cells.

skatole – tryptophan degradation product that may contribute to off-flavor or off-odor in pork.

skeletal muscle – contractile striated, voluntary muscle tissue.

skeletal muscle calpain (Calpain 3 or p94) – calcium-dependent enzyme involved in protein degradation.

small-framed – animal with a relatively small skeleton.

smooth muscle – nonstriated, involuntary muscle tissue.

spicules – elongated, cylindrical structures of bone.

split carcass – carcass that has been separated into halves along the midline.

spongy bone – cancellous bone; bone that is organized into spicules and trabeculae.

spool joint – epichondial of the metacarpals used to identify the skeletal maturity of sheep carcasses.

stratified epithelium – epithelium consisting of multiple layers of cells.

stretch-induced hypertrophy – increased muscle fiber size resulting from muscle stretching.

stromal – connective tissue.

stunning – rendering unconscious.

subcutaneous fat – adipose tissue located underneath the skin.

supplemental fat – addition of fat to a diet to increase energy density.

syncytium – multinucleated mass of protoplasm produced by a merging of cells.

synovial fluid – fluid in a joint cavity.

T

T (transverse) tubules – invaginations of the sarcolemma forming a network of tubules that participate in calcium regulation.

thermogenesis – production of heat.

thick filaments – myosin containing filaments about 14–16 nm in diameter and 1.5 μm long.

thin filaments – actin containing filaments about 6–8 nm in diameter and 1.0 μm in length.

totipotent – capability to develop into any cell type.

trabeculae – long, flattened reinforcements of bone.

triad – T tubule and two associated elements of the terminal cisternae of the longitudinal tubules.

triglyceride – neutral lipid of the adipocyte.

tropoelastin – precursor of elastin.

tropomyosin – filamentous protein lying along the grooves of the thin filament.

troponin – globular protein located along the thin filament at intervals of 7 or 8 G actin molecules.

troponin C – one of the three isoforms of troponin.

troponin I – one of the three isoforms of troponin.

troponin T – one of the three isoforms of troponin.

U

ubiquitin – proteosome proteolytic pathway – one of the proteolytic systems for protein breakdown.

ultimate pH – pH of muscle when postmortem metabolism has ceased.

unilocular – presence of lipid as a single droplet.

V

villi – slender projection of membrane.

W

water-holding capacity – ability of meat to retain its moisture during the application of external forces.

white muscle fibers – anaerobic fibers.

Y

yield grade – grade assigned to segregate carcasses into groups on the basis of expected yield of trimmed cuts.

Z

Z line – dark band bisecting the I band made up of Z filaments that attach to the ends of the thin filaments in a myofibril.

REFERENCES

Aalhus, J. L., A. L. Schaefer, A. C. Murray and S. D. M. Jones. 1992. "The Effect of Ractopamine on Myofibre Distribution and Morphology and Their Relation to Meat Quality in Swine." *Meat Science*. 31, 397–409.

Aalhus, J. L., D. R. Best, F. Costello and A. L. Schaefer. 1997. "The Effects of Porcine Somatotropin on Muscle Fibre Morphology and Meat Quality of Pigs of Known Stress Susceptibility." *Meat Science*. 45, 283–295.

Aberle, E. D., and T. S. Stewart. 1983. "Growth of Fiber Types and Apparent Fiber Number in Skeletal Muscle of Broiler- and Layer-Type Chickens." *Growth*. 47, 135–144.

Aberle, E. D., J. C. Forrest, D. E. Gerrard and E. W. Mills. 2001. "Principles of Meat Science." Fourth Edition. Kendall/Hunt Publishing Co., Dubuque, IA.

Ahima, R. X., and J. S. Flier. 2000. "Adipose Tissue as an Endocrine Organ." *TEM*. 11, 327 332.

Allen, C. E., D. C. Beitz, D. A. Cramer and R. G. Kauffman. 1976. "Biology of Fat in Meat Animals." North Central Regional Research Publication No. 234. Research Division, College of Agricultural and Life Sciences, University of Wisconsin-Madison.

Anderson, P. T. 2000. "Mechanisms by Which Metabolic Modifiers Alter Growth Rate and Carcass Composition of Meat Animals." Pages 31–35 in Proc. 53rd Annual Recip. Meat Conf., Columbus, OH.

Anderson, R. R., R. J. Collier, A. J. Guidry, C. W. Heald, R. Jenness, B. L. Larson, H. A. Tucker. 1985. "Lactation." Iowa State University Press, Ames, IA.

Bechtel, P. J. 1986. "Muscle As Food." Academic Press, Orlando, FL.

Beermann, D. H., V. K. Fishell, K. Roneker, R. D. Boyd, G. Armbruster and L. Souza. 1990. "Dose-Response Relationships Between Procine Somatotropin, Muscle Composition, Muscle Fiber Characteristics and Pork Quality." *Journal of Animal Science*. 68, 2690–2697.

Berg, R. T., R. M. Butterfield. 1978. "New Concepts of Cattle Growth." Sydney University Press, Sydney, Australia.

Berne, R. M., M. N. Levy. 1988. "Physiology." Second Edition. C. V. Mosby Co., St. Louis, MO.

Bernlohr, D. A., N. Ribarik Coe and V. J. LiCata. 1999. "Fatty Acid Trafficking in the Adipocyte." *Cell & Developmental Biology*. 10, 43–49.

Bilezikian, J. P., G. R. Lawrence, G. A. Rodan. 2002. "Principles of Bone Biology." Second Edition. Harcourt Inc., San Diego, CA.

Bloom, W., D. W. Fawcett. 1968. "A Textbook of Histology." Ninth Edition. W. B. Saunders Co., Philadelphia, PA.

Bone, J. F. 1988. "Animal Anatomy and Physiology." Third Edition. Simon & Schuster, Englewood Cliffs, NJ.

Brody, S., 1945. Bioenergetics and Growth, Reinhold Publishing Corp., New York

Brook, C. G. D., N. J. Marshall. 2001. "Endocrinology." Fourth Edition. Blackwell Science, Ltd., Oxford, London.

Buckingham, M. 1994. "Molecular Biology of Muscle Development." *Cell*. 78, 15–21.

Buckingham, M. 2001. "Skeletal Muscle Formation in Vertebrates." *Current Opinion in Genetics & Development*. 11, 440–448.

Bullock, J., J. Boyle, III. and M. B. Wang. 1995. "Physiology." National Medical Series. 3rd Edition. Williams & Wilkins, Malvern, PA.

Butterfield, R. M. 1988. "New Concepts in Sheep Growth." Griffin Press Ltd., Netley, South Australia.

Buttery, P. J., N. B. Haynes and D. B. Lindsay. 1986 "Control and Manipulation of Animal Growth." Robert Hartnoll Ltd., Bodmin, Cornwall, Great Britain.

Campion, D. R., and G. L. Hausman. 1987. "Regulation of Fetal Growth: Importance of Growth and Tissue Deposition in the Pig." Symp. Agric. Research, Beltsville, MD. 12, 293–312.

Capdevila, J., and J. C. I. Belmonte. 2001 "Patterning Mechanisms Controlling Vertebrate Limb Development." *Annu. Rev. Cell Dev. Biol.* 17, 87–132.

Caperna, T. J., I. Fernández-Fígares, N. C. Steele and R. G. Campbell. 2002. "Emerging Metabolic Modifiers." Pages 57–62 in Proc. 53rd Annual Recip. Meat Conf., Columbus, OH.

Castillo, G., S. Hauser, J. K. Rosenfield and B. M. Spiegelman. 1999. "Role and Regulation of PPAR? During Adipogenesis." *Journal of Animal Science.* 77 (Suppl. 3), 9–15.

Christ, B. and C. P. Ordahl. 1995. "Early Stages of Chick Somite Development." *Anatomia, Hitsologia, Embryologia.* 191, 381–396.

Clancy, M. J., J. M. Lester and J. F. Roche. 1986. "The Effects of Anabolic Agents and Breed on the Fibers of the Longissimus Muscle of Male Cattle." *Journal of Animal Science.* 63, 83–91.

Cook, M. E. 2001. "Conjugated Linoleic Acid (CLA) in Growth and Development. Mechanisms Involving Immunity and Eicosanoids." Pages 1–5 in Intl. Anim. Agric. and Food Science Conference, Indianapolis, IN.

Cowherd, R. M., R. E. Lyle and R. E. McGehee Jr. 1999. "Molecular Regulation of Adipocyte Differentiation." *Cell & Developmental Biology.* 10, 3–10.

Dikeman, J. E. 2002. "Effects of Metabolic Modifiers Used in Animal Production on Meat Quality." Pages 36–56 in Proc. 53rd Annual Recip. Meat Conf., Columbus, OH.

Dwyer, C. M., J. M. Fletcher and N. C. Stickland. 1993. "Muscle Cellularity and Postnatal Growth in the Pig." *Journal of Animal Science.* 71, 3339–3343.

Dwyer, C. M., N. C. Stickland and J. M. Fletcher. 1994. "The Influence of Maternal Nutrition on Muscle Fiber Number Development in the Porcine Fetus and on Subsequent Postnatal Growth." *Journal of Animal Science.* 72, 911–917.

Engel, A. G., C. Franzini-Armstrong. 1994. "Myology: Basic and Clinical." Second Edition. McGraw-Hill, Inc., New York, NY.

Evans, H. E., W. O. Sack. 1973. "Prenatal Development of Domestic and Laboratory Mammals: Growth Curves, External Features and Selected References." *Anatomia, Histologia, Embryologia.* 2, 11–45.

Florini, J. R., D. Z. Ewton and S. A. Coolican. 1996. "Growth Hormone and the Insulin-Like Growth Factor System in Myogenesis." *Endocrine Reviews.* 17, 481–517.

Garlick, P. J., G. Caso, D. Mynarcik, M. A. McNurlan and M. E. Gelato. 2001. "Nutritional Regulation of Protein Metabolism." Paper Presented at Triennial Growth Symp., Int. Anim. Agric. and Food Science Conf., Indianapolis, IN.

Gilbert, S. F. 1988. "Early Vertebrate Development: Mesoderm and Endoderm." Pages 196–206 in Developmental Biology. Second Edition. Sinauer Associates, Inc., Sunderland, MA.

Gilbert, S. F. 1991. "Developmental Biology." Third Edition. Sinawer Associates, Inc., Sunderland, MA.

Goll, D. E., V. F. Thompson, R. G. Taylor, and T. Zalewska. 1992. Is calpain activity regulated by membranes and autolysis or by calcium and calpastatin? *Bioessays* 14:549–556.

Greaser, M. L. 1991. "An Overview of the Muscle Cell Cytoskeleton." Pages 1–5 in Proc. 44th Annual Recip. Meat Conf., Chicago, IL.

Gwartney, B. L., S. J. Jones and C. R. Calkins. 1992. "Response Time of Broiler Chickens to Cimaterol: Meat Tenderness, Muscle Composition, Fiber Size, and Carcass Characteristics." *Journal of Animal Science.* 70, 2144–2150.

Hammond, J. Jr., J. C. Bowman and T. J. Robinson. 1983. "Hammond's Farm Animals." Fifth Edition. Edward Arnold Ltd., Bedford Square, London.

Hammond, Sir J. 1960 "Growth in Size and Body Proportions in Farm Animals." Pages 321–334 in Proc. of International Symposium.

Han, S. S., J. O. V. Holmstedt. 1981. "Human Microscopic Anatomy." McGraw-Hill, Inc., New York, NY.

Henckel, P., N. Oksbjerg, E. Erlandsen, P. Barton-Gade and C. Bejerholm. 1997. "Histo- and Biochemical Characteristics of the Longissimus Dorsi Muscle in Pigs and Their Relationships to Performance and Meat Quality." *Meat Science.* 47, 311–321.

Hole Jr., J. W. 1990. "Human Anatomy and Physiology." Fifth Edition. Wm. C. Brown Publishers, Dubuque, IA.

Hollis. G. R. 1993. "Growth of the Pig." Wallingford: CAB Intl., New York, NY.

Hornick, J. L., C. Van Eenaeme, O. Gérard, I. Dufrasne and L. Istasse. 2000. "Mechanisms of Reduced and Compensatory Growth." *Domestic Animal Endocrinology.* 19, 121–132.

Kim, Y. S., and R. D. Sainz. 1991. "Minireview ß-Adrenergic Agonists and Hypertrophy of Skeletal Muscles." *Life Sciences.* 50, 397–407.

Klont, R. E., L. Brocks and G. Eikelenboom. 1998. "Muscle Fibre Type and Meat Quality." *Meat Science.* 49 (Suppl. 1), S219–S229.

Lane, M. D., Q.-Q. Tang and M.-S. Jiang. 1999. "The Adipocyte Diffentiation Program: Repression/Derepression of the C/EBP? Gene." *Journal of Animal Science.* 77 (Suppl. 3), 23–32.

Larzul, C., L. Lefaucheur, P. Ecolan, J. Gogué, A. Talmant, P. Sellier, P. Le Roy and G. Monin. 1997. "Phenotypic and Genetic Parameters for Longissimus Muscle Fiber Characteristics in Relation to Growth, Carcass, and Meat Quality Traits in Large White Pigs." *Journal of Animal Science.* 75, 3126–3137.

Lawrence, T. L. J., and V. R. Fowler. 1997. "*Growth of Farm Animals*." Wallingford: CAB Intl., New York, NY.

Lazar, M. A. 1999. "PPAR8 in Adipocyte Differntiation." *Journal of Animal Science*. 77 (Suppl 3), 16–22.

Lefaucheur, L., A. Missohou, P. Ecolan, G. Monin and M. Bonneau. 1992. "Performance, Plasma Hormones, Histochemical and Biochemical Muscle Traits, and Meat Quality of Pigs Administered Exogenous Somatotropin Between 30 or 60 Kilograms and 100 Kilograms Body Weight." *Journal of Animal Science*. 70, 3401–3411.

Li, Yong, M. F. Seifert, D. M. Ney, M. Grahn, A. L. Grant, K. G. D. Allen and B. A. Watkins. 1999. "Dietary Conjugated Linoleic Acids Alter Serum IGF-I and IGF Binding Protein Concentrations and Reduce Bone Formation in Rats Fed (n-6) or n-3) Fatty Acids." *Journal of Bone and Mineral Research*. 14, 1153–1162.

Lilburn, M. S. 1994. "Skeletal Growth of Commercial Poultry Species." *Poultry Science*. 73, 897–903.

Loveridge, N. 1999. "Bone: More Than a Stick." *Journal of Animal Science*. 77 (Suppl. 2), 190–196. *Dairy Science* 82, (Suppl. 2) 190–196.

MacLennan, D. H., and M. S. Phillips. 1992 "Malignant Hyperthermia." *Science*. 256, 789–794.

Martini, F. 1989. "Fundamentals of Anatomy and Physiology." Prentice Hall, Englewood Cliffs, NJ.

Maynard, L. A., J. K. Loosli, H. F. Hintz and R. G. Warner. 1979. "Animal Nutrition." Seventh Edition. McGraw-Hill Inc., New York, NY.

McMeekan, C.P. and J. Hammond. 1939. Journal of the Ministry of Agriculture, London, vol 46, 232.

McMeekan, C.P. 1940. Growth and development in the pig, with special reference to carcass quality characters; III. Effect of the plane of nutrition on the form and composition of the bacon pig. *Journal of the Agricultural Science,* vol 30, 511.

Mersmann, H. J. 2001. "Mechanisms for Conjugated Linoleic Acid-Medicated Reduction in Fat Deposition." Paper Presented at Intl. Anim. Agric. and Food Science Conf., Indianapolis, IN.

Milan, D., J.-T. Jeon, C. Looft, V. Amarger, A. Robic, M. Thelander, C. Rogel-Gaillard, S. Paul, N. Iannuccelli, L. Rask, H. Ronne, K. Lundstr?m, N. Reinsch, J. Gellin, E. Kalm, P. Le Roy, P. Chardon and L. Andersson. 2000. "A Mutation in PRKAG3 Associated with Excess Glycogen Content in Pig Skeletal Muscle." *Science* 288, 1248–1251.

Nishimura, T., A. Hattori and K. Takahashi. 1999. "Structural Changes in Intramuscular Connective Tissue During the Fattening of Japanese Black Cattle: Effect of Marbling on Beef Tenderization." *Journal of Animal Science*. 77, 93–104.

Oksberg, N., J. S. Petersen, M. T. Sórensen, P. Henckel and N. Agergaard. 1995. "The Influence of Porcine Growth Hormone on Muscle Fibre Characteristics, Metabolic Potential and Meat Quality." *Meat Science*. 39, 375–385.

Oksbjerg, N., P. Henckel and T. Rolph. 1994. "Effects of Salbutamol, a β_2-Adrenergic Agonist, on Muscles of Growing Pigs Fed Different Levels of Dietary Protein." *Acta. Agric. Scand., Sec. A. Animal Science*. 44, 12–19.

Oksbjerg, N., P. Henckel, T. Rolph and E. Erlandsen. 1994. "Effectsof Salbutamol, a β_2-Adrenergic Agonist, on Muscles of Growing Pigs Fed Different Levels of Dietary Protein." *Acta. Agriculture Scand., Sec. A. Animal Science*. 44, 20–24.

Olson, E. N. 1992. "Interplay between Proliferation and Differentiation within the Myogenic Lineage." *Developmental Biology*. 154, 261–272.

Ono, Y., M. B. Solomon, C. M. Evock-Clover, N. C. Steele and K. Maruyama. 1995. "Effects of Porcine Somatotropin Administration on Porcine Muscles Located Within Different Regions of the Body." *Journal of Animal Science*. 73, 2282–2288.

Ono, Y., M. B. Solomon, T. H. Elsasser, T. S. Rumsey and W. M. Moseley. 1996. "Effects of Synovex-S and Recombinant Bovine Growth Hormone (Somavubove) on Growth Responses of Steers: II. Muscle Morphology and Proximate Composition of Muscles." *Journal of Animal Science*. 74, 2929–2934.

Orth M. W. 1999. "The Regulation of Growth Plate Cartilage Turnover." *Journal of Animal Science*. 7(Suppl. 2), 183–189. *Dairy Science*. 82 (Suppl. 2), 183 189.

Owens, F. N., D. R. Gill, D.S. Secrist and S. W. Coleman. 1995. "Review of Some Aspects of Growth and Development of Feedlot Cattle." *Journal of Animal Science*. 73, 3152–3172.

Owens, F. N., P. Dubestk and C. F. Hanson. 1993. "Factors that Alter the Growth and Development of Ruminants." *Journal of Animal Science*. 71, 3138–3150.

Pearson, A. J., and T. R. Dutson. 1991. "Growth Regulation in Farm Animals. Advances in Meat Research. Volume 7." Elsevier Science Publishers, Ltd., Essex, England.

Pearson, A. M., R. B. Young. 1989. "Muscle and Meat Biochemistry." Academic Press, Inc., San Diego, CA.

Perry, J. S. 1981. "The Mammalian Fetal Membranes." *Journal Reproduction Fertility*. 62, 321–335.

Pette, D., and R. S. Staron. 1997. "Mammalian Skeletal Muscle Fiber Type Transitions." *International Review Cytology*. 170, 143–195.

Pourquié, O. 2001. "Vertebrate Somitogenesis." *Annu. Rev. Cell Dev. Biol*. 17, 311–50.

Rahelic, S., and S. Puac. 1980. "Fibre Types in *Longissiums Dorsi* From Wild and Highly Selected Pig Breeds." *Meat Science*. 31, 439–450.

Randall, D. W., Burggren, K. French. 2002. "Eckert Animal Physiology: Mechanisms and Adaptations." W. H. Freeman and Co., New York, Ny.

Rehfeldt, C., I. Fiedler, G. Dietl, and K. Ender. 2000. "Myogenesis and Postnatal Skeletal Muscle Cell Growth as Influenced by Selection." *Livestock Production Science.* 66, 177–188.

Rosen, E. D., and B. M. Spiegelman. 2000. "Molecular Regulation of Adipogenesis." *Annu. Rev. Cell Dev. Biology*. 16, 145–71.

Ruusunen, M., and E. Puolane. 1997. "Comparison of Histochemical Properties of Different Pig breeds." *Meat Science*. 45, 119–125.

Seideman, S. C. and J. D. Crouse. 1986. "The Effects of Sex, Condition, Genotype and Diet on Bovine Muscle Fiber Characteristics." *Meat Science*. 17, 55–72.

Seideman, S. C., J. D. Crouse and H. R. Cross. 1986. "The Effect of Sex, Condition and Growth Implants on Bovine Muscle Fiber Characteristics." *Meat Science*. 17, 79–95.

Smith, S. B., D. R. Smith. 1995. "The Biology of Fat in Meat Animals." American Society of Animal Science, Champaign, IL.

Soest, V. 1982. "Nutrition Ecology of the Ruminant." O. & B. Books Inc., Corvallis, OR.

Solomon, M. B. 1985. "Profile of Fiber Types in Muscles from Wild Pigs Native to the United States." *Meat Science*. 13, 247–254

Solomon, M. B., R. G. Campbell and N. C. Steele. 1990. "Effect of Sex and Exogenous Porcine Somatotropin on Longissimus Muscle Fiber Characteristics of Growing Pigs." *Journal of Animal Science*. 68, 1176–1181.

Solomon, M. B., R. G. Campbell, N. C. Steele and T. J. Caperna. 1991. "Effects of Exogenous Porcine Somatotropin Administration Between 30 and 60 Kilograms on Longissimus Muscle Fiber Morphology and Meat Tenderness of Pigs Grown to 90 Kilograms." *Journal of Animal Science*. 69, 641–645.

Solomon, M. B., R. G. Campbell, N. C. Steele, T. J. Caperna and J. P. McMurtry. 1988. "Effect of Feed Intake and Exogenous Porcine Somatotropin on Longissimus Muscle Fiber Characteristics of Pigs Weighing 55 Kilograms Live Weight." *Journal of Animal Science*. 66, 3279–3284.

Solomon, M. B., T. J. Caperna, R. J. Mroz and N. C. Steele. 1994. "Influence of Dietary Protein and Recombinant Porcine Somatotropin Administration in Young Pigs: III. Muscle Fiber Morphology and Shear Force." *Journal of Animal Science*. 72, 615–621.

Stickland, N. C. 1983. "Growth and Development of Muscle Fibres in the Rainbow Trout (Salmo gairdneri)." *Journal Anat.* 137, 323–333.

Stickland, N. C. 1995. "Muscle Growth: Microstructural Aspects of Skeletal Muscle Growth." 4, 241–245. *Meat Focus International.*

Stockdale, F. E. 1992. "Myogenic Cell Lineages." *Developmental Biology*. 154, 284–298.

Swatland, H. J. 1994. "Structure and Development of Meat Animals and Poultry." Technomic Publishing Co., Lancaster, PA.

Swatland, H. L. 1984. "Structure and Development of Meat Animals." Prentice-Hall, Inc., Englewood Cliffs, NJ.

Vestergaard, M., P. Henckel, N. Oksbjerg and K. Sejrsen. 1994. "The Effect of Cimaterol on Muscle Fiber Characteristics, Capillary Supply, and Metebolic Potentials of Longissimus and Semitendinosus Muscles from Young Friesian Bulls." *Journal of Animal Science.* 72, 2298–2306.

Vestergaard, M., S. Purup, P. Henckel and E. Tonner, D. J. Flint, L. R. Jensen and K. Sejrsen. 1995. "Effects of Growth Hormone and Ovariectomy on Performance, Serum Hormones, Insulin-Like Growth Factor-Binding Proteins, and Muscle Fiber Properties of Prepubertal Friesian Heifers." *Journal of Animal Science*. 73, 3574–3584.

Walbot, V., N. Holder. 1987. "Developmental Biology." First Edition. Random House, Inc., New York, NY.

Wegner, J., E. Albrecht, I. Fiedler, F. Teuscher, H.–J. Papstein and K. Ender. 2000. "Growth- and Breed-related Changes of Muscle Fiber Characteristics in Cattle." *Journal of Animal Science*. 78, 1485–1496.

White, P., D. Cattaneo and M. J. Dauncey. 2000. "Postnatal Regulation of Myosin Heavy Chain Isoform Expression and Metabolic Enzyme Activity by Nutrition." *British Journal Nutrition*. 84, 185–194.

Whittow, G. C. 2000. "Sturkie's Avian Physiology." Fifth Edition. Academic Press, San Diego, CA.

Wu, Z., P. Puigserver and B. M. Spiegelman. 1999. "Transcriptional Activation of Adipogenesis." *Cell Biology*. 11, 689–694.

Yeh, W.-C., and S. L. McKnight. 1995. "Regulation of Adipose maturation and Energy Homeostatis." *Cell Biology*. 7, 885–890.

Young, O. A. 1984. "Effect of Castration on Bovine Muscle Composition." *Meat Science*. 11, 139–156.

INDEX

A

A-band, 30, 241
absolute rate, 93
accretion, 1
  net lipid, rate, 125, 246
  net protein, 91, 246
  protein, rates, 93
  regulation of protein, 213–214
acetate, 128, 241
acetylcholine receptor subunit, 67
actin, 30, 31–32, 39, 67, 105, 241
actomyosin ATPase activity, 105–107, 242
ad libitum access, 156, 203, 241
adherins, 74, 241
adipoblasts, 121, 241
adipocytes, 15, 122, 124, 128–129, 241
adipogenesis, 121, 123–125, 241
adipose tissue, 241. *See also* fat
  age affect on, 129
  anatomical location of, 119, 129, 131
  cellularity, 129
  composition, 129–135
  as connective tissue, 15–16, 119, 160
  contribution to muscle fiber size, 100
  depots, 119–121, 129, 131, 132, 133, 134, 154, 161, 168, 169, 201, 212, 225–226
  environmental temperature influence on, 135
  genetics affect on, 131
  growth/development, 119–135
  hormones affect on, 132, 133
  mass, 193
  metabolism, 125
  nutrition affect on, 132–134
  in pigs, 121, 123, 124, 133
  sex affect on, 131–132
  species differences in, 122, 129–130
adiposity, 122, 241
ADP levels, 228
adrenaline, 194
aerobic metabolism, 228, 241
age
  adipose affect tissue, 129
  chronological v. physiological, 151–152
  conceptus, 61–62
  meat quality and, 234
  muscle fiber number affect on, 81
  muscle fiber size affect on, 99
ageing, affect of, 232
aitch bone, 154
allantois, 58, 241
alpha-actinin, 27, 241
amino acid(s), 241. *See also specific amino acids*
  deaminated, 210
  deficiencies, 134, 210
  imbalance, 210
  supplementation, 210, 211
amnion, 58, 241
anaerobic glycolysis, 228
anaerobic metabolism, 110, 228, 241
androgens, 98, 184–188
androstenone, 184
animal growth/development, study of, 1–5
animal industry, 1–4. *See also* meat animal industry
animals. *See also* meat animals
  early maturing, 170, 172, 243
  large-framed, 168, 245
  late maturing, 170, 172, 245
  lean growth v. fat, 3–4
  market, performance, 1–4
  small-framed, 168, 248
  transgenic, 5
antemortem handling, 228, 241
antibacterial feed additives, 212–213
antibiotics, 213, 241
antidiuretic hormone (ADH), 183
apical, 51, 241
apical ectodermal ridge, 53, 241
aponeurosis, 24, 241
articulations, 10, 24, 242
astrocytes, 21, 242
ATP, 105–107, 228, 230, 236, 242
autocrine action, 179, 242
autolysis, 92, 242
average growth impetus, 161
avian early development, 50, 51, 52

B

backfat, 120, 226, 242
basement membrane, 26, 242
beef
  carcass maturity, 141
  quality grades, 223
beta-adrenergic agonists, 117, 195–196, 197, 239
biological value, 208, 242
biotechnological advances, 4
birds. *See also* avian early development
  muscle fiber characteristics of, 109
  myogenic lineages in, 96
bison, 165
blastocyst, 48, 58, 242
blastomeres, 47, 242
blastulation (blastula phase), 49–50, 242
blood, as connective tissue, 16, 242
bloodstream conveyance, 179, 242
body
  components, 156–158
  composition, 199
  form, 153–154
  measurement, 154–156
  polar, 47
  type changes, 2, 3

weight, 156–160
bone
aitch, 154
calcium salts in, 19–20
cancellous, 18, 242
cells, 20, 137–138
collagen fibers in, 19–20
compact, 18
as connective tissue, 18–20
cortical, 137
diameter of growing, 141–142
formation, 137–138, 185, 186
hormones and, 147
lining cells, 138, 242
long, 18, 137, 139, 141, 142, 145, 245
marrow, 18, 139, 142, 242
matrix, 137–138
measuring methods, 144–145
medullary, 137
nutrition requirements of, 146
osteons/ Haversian system in, 20, 141, 244
ratio of muscle to, 173–174, 175, 177, 222
remodeling, 144–146, 242
resorption, 138, 142–144, 185, 186, 247
spongy, 18, 139, 142, 144, 248
tissue, 18–20
types of, 20
break joint, 141, 219, 242
breed
effect on fat distribution, 174
frame size and, 168
muscle fiber number and, 81–84, 85
muscle fiber size and, 98
broiler performance, 3
buttons, 217, 242
butyrate, 128, 242
bypass protein, 209, 242

C

calcitonin, 145, 196–198
calcium carbonate, 20
calcium phosphate, 20
calcium pumps, 36
calcium salts, 19–20
Callipyge sheep, 89, 100, 114, 242
μ-calpain (Calpain 1), 92, 241
calpain system, 92, 93, 242
calpastatin, 92, 242
calsequestrin, 36, 242
canaliculi, 20, 242
cap Z, 32, 242
carbohydrates, low v. high, 213
carcass
components, 158–159
composition, 163–178
growth hormone (GH) effect on, 190–191
length, 154–156
maturity, 141, 216–219
pork, 225, 226
ribbed, 220
size, 226
split, 217, 248
weight, 157–158, 226
weight/length of pig, 154, 155
yields of meat animals, 147
cardiac muscle, 37–38, 242
carnucles, 60, 242
cartilage
articular, 140, 141, 241
as connective tissue, 16–18
elastic, 18, 243
fibrocartilage as, 18, 19, 244
ground substance of, 16–17
growth (epiphyseal) plate, 18, 20, 140–141, 145–146
hyaline, 17–18, 244
matrix, 16–17, 139
cartilaginous structure, 137, 138–139
castration, 242. *See also* males, intact v. castrate
elimination of, 233
growth/composition and, 176–178
hormones and, 185, 186
skeletal growth and, 146
catecholamines, 194–195
cathepsins, 92, 242
cattle. *See also* beef; steer, market
bovine fetus weight of, 150
Cimaterol effect on, 117
cotyledon structure in, 61
double muscled, 5, 85–86, 87, 99–100, 114, 236–237, 243
endocrine glands, 183
fat in, 167–168
growth hormone (GH) administration in, 5
milk v. meat type, 170–175
muscle fiber type in, 107–108, 113, 114–116
muscle growth in, 162
muscle to bone ratio, 177
muscle weight distribution in, 178
prenatal development in "continental" v. "British," 43
semitendinosus muscle, 80, 85
steroid use in, 186–188
twin, 204–206
cell(s). *See also specific cells*
action of growth factors on, 65
ameboid, 10, 241
bone, 20, 137–138
bone lining, 138, 242
columnar, 8
cuboidal, 8
cycle, 63–66, 242
daughter, 63–64, 243
definition of, 7
determination, 66–67, 243
determined/committed, 66, 242
division, 43, 63, 64, 242
dormant *(quiescent),* 66
ependymal, 21
epithelial, 137
fat, 15, 17, 121, 123–125, 244
fixed, 9, 10
glial, 21
lineages, 43, 242
mast, 10, 245
maturation, 68
mesenchymal, 137
miroglia, 21
mitosis, 64
multipotent, 43, 246
muscle, 26, 92
muscle fiber precursor, 63
muscle satellite, 80
myogenic precursor, 72
nerve, 21, 22
oligodendrocytic, 21, 23
postmitotic, 63
red blood, 16
replication competent, 67
satellite, 66, 93–98, 248
Schwann, 21
scleratomal, 56, 137
secretory, 7
squamous, 8
structural proteins in, 74
totipotent, 43
types, 10
wandering, 9, 10
white blood, 16
cell surface receptors, 65, 242
cell-adhesion molecules, 74, 242
cellular autonomy, 50
cellular replications, 43, 242
central nervous system, 21, 242
chemical composition, changes in, 159–160
chemotaxis, 52, 242
chickens
fat in, 167
growth/muscle fiber diameter in, 90
chondroblasts, 138, 139, 242

chondrocytes, 16, 18, 138, 140, 242
chondroitin sulfates, 10, 16, 242
chorion, 58, 242
chronic stimulation, 116
chronological age, 216
ciliation, 8
Cimaterol, effect on cattle, 117
class, assignment of, 216
cleavage phase, 47, 48, 49
cold-shortening, 234
collagen
  composition of, 10–11, 12–13
  fibers, 10–13, 19–20
  ground substance of, 17–18
  molecule, 10–11
  pentafibrils of, 11
  procollagen and, 11
  synthesis/maturation, 12
  types, 13
color, of lean, 221
compaction, 47–48, 242
compartment, 65
conceptus, 59, 60, 61–62
conformation, 221–222, 243
conjugated linoleic acids (CLAs), 134
connectin, 32
connective tissue, 7, 9–20, 160–161, 243
  adipose tissue as, 15–16, 119, 160
  blood as, 16, 242
  bone as, 18–20
  cartilage as, 16–18
  cell types, 10
  collagen fibers as, 10
  contribution to muscle size, 101
  dense regular/irregular, 14–15, 16, 243
  elastin as, 13
  endomysial, 25–26
  epimysial, 24–25
  fibers, 10–13
  ground substance of, 9, 10
  hematopoietic, 16
  loose, 13–14, 16, 245
  lymph as, 16, 242
  meat tenderness and, 231
  muscle, 24–26
  perimysial, 25
  proper, 9–10, 138, 243
  reticular fibers as, 13
  specialized, 15–16
  structure of, 10
  supportive, 16–20
  types of, 13–15
contraction, muscle, 100, 105–106, 109
costameres, 33, 34, 243
cotyledons, 60, 61
crown-rump length (CRL), 61–62, 243
Culard, 85, 243. *See also* cattle, double muscled
cutability, 223, 243
cystatin, 92
cytochromes, 109–110
cytokinesis, 47, 63, 64, 243
cytoplasmic calcium concentrations, 110, 111
cytoskeleton, 243
  muscle, 26–27
  stress fibers in, 74

D

daily gain, average, 152, 243
dairy type, 168, 243
deer, 165
defects, 222
degenerative joint diseases, 141
delamination, 49, 52, 243
deletions, 86
dendrites, 21, 243
dense zones/bodies, 39, 243
depots. *See* adipose tissue
dermamyotome, 51, 68, 69, 243
dermatome, 52, 55, 56, 243
desmin, 33
desmosine, 13, 243
determination, during myogenesis, 66–67
development, 1. *See also* growth; *specific types of development*
diabetes, 132, 243
diameter
  of growing bone, 141–142
  muscle fiber, 90
diaphysis, 18, 139, 243
diet
  composition, 132–134
  fat hardening, 212, 244
dietary energy, 199–202, 210–212
dietary protein, 208–210
differentiation, during myogenesis, 67, 243
diffuse attachment, 60
dihydropyridine receptor, 36
distal light chains, 31
DNA, 243
  accumulation in postnatal muscle, 97
  mutation during replication of, 85
  in ovum phase, 47
  ratio/unit, 93
  replication during mitosis, 64
  RNA:DNA ratio of, 96
domestication, 164–167
Dopplelander, 85, 243
double muscled cattle, 5, 85–86, 87, 99–100, 114, 236–237, 243
dressing percentage, 156–157, 170
dyschondroplasia, in poultry, 141

E

early maturing animal, 170, 243
ectoderm, 50, 243
elastic fibers, 10
elastin, 13, 14, 243
embryo
  ectoderm of, 50
  splitting, 47
embryonic disc, 50
embryonic growth/development, 43–62, 243. *See also* prenatal growth/development
  anatomical indications during, 62
  cell number changes during, 47
  in "continental" v. "British" cattle, 43
  differentiation of tissues during, 45
  embryonic disc and, 50
  fetal phase of, 45, 53–58
  germ layers in, 50–51
  human, 46, 48
  impact on postnatal growth, 43, 46
  measurements of, 60–62
  muscle fiber development during, 71, 72–73
  myogenesis during, 111, 113
  ovum phase of, 45–49
  primitive streak in, 50, 51, 52
  somite formation during, 60
  splitting during, 47–49
endocrine glands, 183, 198
endocrine system, 5, 199, 201, 243
endoderm, 51, 243
endomysium, 25, 243
endoneurium, 21, 243
endosteum, 20, 243
endotheliochorial condition, 59, 60
endotheliochorial connection, 59
endothelium, 7, 243. *See also* epithelial tissue
endpoint, 152, 243
energy balance, 201–202, 243
energy, dietary, 199–202, 210–212
energy metabolism, 105, 243
environment
  impact of, 163

environmental factors, affect on partitioning of nutrients, 208
environmental temperature
  influence on adipose tissue, 135
  influence on muscle fiber types, 116
enzymes. *See specific enzymes*
epiblast, layer of embryonic disc, 50
epiboly, 49, 243
epimysium, 24, 243
epinephrine (adrenalin), 194
epineurium, 21, 243
epiphyseal closure, 141, 243
epiphyseal plates, 18, 20, 140–141, 145–146, 219, 243
epithelial tissue, 7–9, 243
epitheliochorial condition, 59
esterification, 125
estrogens, 145–146, 186, 198
evolution, 164
extended growth period, 204, 244
extracellular molecules, 74, 244
extraembryonic tissues, 58

F

F-actin, 31, 32, 244
fascia, 24, 244
fasiculi, 25
fat. *See also* adipoblasts; adipocytes; adipose tissue; backfat
  absence of, 234
  amount of, 225–226
  breed type and, 174
  brown, 125, 203
  in cattle, 167–168
  caul, 119, 242
  cells, 15, 17, 121, 123–125, 244
  in chickens, 167
  content of newborns, 203
  deposition by species, 167–168
  depots, 119–121, 129, 131, 132, 133, 134, 154, 161, 168, 169, 201, 212, 225–226
  depth, 226, 244
  dietary, 127
  endpoint, 170, 244
  feed intake effect on, 203
  growth impetus, 162
  hardening diet, 212, 244
  intermuscular (seam), 120, 121, 245
  intramuscular, 25, 101, 238, 245
  intraosseus, 161
  kidney, 119, 245
  lace, 119
  leaf, 119, 245
  messenteric, 119
  neutral, 125, 246
  perirenal, 119, 247
  in pigs, 167
  in sheep, 167
  subcutaneous, 120, 156, 248
  supplemental, 212, 248
  thickness, 226, 244
  visceral, 119–120
fat animals v. lean growth, 3–4
fat free lean, 228, 244
fatback, 120, 244
fatty acid synthesis, 125–128, 147, 212
feathering, 220, 244
feed
  antibacterial additives, 212–213
  efficiency, 199
  intake effect on muscle/fat, 203
  withdrawal, 238, 244
feed:gain ratios, 201
fetal phase, 53–58, 244
fetal tissues, 59
fetus
  transfer of nutrients to, 202
  weight of bovine, 150
fiber. *See* muscle fiber; muscle fiber number; muscle fiber size; muscle fiber types; *specific fibers*
fibroblasts, 10, 65, 138, 244
filamentous actin, 31, 244
filamentous myosin, 31
filamin, 33
fill, 156, 244
finish, 220, 244
firmness, 220–221
fish
  hyperplasia in, 81, 83
  muscle fiber size/number, 82
flank streaking, 219–220, 244
fleshing, 221–222, 244
fractional rate, 93, 244
frame scores, 168–170, 244
frame size, 168–173, 244
free choice access, 156, 244
fresh meat quality, 216
fusion, during myogenesis, 67–68

G

G-actin, 31, 244
gap junctions, 72, 244
gastrulation, 49, 50–51, 244
gene. *See also specific genes*
  product, 86
  "stress," 234–236
genetics, 244
  affect on adipose tissue, 131
  affect on carcass composition, 163
  affect on muscle fiber number, 85–86
  affect on muscle fiber size, 99–100
  influence on muscle fiber types, 114
genome, 64
genotype, 163, 244
germ layers, 50–51
girth, 100, 244
glial cells, 21
globular head, 105, 244
glossary, 241–249
glucagon, 192
glucocorticoids, 194
glucose residues, 228
glycogen, 228
glycogenolysis, 228, 230
glycolytic potential values, 234, 244
gonads, 45, 244
grading system, meat, 215–232, 244
ground substance, 244
  of cartilage, 16–17
  of collagen, 17–18
  of connective tissue, 9, 10
growth. *See also* animal growth/development, study of; embryonic growth/development; muscle; prenatal growth/development
  allometric differential, 161–163
  appositional, 143
  augmenting, 233–240
  catch-up, 204, 242
  compensatory, 204–208, 243
  definition of, 1, 149
  determinate v. indertiminate, 81
  diphasic, 163, 243
  hormonal regulation of, 179–198
  interstitial, 17, 20, 245
  meat animal, 77
  meat quality and, 233–240
  monophasic, 162, 163, 245
  nutritional regulation of, 199–214
  perinatal stages of, 203
  permanent stunting of, 205
  protein changes during, 160
  quality v. quantity in, 233
  skeletal, 146
  true v. fattening, 1
  in utero, 45
  whole body animal, 149–178
growth biology, 4
growth coefficient, 161
growth curves, 149–153, 159
  absolute rate, 152, 208, 241
  compensatory growth, 204
  early v. late maturing animals, 172
  postnatal, 151, 158

relative rate, 152–153
sigmoidal, 149, 158
species/sex, 176
growth factors, 65, 147, 244
growth hormone (GH), 189–192. *See also* somatotropin
administration in cattle, 5
administration in pigs, 99, 100
effect on carcasses, 190–191
effect on muscle fiber type, 117
induced muscle growth, 190
growth hormone releasing hormone (GHRH), 183, 189
growth plate, 18, 20, 140–141, 145–146
growth promotants
affect on muscle fiber size, 99
effect on muscle fiber types, 117–118
growth rate
improvement of, 2–3
relative, 152–153, 247

H

H zone, 30, 244, 247
halothane (anesthetic), 234, 244
halothane gene, 88, 234–236, 244
Hammond, Sir John, 4, 205
Hampshire effect, 234, 244
hatching, 47
Haversian system, in bone, 20, 141, 244
head size, 153
heavy meromyosin, 31, 244
helix-loop-helix proteins, 66–67
hemochorial interface, 60
Henson's Node, 50
heredity, 163, 164
heritability estimates, 163
heterodimers, 66
high frequency/low amplitude stimulation, 116
high growth impetus, 161
histamines, 10, 244
hogs. *See* pigs
homeorhesis, 192, 201, 244
homeostasis, 228, 244
homodimers, 66
hormonal profiles, 5
hormonal regulation
of growth, 179–198
of osteolysis, 142
hormone(s). *See also* growth hormone (GH); growth hormone releasing hormone (GHRH); somatotropin; *specific hormones*; steroids
action, 179–180, 181
affect on adipose tissue, 132, 133
bone growth and, 147
castration and, 185, 186
chemical nature of, 180
combinations of synthetic, 186–188
definition of, 179
effect on muscle fiber types, 117
hypothalamic-pituitary -peripheral gland axes and, 180–184
implants, 188, 238–240
inhibiting, 183
negative feedback loop regulation of, 180–181, 183–184
receptors, 179
regulation of growth, 179–198
releasing, 183, 247
sex, 146
tropic, 183
horses, microcotyledon structure in, 60, 61
hyaluronic acid, 10, 16, 244
hydroxyapatite, 20, 244
hyopoblast formation, avian, 50
hyperphagia, 132, 244
hyperplasia, 1, 85, 244
adipocyte, 121–123
effect on muscle hypertrophy, 78
fish, 81, 83
muscle cell, 78
muscle fiber, 73, 81, 83, 103
hypertrophy, 1, 78, 244
adipocyte, 121–123
v. atrophy in muscle fiber, 91–92
hyperplasia effect on, 78
muscle, 77, 78
muscle cell, 77, 78
stretch-induced, 90, 248
work/exercised-induced, 88, 243
hypoblast, 50, 244
hypoglycemia, 192
hypothalamic-pituitary-peripheral gland axes, 180–184
hypothalamus, 180, 182, 184, 244

I

I-band, 30, 245
IGF family, 190
implantation, 245
in domestic farm animals, 44
interstitial v. superficial, 59
in ovo, 43, 80, 245
in utero, 43, 45, 80, 245
in vivo, 72
ingression, 49, 245
inhibiting hormones, 183
inner cell mass, 49, 245
insertion, of articulation, 24, 245
insulin, 192
insulin-like growth factors, 65
integrins, 74
intercalated discs, 38, 245
intermediate filaments, 26
interstitial fluid, 16, 245
interstitial lamellae, 20
intracrine action, 179, 245
invagination, 49, 245
involution, 49, 245
ionophores, 212–213, 245
isocitrate dehydrogenase, 106
isodesmosine, 13, 245
isoforms, 31

J

joints, 24

K

keratan sulfate, 16, 245
kind, assignment of, 216
Koller's Sickle, 49, 245
Kreb's cycle, 228

L

lactate dehydrogenase, 106
lactation, utilization of energy for, 201
lacunae, 17, 18, 20, 245
lambs. *See also* sheep
Callipyge v. normal genotype, 89
carcass maturity, 141
grades, 219
performance, 3
lamellae, 20, 245
large-framed animals, 168, 245
late maturing animal, 170, 245
lateral plate mesoderm, 51, 137
lathyrism, 11–12
lean growth v. fat animals, 3–4
lean tissue color, 219
length, conceptus age and, 61–62
leptin, 5, 132, 192–193, 245
light meromyosin, 31, 245
limb bud formation, 53, 56, 57
limit feed basis, 203, 245
linoleic acids, 134
lipases, 128, 245
lipid metabolism, 132–135
lipogenesis, 125–128, 245
lipolysis, 125, 128–129, 245
lipoprotein lipase, 125

livestock, marketing, 177
lobules, 121
local conveyance, 179, 245
loin-eye area, 225
longissimus dorsi muscle
  appraisal of, 220
  area, 154–156
  in cattle, 85, 104, 113–114
  measurement of, 225
  muscle fiber number in, 77–79, 85
  in pigs, 90, 108, 113–114, 115, 117
low frequency/high amplitude stimulation, 116
low growth impetus, 161
lumen, 47
lymph, as connective tissue, 16, 242
lymphocytes, 15
lysl oxidase, 12, 245
lysosomal system, 92, 245

M

macrophages, 10, 15
males, intact v. castrate, 146, 176–178, 185, 186, 233
malignant hyperthermia, 234, 245
mammalian development, 44
marbling, 25, 101, 121, 223, 233, 238, 245
  flecks, 219–220
  scores, 220
marrow, bone, 18, 242
  cavity, 139, 142
mastication, 231, 245
maternal membrane development, 58–60
maternal tissues, 59
maturation, 245
  during myogenesis, 68
maturity, 245
  beef carcass, 141
  carcass, 216–219
  chemical, 152, 242
  compositional, 152, 243
  estimating, 141
  lamb carcass, 141
  measuring, 152
  physiological, 152, 247
  sexual, 152, 248
m-calpain (calpain 2), 92, 245
McMeeken, C.P., 4, 205
mean packing density, 80, 245
measurement, body, 154–156
meat
  acid, 231, 241
  cuts of, 77, 79
  dark cutting, 228–229
  grading, 215–232
  pale, soft, exudative (PSE), 230–231
  tenderness, 231, 232, 237–238
  transformation of muscle to, 221
  type, 168, 245
meat animal industry. *See also* animal industry
  estimating maturity by, 141
  goal of, 215
  importance of bone growth to, 4, 5
  optimizing of muscle growth, 23
meat animals, 170–175
  carcass yields of, 147
  growth in, 77
  muscle fiber characteristics of, 109
meat quality, 111, 215–240
  age and, 234
  grading system for, 215–232
  hormone implants effect on, 238–240
  muscle fiber type and, 237–238
  relationship with growth, 233–240
medullary cavity, 18, 245
meiosis, 47, 245
melatonin, 198
melengestrol acetate (MGA), 186
mesoderm, 51, 245
metabolism
  adipose tissue, 125
  aerobic, 228, 241
  anaerobic, 110, 228, 241
  energy, 105, 243
  glycolytic, 116, 244
  lipid, 132–134
  modifiers, 238
  muscle protein, 184–185
  oxidative v. glycolytic, 109–110, 244, 246
  postmortem, 228, 230, 236, 247
  steroid, 233
  stress-induced changes in, 195
  whole body, 5
mice
  leptin hormone in, 132
  muscle growth in, 98
microcontyledonary, 60
microcotyledon structure, in horses, 60, 61
microfilaments, 26, 245
microphages, 15, 245
microtubules, 26, 245
microvilli, 9, 245
milk production, 201–202
minerals, ash, 161
miosin heavy chain molecules, 30
miosin light chain molecules, 30
mitosis, 63, 64, 65, 245
morphogenesis, 49
morphogens, 52
motor neurons, 21
M-phase, 65
m-protein, 31, 245
MRF-4, 67, 246
mucous, 7, 246
muscle. *See also* myogenesis; *specific muscles*
  amount of, 223–225
  atrophy, 93, 242
  cardiac, 37–38
  cells, 92
  chemical composition of, 39–41
  comparison, 165
  connective tissue, 24–26
  contraction, 100, 105–106, 109
  cytoskeleton, 26–27
  development, proteins coordinating, 85
  different types of, 78
  extraocular, 110
  growth, 23, 77–101
  growth hormone (GH) and, 190
  growth impetus, 162
  growth in cattle, 162
  intercostal, 220
  major groupings of, 162
  into meat, 228
  in meat cuts, 79
  myoblasts in, 96
  myofibrillar splitting in, 91
  plasticity, 103, 104
  precursor cells, 69
  protein synthesis in, 91–93, 95, 96, 247
  proteins effect on, 99
  proteolytic systems in, 92–93
  ratio to muscle, 173–174, 175, 177, 222
  red v. white, 103, 111
  sarcomere in, 27, 29, 30, 31
  satellite cells, 80
  scores, 228
  semimembranosus, 23, 80
  semitendinosus, 23, 79, 80, 85, 103, 104
  size, 101
  skeletal, 68
  smooth, 38–39, 248
  species differences in, 166–167
  texture, 219
  water to protein ratio in, 160
  weight, 77, 111
  weight distribution in cattle, 178

muscle cell hyperplasia, 78
muscle cell hypertrophy, 77, 78
muscle cell lineage, 66, 246
muscle creatine kinase, 67
muscle fiber(s), 26–37, 77–101, 246
  arrangement of, 78
  characteristics in meat animals/birds, 109
  composition, 113
  development, 72–73
  diameter, 90
  diversification, 104
  embryonic/fetal development of, 71
  fish, 82
  giant, 88
  hyperplasia, 73, 81, 83, 103
  hypertrophy v. atrophy in, 91–92
  maturation during myogenesis, 68
  mitochondria in, 36–37
  nuclei in, 36
  organization of, 79
  parameters, 79
  pig skeletal, 26
  plasticity, 103, 104, 116
  precursor cells, 63
  primary/secondary, 70, 247
  regeneration, 97
  sarcoplasm in, 37
  secondary, 81, 248
  size changes over time, 70
  typing, 105
muscle fiber number, 77–86
  absolute, 79–80, 241
  age affect on, 81
  animal variation affect on, 80
  apparent, 79–80, 241
  breed and, 81–84, 85
  in double muscled cattle, 85–86, 87
  in fish, 82
  genetics affect on, 85–86
  intrafasicularly terminating fibers in increased, 80
  in longissimus dorsi muscle, 77–79, 85
  mean packing density in, 79–80
  mechanisms for increasing, 80
  muscle affect on, 80
  muscle fiber orientation and, 79
  v. muscle fiber size, 77, 86
  muscle regeneration and, 80
  nutrition affect on, 80–81
  performance traits relationship with, 84
  in pigs, 83–84, 85
  sex affect on, 84–85
  species affect on, 80
muscle fiber size, 77, 86–101
  adipose tissue contribution to, 100
  age affect on, 99
  beta-adrenergic agonists effect on, 99
  breed effect on, 98
  in Callipyge sheep, 100
  feed intake effect on, 99
  in fish, 82
  genetic anomalies affect on, 99–100
  growth promotants affect on, 99
  intrafasicularly terminating muscle fibers and, 100
  mechanisms for increasing, 88–98
  muscle fiber number v., 77, 86
  nutrition affect on, 99
  orientation of muscle fibers in, 100
  in pigs treated with somatotropin, 100
  postnatal growth, 77
  protein synthesis/degradation and, 91–93, 95
  satellite cell recruitment and, 93–98
  sex effect on, 98–101
  species affect on, 100
  structural changes in, 88–91
muscle fiber types, 246
  actomyosin ATPase activity in, 105–107, 236
  beta-adrenergic agonists effect on, 117
  in cattle, 107–108, 113, 114–116
  changing, 237
  classification of, 105–109
  composition, 111–118
  confusion in, 107
  development of, 110–111
  diversification/specialization, 110
  environment temperature influence on, 116
  enzyme abundance in, 106–108
  exercise influence on, 116–117
  fast v. slow, 109
  features of, 109–110
  genetics/genes influence on, 114
  growth performance and, 111
  growth promotants effect on, 117–118
  hormones effect on, 117
  innervation and, 116
  meat quality and, 237–238
  myosin heavy chain isoform in, 108, 109, 110, 112, 113
  nutrition influence on, 116
  organization, 103–105
  oxidative v. glycolytic metabolism and, 109–110
  PH sensitivity in, 107
  physiological function and, 113
  in pigs, 105, 107, 110, 112, 114
  in poultry, 103
  red v. white, 103–105, 108, 109, 110, 241, 247, 249
  sex and, 116
  time and, 114–116
  variation with species, 113–114
muscle hypertrophy, 77, 78
muscle regulatory factor genes (MRFs), 66–67, 246
muscle tissue, 7, 21–41, 246
  cardiac muscle as, 23
  non-striated, 23
  proteins in, 30–35
  skeletal muscle as, 23–37, 63
  smooth muscle as, 23, 248
  striated, 24, 28, 29, 37
muscle to bone ratio, 173–174, 222
  cattle, 177
  pig, 175
mutation, during DNA replication, 85
myelin sheath, 21, 246
Myf-5, 67, 246
myoblasts, 63, 67–66, 71, 96
myoctyes, 26, 246
Myo-D, 67, 246
myofibers, 26, 85, 99, 246
myofibrillar proteins, 41, 232
myofibrillogenesis, 88, 246
  myogenesis and, 73–75
  stress fiber v. premyofibril model, 74
myofibrils, 28–29, 33, 246
  banding patterns in, 30
  nascent, 74, 75
  splitting, 91
myofilamints, 30, 31–32, 39, 246
myogenesis, 63–75, 85, 246
  cell cycle and, 63–66
  determination during, 66–67
  differentiation during, 67
  embryonic/fetal myosins during, 111, 113
  fusion during, 67–68
  maturation during, 68
  molecular aspects of, 63–68
  morphological apsects of, 68–73
  muscle fiber maturation during, 68
  myofibrillogenesis and, 73–75
  proliferation/replication and, 63
  somite, 68
myogenic lineages, in birds, 96

myogenic precursor cells, 72
myogenin, 67, 246
myoglobin, 109, 246
myosin, 30–31, 39, 67, 105, 246
myosin heavy chain isoform, 108, 109, 110, 112, 113
myostatin gene, 86, 246
myotendinal junction, 24, 246
myotome, 52, 55, 56, 246
myotubes, 67, 68–73, 110, 246

N

NCAMs (neural cell adhesion molecules), 74
nebulin, 33
negative feedback loop, 180–181, 183–184, 246
neonatal phase, 45, 203, 246
nerve bundles, 21, 23
nervous system, 21
nervous tissue, 7, 21, 246
net lipid accretion rate, 125, 246
net protein accretion, 91, 246
net protein deposition, 91, 246
neural crest, 51, 246
neural ectoderm, 51, 246
neural tubes, 51, 53, 55, 246
neuralation, 51, 52, 246
neurocrine, 179, 246
neuroendocrine system, 179, 180, 246
neuroglia, 21, 246
neurohormone, 179, 246
neurons (nerve cells), 21, 22
nicotinamide adenine dinucleotide (NADH), 106
nodes of Ranvier, 21, 246
non-protein nitrogen (NPN), 41, 208, 246
norepinephrine, 194
notochord, 51, 246
nuceloproteins, 161
nuclei, 36
nutrient partitioning, 199–202, 208
nutrition
  affect on adipose tissue, 132–134
  affect on muscle fiber number, 80–81
  affect on muscle fiber size, 99
  distribution, 199–200
  environmental factors affect on, 208
  fetus and, 202
  growth and, 199–214
  influence on muscle fiber types, 116
  meat tenderness and, 237–238
  plane, 199, 205–208
  during postnatal stage, 203
  during prenatal growth/development, 202–203
  requirements of bone, 146
  restriction, 204
  undernutrition and, 202–203
nutritional regulation, 199–214

O

obesity, 132, 246
oligodendrocytes, 21
oocytes, 45, 47, 246
organizing centers, 53
origin, 24, 246
ossification
  endochondral, 137, 138–141, 243
  intramembranous, 137, 141, 142
osteoblasts, 20, 137–144, 246
osteochondrosis, 141, 147
osteoclasts, 138, 139, 142–143, 144, 246
osteocytes, 20, 137–138, 142, 246
osteocytic osteolysis, 145, 246
osteogenesis, 137
osteolysis, 142
osteons, 20, 141, 246
ovaries, 45, 246
ovum (ova), fertilization of, 45–47, 246

P

papain, 31, 247
paracrine action, 179, 247
paranemin, 33
Parathyroid (PTH), 145, 198
paraxial mesoderm, 51, 68
parenchyma, 13
partitioning of nutrients, 208
Paylean®, pigs fed, 4, 117, 195
pentafibrils, 11
peptidases, 11–12, 247
performance, market animal, 1–4
perichondrium, 17, 139, 247
perimysium, 25, 247
perinatal stages, of growth, 203, 247
perinerium, 21, 247
periosteum, 20, 247
peripheral nervous system, 21, 247
permanent stunting, 205, 247
PH sensitivity, in muscle fiber types, 107
phasic, 110, 247
phenotype, 163, 247
phospholipids, 125, 247
physiological age, 217, 247
pigs. *See also* pork
  adipose tissue in, 121, 123, 124, 133
  body type changes in, 2
  carcass weight/length, 154, 155
  changes in muscle weight/nuclei acid content, 98
  energy partitioning in, 200
  fat in, 167
  fed Paylean®, 4, 117, 195
  fetal membrane organization, 58
  growth hormone administration in, 99, 100
  halothane gene in, 114
  halothane positive, 234–236
  longissimus dorsi muscle in, 90, 108, 113–114, 115, 117
  maternal and fetal membranes in, 60
  muscle fiber number in, 83–84, 85
  muscle fiber size in, 100
  muscle fiber types in, 105, 107, 110, 112, 114
  muscle to bone ratio of, 175
  performance, 2
  plane of nutrition effects on, 205–208
  porcine stress syndrome in, 234
  porcine v. bovine semitendinosus muscle, 80
  Rendement Napole (RN) gene in, 114
  runt, 80–81, 116, 202–203, 248
  semitendinosus muscle in, 103, 104
  skeletal muscle fibers of, 26
  "stress" gene in, 88, 234
  treated with somatotropin, 92, 100, 183, 210, 211, 212, 239
pineal gland, 198
pituitary gland, 181–184, 247
placenta, 202
placental lactogen, 198
placental tissues, 58
placentomes. *See* cotyledons structure
plane of nutrition, 199
  effects on pigs, 205–208
plasma, 16, 247
plasma calcium concentration, 198
plasticity, muscle/muscle fiber, 103, 104, 116
platelets, 16, 247
point mutation, 86
polar body, 47
polarizing zone, 53

polymerization, 10
porcine stress syndrome, 234, 247
pork
  carcass, 225, 226
  quality grades, 223
posterior marginal zone, 49–50
postmitotic, 63, 247
postmortem metabolism, 228, 230, 236, 247
postnatal growth/development
  DNA accumulation and, 97
  embryonic growth/development impact on, 43, 46
  growth curves and, 151, 158
  muscle fiber size, 77
  nutrition during, 203
  regulation of body composition, 199
postweaning lag, 203, 247
poultry. *See also* chickens
  dyschondroplasia in, 141
  grades, 222
  muscle fiber types in, 103
  quality grades, 223, 224
  selection, 175
prenatal growth/development, 43–62
  bovine fetus weight and, 150
  nutrition during, 202–203
  regulation of body composition, 199
primary bundle, 25, 247
primary myotubes, 68, 247
primary ossification center, 139, 247
primitive groove, 50, 247
primitive knot, 50
primitive streak, 50, 51, 52, 247
primordial, 53, 247
procollagen, 11
progenitor, 52, 247
progestins, 186
proliferation, *63,* 247. *See also* myogenesis
proliferative zone, 140, 247
pronucleus, 47
propionate, 128, 247
protein(s). *See also specific proteins*
  accretion, net, 91, 246
  accretion rates, 93
  accretion, regulation of, 213–214
  adherins, 74
  changes during growth, 160
  coordinating muscle development, 85
  degradation, 91–93, 95, 232, 247
  deposition, net, 91, 246
  dietary, 208–210
  effect on muscle, 99
  helix-loop-helix, 66–67
  insufficient, 208–209
  intersarcomeric, 33
  intrasarcomeric, 32–33
  isoform, 108
  muscle fiber size and, 91–93, 95
  muscle, metabolism, 184–185
  in muscle tissue, 30–35
  myofibrillar, 41, 232
  NCAMs (neural cell adhesion molecules), 74
  ratio of water to, 160
  regulatory, 32
  rumen degradable v. undegradable, 210, 248
  rumen escape, 209, 248
  structural, 74
  synthesis, 91–93, 95, 96, 247
  turnover, 91, 93, 111, 213–214, 247
proteinaceous substance, 7, 247
protein:RNA ratio, 93
proteoglycans, 10, 247
proteolytic systems, in muscle, 92–93
protracted G1-phase, 65
proximal light chains, 31

Q

quality grades, 216–223
quality, of meat. *See* meat, quality
quiescent cells, 66, 247

R

rate-limiting enzyme, 125
receptors
  adrenergic, 194–195
  hormone, 179
recovery weight for age, 204, 247
rectus abdominus, 220
red blood cells, 16, 247
releasing hormones, 183, 247
Rendement Napole (RN), 114, 234, 247
repartitioning agents, 238
replication, 63, 247
reproduction, process of, 43
reserve zone, 140, 247
resorption, 247
resting tension, 33
resting zone, 140, 247
restriction, 116
restriction point, 65
retail yield, 223, 225, 248
reticular fibers, 13, 248
rib-eye area, 225, 248
RNA:DNA ratio, 96
ruffled border, 142, 248
rumen degradable v. undegradable protein, 210, 248
rumen escape protein, 209, 248
ruminants, 167
  dietary protein and, 208–210
  v. nonruminants fat depots, 212
runt, pigs, 80–81, 116, 202–203, 248
ryanodine receptor, 36, 236, 248

S

sac, 58
sacral vertebrae, 217–219
sarcolemma, 26, 248
sarcomere, 90–91, 248
  addition, 90–91, 248
  length, 232, 248
  in muscle, 27, 29, 30, 31
sarcoplasmic calcium channel, 36
sarcoplasmic reticulum, 35–36, 37, 248
scaffolding, 68
Schwann cell, 21, 248
scleratomal cells, 56
scleratome, 51, 248
seam fat, 120, 248
secondary center of ossification, 139, 248
selection, 164–165, 170–175
  poultry, 175
  pressure, 164, 168
  programs/strategies, 3–4, 169
semimembranosus muscle, 80
senescence, 81, 248
serotonin, 198
sex
  affect on adipose tissue, 131–132
  affect on carcass composition, 176–178
  affect on muscle fiber number, 84–85
  growth curves, 176
  hormones, 146
  improved efficiency based on, 233–234
  muscle fiber size effect on, 98–101
  muscle fiber types and, 116
sheep. *See also* lambs
  body type changes in, 3
  Callipyge, 89, 100, 114, 242
  cotyledon structure in, 61
  fat depots in, 133, 134
  fat in, 167
  growth, 150
  wool v. meat type, 170–175

signal transduction, 110, 248
skatole, 233, 248
skelemin, 31
skeletal growth/development
  castration and, 146
  phases of, 138
  problems, 147
skeletal muscle, 68, 77, 248
skeletal muscle calpain (Calpain 3 or p94), 92, 248
skeleton, axial v. appendicular, 137
small-framed animals, 168, 248
soft lean, 221
soma, 21
somatocoele, 51, 55
somatotropin, pigs treated with, 92, 100, 183, 210, 211, 212, 239
somite(s), 68
  formation, 60
  maturation, 55
  morphology, 54
  myogenesis in, 68
  newly formed, 55
somitogenesis, 51
speciation, 164
species
  affect on carcass composition, 163–168
  affect on muscle fiber number, 80
  affect on muscle fiber size, 100
  differences in adipose tissue, 122
  differences in muscle distributions, 166–167
  fat deposition by, 167–168
  growth curves, 176
  variation with muscle fiber types, 113–114
speed of contraction, 105
S-phase, 65
spicules, 18, 248
spool joint, 219, 248
static stimulation, 116
steer, market
  body type changes in, 2
  performance, 3
stem cell zone, 140, 248
sternomadibularis, 24
steroid(s), 186–188
  hormones, 180
  implants, 118
  metabolism, 233
  use in cattle, 186–188
strain. *See* breed
stress fibers, 74
"stress" gene, in pigs, 88, 234
stressors, 228–229
stromal, 41
structural, 74
stunning, 228, 248
succinate dehydrogenase (SDH), 106
swine. *See* pigs
synapatic junctions, 21
syncytium, 38, 248
syndesonchorial interface, 60
synemin, 33
synovial fluid, 10, 248
synthesis
  collagen, 12
  fatty acid, 125–128, 147, 212
  muscle fiber size and, 91–93, 95
  protein, 91–93, 95, 96, 247

T

talin, 33
TCA, 228
tenderness, meat, 231, 232, 237–238
tendons, 24
terminal cisternae, 35
testosterone, 65, 98, 145–146, 184
tetrazolium reductase, 106
thermogenesis, 116, 248
thymus, 198
thyroid, 194
  gland, 145
  hormones, 116, 117, 145, 194, 196–198
thyroxine (T4), 196
tissues. *See also specific tissues*
  early/late developing, 162
  maternal, 59
  structure/organization, 7–41
tissue-specificity, 179
titin, 32
tonic, 110
trabeculae, 18, 248
traits, 168
transformation, of muscle to meat, 221
transforming growth factor, 65
transgenic animals
  advent of, 5
transverse tubules. *See* T-tubules
triad, 35, 248
triglycerides, 125, 248
triiodothyronine (T3), 117, 196
tropoelastin, 13, 249
tropomyosin, 32, 249
troponin, 32, 249
T-tubules, 35–36, 37, 248
turkey, tom, 3
turnover, protein, 91, 93, 111, 213–214, 247
26S proteosome, 92
twin cattle, 204–206
type 2C fibers, 113

U

ubiquitin-proteosome proteolytic pathway, 92–93, 94, 249k
ultimate pH, 228, 249
undernutrition, 202–203, 205
United States Department of Agriculture (USDA), 216–217, 223

V

villi, 60
viniculin, 33

W

water holding capacity, 231, 249
water to protein ratio, in muscle, 160
weaning, 203
weight. *See also specific animals*
  average daily gain in, 1–3, 243
  body, 156–160
  carcass, 157–158
  conceptus age and, 61
  low birth, 202
  muscle, 77, 111
wethers, market. *See* sheep
white blood cells, 16, 249
whole body animal growth. *See* growth

Y

yield grade, 223, 227–228, 233, 249
yolk, 58

Z

Z-line, 27, 249
zona/area pellucida, 47, 241
zone of polarizing activity, 53